Practical Micromechanics
of Composite Materials

Practical Micromechanics of Composite Materials

Jacob Aboudi

Professor Emeritus, School of Mechanical Engineering,
Tel Aviv University, Tel Aviv, Israel

Steven M. Arnold

Technical Lead for Multiscale Modeling,
Materials and Structures Division,
NASA Glenn Research Center, Cleveland, Ohio, USA

Brett A. Bednarcyk

Senior Research Engineer, Materials and Structures Division,
NASA Glenn Research Center, Cleveland, Ohio, USA

ELSEVIER

Butterworth-Heinemann
An imprint of Elsevier
elsevier.com/books-and-journals

Butterworth-Heinemann is an imprint of Elsevier
The Boulevard, Langford Lane, Kidlington, Oxford OX5 1GB, United Kingdom
50 Hampshire Street, 5th Floor, Cambridge, MA 02139, United States

British Library Cataloguing-in-Publication Data
A catalogue record for this book is available from the British Library

Library of Congress Cataloging-in-Publication Data
A catalog record for this book is available from the Library of Congress

ISBN: 978-0-12-820637-9

For Information on all Elsevier publications visit our website at
https://www.elsevier.com/books-and-journals

Publisher: Matthew Deans
Acquisitions Editor: Dennis McGonagle
Editorial Project Manager: Chiara Giglio
Cover Designer: Miles Hitchen
Production Project Manager: Prasanna Kalyanaraman

Working together
to grow libraries in
developing countries
www.elsevier.com • www.bookaid.org

Typeset by Aptara, New Delhi, India

This book is dedicated to
the next generation of composites professionals
and
our families, with all of our Love

Jacob Aboudi,
Steven M. Arnold,
and
Brett A. Bednarcyk

Contents

Please visit the book's GitHub site hosted by NASA for the companion Matlab code:
https://github.com/nasa/Practical-Micromechanics
Please visit the book's instructor site for additional materials:
https://textbooks.elsevier.com/web/Manuals.aspx?isbn=9780128206379

Preface

Composite materials comprising two or more constituent materials are designed and manufactured to obtain better performance than the individual constituents. Micromechanical theories are developed to predict the effective properties of the resulting composite (e.g., stiffness tensor, coefficients of thermal expansion, etc.), as well as the stress and strain distributions obtained by applying a prescribed loading. These distributions provide the designer/analyst information about high-stress locations where failure may initiate. Sophisticated micromechanics analyses provide more useful information about the composite's performance (e.g., properties and behavior) than simple analyses.

There are numerous books dealing with the micromechanical analyses of composites, but most of them are quite theoretical. This includes the authors' previous book entitled "Micromechanics of Composite Materials: A Generalized Multiscale Analysis Approach," which presents various types of micromechanical theories that describe the behavior of elastic, viscoelastic, thermoelastic, inelastic, and smart composites. It also includes the modeling of functionally graded materials, composites behavior undergoing large deformation, and elastic wave propagation in composites. The previous book provided a collection of the authors' work over several decades. Thus, the book was essentially written for documentation rather than instruction.

In contrast, the present book was written with the express intent of providing the reader with an accessible treatment of the underlying theory and *its implementation* to enable *practical use* of the micromechanics theories. The presented theories can be used on their own and in conjunction with the classical lamination theory (CLT). In particular, the simple and easy to use Mori–Tanaka (MT) method, which is followed by the method of cells (MOC), generalized method of cells (GMC), and the more sophisticated high-fidelity generalized method of cells (HFGMC), is derived in the context of thermoelastic material behavior and implemented within the MATLAB environment. Both the theory and implementation show that the MT method requires only matrix multiplication and inversion. The MOC, on the other hand, requires the solution of a system of 24 algebraic equations, and more sophisticated programming is required for GMC and HFGMC. All of these micromechanics theories, as well as CLT, are provided as complementary open-source MATLAB functions and scripts at https://github.com/nasa/Practical-Micromechanics for immediate use. An additional unique feature of this book is its treatment of "failure" in which four commonly used failure criteria are provided, along with a unified treatment of calculating margins of safety, for both unidirectional and laminated composites under general thermomechanical loading. Lastly, an advanced chapter, Chapter 7, deals with the ability to perform a progressive, micromechanics-based, damage analysis of composites.

Throughout the book, example problems and exercises highlight lessons learned in developing and applying these theories over the past two decades. It is believed that a reader who understands this "practical" book will be able to adapt the current implementation to their problems of interest. Users are encouraged to upload such adaptations to the above open-source website. Consequently, it is our sincere hope that this unified multiscale approach as well as its companion open-source implementation will help materials scientists, researchers, engineers, and structural designers (both students and practitioners) in developing a better understanding of composite mechanics at all scales, and thereby contribute to composites reaching their full potential. An exercise solution manual is available for instructors.

Jacob Aboudi
Steve Arnold
Brett Bednarcyk
December 2020

Acknowledgments

We extend a tremendous "Thank You" to everyone who contributed to and helped us complete this book. Specifically, we would like to thank our NASA colleagues Trent M. Ricks, Evan J. Pineda, and Pappu L. N. Murthy for their help with getting us up to speed on MATLAB coding and their advice. Thank you to Robert Earp (GRC legal counsel) for making sure we are not exposed and that we followed all the rules regarding release of the open source MATLAB. Special thanks to Jennifer Bednarcyk for sending years of weekly update emails to Jacob, forcing us to stay on schedule.

Special thanks to NASA for sponsoring the writing of the book and the development of the associated MATLAB scripts and functions that accompany this book. Specific thanks go to the Transformational Tools and Technologies Project (Dale Hopkins and Mike Rogers) within the Transformative Aeronautics Concepts Program and to Jim Zakrajsek and Joyce Dever, Materials and Structures Division Chief and Deputy Chief, and George Stefko, Multiscale and Multiphysics Modeling Branch Chief, for their encouragement and support of this effort.

I, Jacob Aboudi, would like to thank my friend and engineer, Mr. Rami Eliasy, whose help with computer issues including system management, program compilations, execution commands and files has been indispensable.

I, Brett Bednarcyk, would like to thank my daughter Adia Bednarcyk for assistance with graphics and presentation aesthetics.

Finally, I, Steve Arnold, would like to thank my Lord and Savior Jesus Christ, for providing me an opportunity to embark on a second journey with two such excellent gentlemen as Jacob and Brett of whom I am proud to have been associated with and to call my friends for so many years. I specifically want to thank Brett for his attention to details, resilience, and steadfastness to see this second book through to completion. I consider myself, once again, blessed to have been able to make this second journey with them both.

Introduction

1

Chapter outline

1.1 Introduction

The intent of this book is to provide both students and practitioners with a solid understanding of micromechanics concepts and theories, which can be used in practice, and provide their theoretical underpinnings in a clear and concise manner. A MATLAB implementation of the discussed theories is provided to illustrate their utility for the design and analysis of advanced composites. Throughout the book, emphasis is placed on the determination and application of the local (stress and strain) fields within composites. Multiple micromechanics theories are presented in terms of the unifying concept of strain concentration tensors. These tensors provide the local strains throughout the composite given applied global strains. They are extremely powerful because, as demonstrated herein, once obtained, the composite effective mechanical properties and local fields can be easily calculated. Similarly, thermal strain concentration tensors are developed to account for thermal effects due to a global temperature change.

The utility of micromechanics theories becomes particularly clear with their application in multiscale modeling of composites (e.g., Kanoute et al., 2009; Aboudi et al., 2013; Fish, 2014). Because they provide effective anisotropic constitutive equations for composite materials, these theories represent a material point. As such they can be used to represent the material behavior at a point in a composite structure that is being analyzed using a higher scale model such as lamination theory (used herein) or even finite element analysis. The transition between scales (via homogenization and localization) involved in a continuum-based analysis of composite materials/structures is depicted in Fig. 1.1. For example, in the context of multiscale lamination theory, the constituent deformation and damage behavior affects the behavior of the individual plies, which in turn affect the overall laminate response. Furthermore, for multiscale problems to remain tractable, the micromechanics methods used must be very efficient. Efficiency is a hallmark of the micromechanics theories presented herein as they are closed form or semiclosed form.

This introductory chapter provides some fundamental information about composites and then focuses on introducing modeling of composites, particularly micromechanics

Practical Micromechanics of Composite Materials. DOI: 10.1016/C2019-0-04193-3

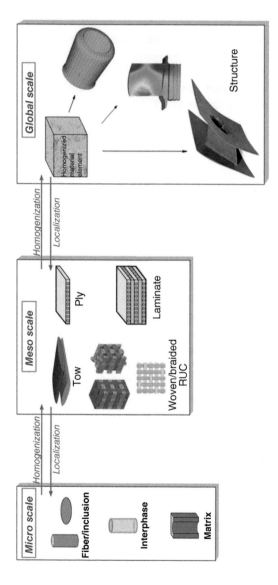

Figure 1.1 Illustration of the relevant levels of scale for continuum-base multiscale composite analysis.

and multiscale modeling. There are many excellent texts, however, that go into much greater detail regarding the how and the why of composite materials and structures. Rather than repeat this information, the reader is referred to Jones (1975), Christensen (1979), Carlsson and Gillespie (1990), Daniel and Ishai (1994), Herakovich (1998), Hyer (1998), Zweben and Kelly (2000), Miracle and Donaldson (2001), Barbero (2011), Herakovich (2012), and Aboudi et al. (2013). It should also be noted that this book, in contrast to the expansive text on micromechanics by Aboudi et al. (2013), is limited to thermoelastic behavior and has been made much more accessible so as to introduce students and practitioners to the art and science of micromechanics-based composite analysis. As such, an accompanying open source MATLAB (2018) code is provided and discussed in detail throughout this text, see Section 1.4.

1.2 Fundamentals of composite materials and structures

In the fields of Structural Engineering and Materials Science and Engineering, the difference between a structure and a material comes down to the presence of a boundary. A material is the substance of which an object is composed. The material itself has no boundaries. Rather, it is what is present at a point in the object. Scientists and engineers have developed ways to represent materials through properties that describe how the material behaves at a point in an object. These properties include Young's modulus, thermal conductivity, density, yield stress, Poisson's ratio, and coefficient of thermal expansion. The object itself, on the other hand, is a structure. It has boundaries, and its behavior is dependent on the conditions at these boundaries. For example, given a steel beam, the beam itself is a structure, while the material is steel. This distinction between materials and structures is natural and extremely convenient for Structural Engineers and Materials Scientists. Imagine attempting to combine properties of materials and structures in the case of the aforementioned beam. A beam's bending characteristic is dictated by its flexural rigidity (the Young's modulus E times the cross-section moment of inertia I, or EI, and not just its Young's modulus). If this were not separated into a material property (E) and a structural property (I) but rather bookkept as a combined property, one would need to look up a value for every combination of beam shape and material.

The above discussion implies that a material is a continuum, meaning it is continuous and completely fills the region of space it occupies. The material can thus be modeled using continuum mechanics, which considers the material to be amorphous and does not explicitly account for any internal details within the material, such as the presence of inclusions, grain orientation, or molecular arrangement. To account for such internal details, some additional theory beyond standard continuum mechanics is needed.

In its broadest context, a composite is anything comprised of two or more entities. A composite structure would then be any object made up of two or more parts or two or more materials. Likewise, a composite material is a material composed of two or more materials with a recognizable interface between them. Because it is a material, it has no external boundaries; once an external boundary is introduced, it becomes a structure composed of composite materials. Clearly, however, a composite material does have distinct internal boundaries. If these internal boundaries are ignored, continuum mechanics can be used to model composite materials as pseudo-homogenous anisotropic materials with directionally dependent "effective," "homogenized," or "smeared" material properties. This approach is called macromechanics. Micromechanics, on the other hand, attempts to account for the internal boundaries within a composite material and to capture the effects of the composite's internal arrangement. In micromechanics, the individual materials (typically referred to as constituents or phases) that make up a composite are each treated as continua via continuum mechanics, with their own individual representative properties and arrangement dictating the overall behavior of the composite material.

In many cases, especially with composite materials used in structural engineering, the geometric arrangement of one phase is continuous and serves to hold the other constituent(s) together. This constituent is referred to as the matrix material. Whereas, the other constituent(s), often referred to as inclusion(s) and/or reinforcement(s), are materials that can be either continuous or discontinuous and are held together by the matrix. There may also be interface materials, or interphases, present between the matrix and inclusion(s). A fundamental descriptor of composites that should always be indicated when denoting a given system (since it strongly influences the effective behavior) is the volume fraction of the phases present. Typically, only the reinforcement phase is indicated unless multiple phases are present since the sum of all phases must equal 100%; for example, in a two-phase fiber-reinforced composite v_f is the volume fraction of fibers and $v_m = 1 - v_f$ is that of the matrix.

Composites are typically classified at two distinct levels. The first level of designation is usually made with respect to the matrix constituent. This divides composites into three main categories: polymer matrix composites (PMCs), metal matrix composites (MMCs), and ceramic matrix composites (CMCs). The second level of classification refers to the form of the reinforcement: discontinuous (particulate or whisker), continuous fiber, or woven/textile (braided or knitted fiber architectures are included in this classification). In the case of woven and braided composites, the weave or braiding pattern (e.g., plane weave, triaxially braided) is also often indicated. Examples of some of these types of composites are shown in Fig. 1.2. Note that particulate composites are typically isotropic whereas most other composite forms have some level of anisotropy (e.g., a unidirectional continuous fiber reinforced composite is usually transversely isotropic).

Composites, particularly PMCs, are often manufactured as an assembly of thin layers joined together to form a laminate (see Fig. 1.2B). Each layer is referred to as a lamina or ply. By orienting the reinforcement direction of each ply, the properties and behavior of the resulting laminate can be controlled. A quasi-isotropic laminate can be formed by balancing the orientations of the plies such that the extensional stiffness of the laminate is constant in all in-plane directions. Quasi-isotropic laminates have thus been very popular as—under in-plane elastic extension—they behave like isotropic metals, with which most engineers are familiar. However, this has also led to engineers attempting to simply replace metals with quasi-isotropic laminates in structures that were designed based on isotropic metallic properties. This is the origin of the expression "black aluminum," which refers to a black quasi-isotropic carbon/epoxy

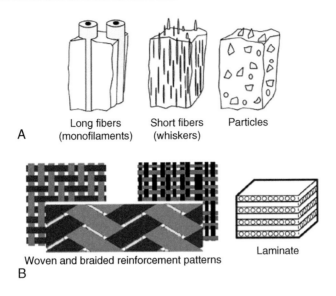

Figure 1.2 Composite systems. (A) Reinforcement types. (B) Laminate and woven constructions.

laminate, whose in-plane effective elastic properties are often very close to those of aerospace aluminum alloys. The tremendous pitfall of this approach, which has in many ways slowed the realization of the full potential of composites, is that quasi-isotropic carbon/epoxy laminates are not even close to isotropic in terms of their out-of-plane behavior. They are also prone to delamination and interlaminar failure; failure modes which do not afflict isotropic metals. The "black aluminum" design approach, while simple, is therefore typically very inefficient. Some pros and cons associated with employing composite materials are listed in Table 1.1.

Table 1.1 Pros and Cons associated with using composite materials

Advantages	Disadvantages
Weight reduction (high strength/stiffness to weight ratio)	Higher recurring and nonrecurring costs
Enhanced fatigue and corrosion resistance	Isolation requirements for some materials (e.g., to avoid galvanic corrosion)
High toughness	Lack of electrical conductivity
Lower part count	Anisotropy
Tailorable strength and stiffness (via optimization)	Impact damage is difficult to detect and repair (bird strike, tool drops) – delamination
More efficient use of materials (lower "buy-to-fly" ratio = material weight used to fabricated/final machined weight, Boyer et al. (2015))	Moisture absorption
Better durability and lower maintenance cost	Higher possibility of manufacturing defects

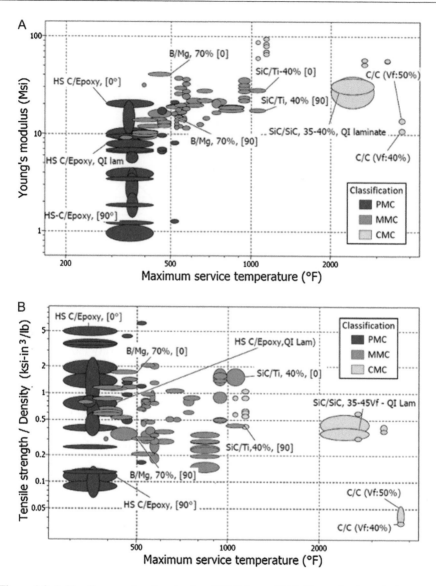

Figure 1.3 Ashby diagrams [produced using CES Selector 2012 (Granta Design Limited, 2012)] for various PMCs, MMCs, and CMCs. (A) Young's modulus versus maximum service temperature. (B) Specific strength (tensile strength divided by density) versus maximum service temperature.

A key distinction among PMCs, MMCs, and CMCs is their maximum service temperature. As shown in Fig. 1.3, most PMCs are limited to an operating temperature under 450°F (250°C). Metal matrix composites extend this range to approximately 1200°F (649°C), depending upon the capability of the chosen matrix, and typical CMCs can remain functional to over 2000°F (1093°C). Obviously, the temperature limitations are dependent on the limitations of the composite constituent materials. Indeed, the groups of small yellow ellipses in Fig. 1.3 representing CMCs with maximum service temperatures of approximately 1100°F (593°C), and high Young's modulus, are tungsten carbide ceramic matrix materials with particulate metallic cobalt inclusions. Thus, the lower operating temperature is due to the metallic reinforcement; most CMCs are composed of ceramic matrices and ceramic reinforcements.

The vertical axes in Fig. 1.3 represent the composite's (A) effective Young's modulus and (B) specific strength (strength divided by density). The wide spread in properties, especially in the case of PMCs, is indicative of the anisotropy present in continuously reinforced composites. For the carbon/epoxy composite labeled in Fig. 1.3, there is a factor of 20 between the Young's moduli in the longitudinal direction (i.e, along the continuous carbon fibers, 0°) and the transverse direction (i.e., perpendicular to the fiber direction, 90°). For the specific strength, the corresponding factor is close to 40. The composite labeled "HS C/epoxy, QI lam" represents the effective in-plane properties of a quasi-isotropic laminate composed of the previously discussed high strength carbon/epoxy material. As would be expected, this laminate's effective properties are intermediate to those of its plies in each direction. It is also noteworthy that this quasi-isotropic laminate, which is actually a structure with external boundaries, is compared here to unidirectional composite materials. Such a laminate would only behave like a material if appropriate extensional in-plane boundary conditions were applied. If it were subjected to bending, it would behave like an anisotropic plate, and its properties would be dependent on its thickness and ply stacking sequence, which are structural rather than material properties.

Without differences in properties between the constituents, a composite would cease to be a composite. That is, it is the difference in properties between the constituents that makes a composite behave differently than a monolithic material and enables its tailoring for specific purposes. PMCs are characterized by a large property mismatch between the constituents. As shown in Fig. 1.4, the extremely low stiffness of the polymer matrix (epoxy shown here) results in a stiffness mismatch on the order of 80:1 in the case of carbon fibers in the longitudinal direction (L), and 20:1 in the case of glass fibers. Carbon fibers are transversely isotropic, so the mismatch in the transverse direction (T) for carbon fiber reinforced polymers (CFRP) is typically much lower. Alternatively, MMCs typically have constituent stiffness mismatches of 3:1 to 4:1. The longitudinal stiffness mismatch between the fiber and matrix in CMCs is much lower: close to 1:1 in the case of SiC/SiC composites (see Dunn, 2010) and approximately 1:1.6 in the case of C/SiC. In the transverse direction, however, there is a large mismatch in stiffness in the case of C/SiC (on the order of 1:17). In SiC/SiC, the SiC fibers are coated with a very compliant material, such as BN, to present a barrier to matrix crack growth. This then sets up a large mismatch in stiffness at the composite level for SiC/SiC between the longitudinal and transverse directions (on the order of 1:18). C/SiC composites also typically have a compliant pyrolytic carbon interface material that serves the same purpose. The extent of property mismatch is a key feature of composites that impacts the efficacy of composite models. For a model

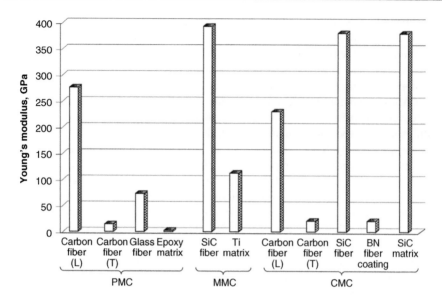

Figure 1.4 Comparison of the Young's moduli of typical PMC, MMC, and CMC constituent materials, with (L) indicating the longitudinal direction and (T) indicating the transverse direction in the case of transversely isotropic carbon fibers. The other constituents are considered to be isotropic.

to be applicable to all types of composites, it must properly handle widely varying degrees of property mismatch in each direction.

The development of high-performance composite materials started in the 1940s with the introduction of glass-fiber-reinforced polymer matrix composites. It has continued to grow with the introduction of additional polymeric, metallic, and ceramic composite systems to become a major force in the world materials market. Composites have penetrated such key industries as aerospace, automotive, building and construction, sports and leisure, and most recently, wind energy. The global composite market can be broken down into nine distinct market sectors (see Fig. 1.5), wherein electrical and electronics (E&E), construction, and aerospace markets appear to have the largest market share, followed by the transportation, wind energy, and pipe and tank markets. Furthermore, in the first decade of the 21st century, the composite market as a whole saw an annual growth of approximately 4% to 5% per year in value and 3% in volume, with emerging countries seeing approximately twice as much growth per year in value as compared to developed nations (JEC Composites, 2011).

Arguably the most aggressive industry utilizing composites is aerospace, as illustrated in Fig. 1.6, because of their attractive (weight-saving) properties and lower part counts, which translate directly to cost savings. The most recently publicized commercial aircraft is the Boeing 787 with over 50% by weight of its materials being composites, as shown in Fig. 1.6. This figure illustrates the significant increase in the use of composites in commercial aircraft over the past half century as manufacturing, joining, and analysis methods have improved and performance demands have

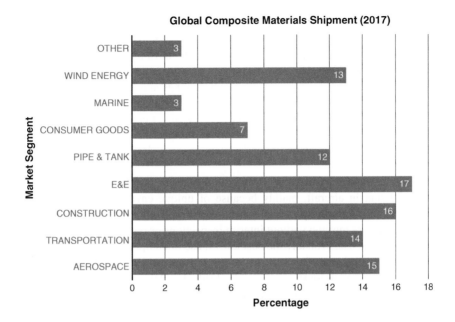

Figure 1.5 The global composite market divided into market segments. Percentage distributions are as of 2017 as reported by TextileToday (2020).

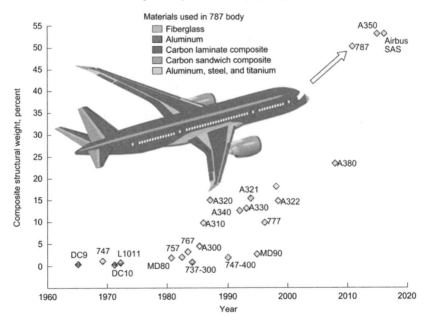

Figure 1.6 Composite material usage by weight in commercial aircraft (updated from Harris et al., 2001).

increased. Prior to the mid-1990s, composites were mainly limited to use in secondary structures (i.e., those that do not cause immediate danger upon failure). However, with the development of the Airbus A380, Boeing 787, and most recently the Airbus A350, composites have seen step changes (as a percent of structural weight) as they are being extensively utilized in primary structures, such as wings and fuselage components, as well. This increased use of CFRP composites has also resulted in an increase use of titanium (almost 2X in 787 as compared to 777, see Boyer et al., 2015), due to its low coefficient of thermal expansion, density, and its compatibility with the carbon fibers in the presence of moisture. Consequently, titanium is an ideal material for interfacing with composites.

Usage of composites (primarily PMCs) in spacecraft and launch vehicles is lagging behind that in aircraft, mainly because of unique environmental and loading requirements. Furthermore spacecraft structures are typically designed and certified to perform their entire mission without interim inspections or repairs, while most launch vehicles are single use. Probably the best known use of composites in spacecraft is the 60-ft long payload doors of the space shuttle; however, other launch vehicles using carbon fiber reinforced composites are the Delta IV, Atlas V, EELV, and Pegasus. NASA and its industrial partners are actively pursuing the development and use of composites in large structures (e.g., composite crew module, space launch systems (SLS), antennas and solar arrays, and propellant tanks, to name a few) for future space missions.

1.3   Modeling of composites

In this text, the phrase "modeling of composites" is intended to refer to simulating or analyzing the behavior of a fully consolidated composite material or structure. Process modeling, the simulation of the manufacturing, and forming of composite materials and parts, is not addressed. In this context, there are two basic approaches to modeling composites: the macromechanical approach and the micromechanical approach. The macromechanical approach involves constructing models strictly at the ply scale or higher (see Fig. 1.1), wherein the composite is viewed as an anisotropic material, and the details of the underlying arrangement of the constituent materials is ignored. In the linear elastic regime, this approach is straightforward; it involves only determining (usually experimentally) the anisotropic elastic properties of the composite material. These can then be entered into a structural analysis code, such as lamination theory or finite element analysis (FEA), to determine structural performance. In fact, this is the current standard design procedure for composite structures and is illustrate in Chapter 2. In addition to the elastic properties, statistically meaningful design allowable stresses are determined for the composite material through extensive testing. Then, the part is designed such that the stresses never exceed these allowables, with a sufficient margin of safety. This is illustrated in Chapter 4. Of course, while computationally efficient and straightforward, this approach is heavily reliant on costly experimental data (both coupon and structural), which must be generated for each variation of the composite (e.g., change in fiber volume fraction).

In the nonlinear regime (e.g., high temperature applications), and when trying to predict damage and failure, the macromechanical approach becomes somewhat problematic. Anisotropic constitutive, damage, and failure models must be constructed to account for the widely varying behavior and failure mechanisms of the composite in the various directions and characterized through extensive composite testing. A benefit is that the intrinsic, history-dependent, interactive effects of the composite constituents are embedded in the experimental results, and their in-situ behavior is automatically captured. However, such models are hampered by the fact that, physically, the deformation, damage, and failure occur in the actual constituent materials of the composite, not within an idealized effective anisotropic material. Thus, while many have attempted to model the nonlinear behavior of composites with macromechanics, the approach typically will be highly phenomenological since, by definition, it does not consider what is happening in (or between) each constituent at the appropriate physical scale.

In contrast, the micromechanical approach to modeling composites explicitly considers the constituent materials and how they are arranged within the composite. The goal of micromechanics is to predict the effective behavior of a heterogeneous material based on the behavior of the constituent materials and their geometric arrangement. By determining a composite's effective behavior via micromechanics, it can then be treated as a material in higher-scale analyses (similar to the macromechanical approach). For example, the effective material properties of the composite determined via micromechanics can be used in a laminate analysis to represent the ply materials (as done in Chapter 3 herein), or in FEA of a composite structure to represent the materials in different regions. One benefit of micromechanics is that composite properties can be determined, in any direction, for any fiber volume fraction or reinforcement architecture, even if the composite has never been manufactured. It can therefore assist in designing the composite materials themselves as well as the structures comprised of them. Additionally, micromechanics enables "virtual testing," and if the micromechanics approach is sufficiently validated, it has the potential to drastically reduce development and certification costs. Furthermore, since design engineers typically use stress/strain allowables that are several standard deviations worse than average to account for processing variations, micromechanics offers a methodology to calculate distributions of properties and allowables that account for manufacturing defects and flaws (as well as other variabilities).

In terms of nonlinearity associated with inelasticity, damage, and failure (addressed in Aboudi et al., 2013), micromechanics allows the physics of these mechanisms to be captured at the constituent scale, where they are actually occurring. There are many (molecular dynamicists, for example) who argue that even this scale is too high to properly account for these mechanisms. It is clear, however, that micromechanics allows the physics to be captured at a more fundamental scale compared to macromechanics. Further, if the interface between the composite constituents contributes significantly to the overall composite behavior (as in MMCs and CMCs), it can be addressed through micromechanics, wherein the interface can be explicitly modeled.

As mentioned previously, multiscale modeling of composites refers to simulating their behavior through multiple time and/or length scales. Although the nomenclature in the literature varies, typically a multiscale modeling analysis will involve the length scales shown in Fig. 1.1 for continuum-based modeling. These scales, progressing from left to right in Fig. 1.1, are the microscale (constituent level; fiber, matrix, and interface), the mesoscale [composite level; tow/yarn and woven repeating unit cell (RUC), ply, laminate], and the macroscale (structural level; composite parts). Traditionally, one ascends (moves right) or descends (moves left) these scales via homogenization and localization techniques, respectively. A homogenization technique provides the properties or response of a "structure" (higher scale) given the properties or response of the structure's "constituents" (lower scale). Conversely, localization techniques provide the response of the constituents given the response of the structure.

Herein, the micromechanics theories used to localize and homogenize are the Voigt model, Reuss model, Mori-Tanaka (MT) Method, and the family of Method of Cells theories: Method of Cells (MOC), Generalized Method of Cell (GMC), and High-Fidelity Generalized Method of Cells (HFGMC). A significant distinction divides these theories into those that consider only the average stresses and strains in the constituent materials (Voigt, Reuss, and MT) and those that account for stress and strain gradients through further subdividing the composite and consider a RUC representing the composite microstructure (MOC, GMC, HFGMC). All of these theories consider a unidirectional composite with long continuous fibers but are capable of considering discontinuously reinforced composites as well. The MOC, GMC, and HFGMC theories for discontinuous and woven composites are not presented herein as they require triply-periodic (as opposed to doubly-periodic) implementations, and, as such, these versions are more involved mathematically. The reader is therefore referred to Aboudi et al. (2013) for the triply-periodic versions of the MOC, GMC, and HFGMC, along with relevant applications. However, the equations for the MT Method for discontinuously reinforced composites are given in Chapter 3. All of the theories presented in this text can be used to represent plies within a multiscale composite laminate analysis using classical lamination theory, which is introduced in Chapter 2.

A measure of the usage of the four primary micromechanics theories (MT, MOC, GMC, HFGMC) presented in this book is shown in Fig. 1.7. The data in this figure were obtained from Engineering Village (2020), which searches the titles and abstracts of 38740 journals and conference proceedings dating back to 1884. The search terms were "micromechanics" plus the name of each of the four theories. Clearly, MT has been extensively used in the literature as compared with the MOC family of theories, likely due to its simplicity and efficiency. The MT theory was developed in the early 1970s, but was popularized by Benveniste's (1987) reformulation. The MOC was developed in the 1980s, GMC in the 1990s, and HFGMC in the 2000s, which is reflected in Fig. 1.7. Over the last five years the MOC family has good representation in the literature with approximately a third of the publications of the simpler MT theory. Although not shown, it should be noted that the finite element micromechanics approach (wherein the composite microstructure is represented using standard finite element analysis (FEA) techniques) is the most popular in the literature with an average of over 100 publications per year since 1988. A brief discussion of the finite element micromechanics approach is given in Chapter 3.

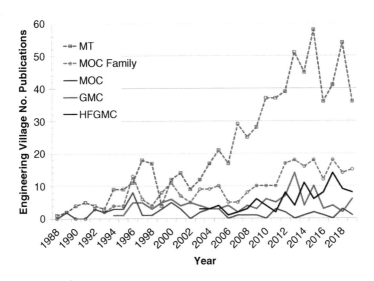

Figure 1.7 Number of publications per year for several micromechanics theories according to Engineering Village (2020).

1.4 MATLAB code

An extensive MATLAB (2018) code is provided with this book to solve micromechanic problems dealing with effective unidirectional composite material behavior and the behavior of laminates. The plies within the laminates can be represented using either effective anisotropic ply properties (macromechanics) or using micromechanics. Many of the MATLAB functions and scripts are provided in the text in the appropriate · chapters, with flowcharts shown for the critical scripts and functions. A list and description of the MATLAB scripts and functions are given in Table 1.2.

The entirety of the open source MATLAB code is available from:

https://github.com/nasa/Practical-Micromechanics

The MATLAB code solves three classes of problems, all providing effective properties and, given applied loading, the local stress and strain distributions and margins of safety. These are: (1) classical lamination theory using ply-level properties, (2) micromechanics solutions based on constituent properties, and (3) micromechanics-based classical lamination theory using constituent-level properties. Problems are defined (mainly) using the scripts and functions located in the `root` directory (see Table 1.2). The script `LamAnalysis.m` analyzes linearly thermoelastic laminates using classical lamination theory, and associated problems are defined in the function `LamProblemDef.m`. If using the macro approach, the effective ply material properties are specified in the function `GetPlyProps.m`. Likewise, the script `MicroAnalysis.m` analyzes linearly thermoelastic unidirectional composite materials, and associated problems are defined in the function `MicroProblemDef.m`. For these micro analyses,

Table 1.2 List and description of MATLAB scripts and functions.

Path/Name	Description	Chapter Introduced
root/		
GetConsitProps.m	Defines constituent material properties and allowables	3
GetEffProps.m	Calculates effective composite properties from micromechanics	3
GetPlyProps.m	Defines effective ply properties and allowables	2
LamAnalysis.m	Performs laminate analysis using classical lamination theory	2
LamProblemDef.m	Defines laminate problems	2
LamSimDamage.m	Performs micromechanics-based progressive damage on laminates	7
MicroAnalysis.m	Performs micromechanics-based RUC analysis	3
MicroProblemDef.m	Defines micromechanics-based RUC problems	3
MicroSimDamage.m	Performs micromechanics-based progressive damage RUC analysis	7
RandomRUCHist.m	Runs multiple micromechanics-based problems and plots histograms of the effective properties	6
Functions/CLT/		
CLT.m	Classical lamination theory analysis	2
GetABD.m	Calculates laminate ABD matrix	2
GetT.m	Obtains ply transformation matrices	2
GetTherm LoadsNM.m	Calculates the laminate thermal force and moment resultants	2
Qbar.m	Calculates ply rotated reduced stiffness matrices	2
Functions/Damage/		
DamageConstit.m	Calculates damaged (failed) material properties	7
DamageLam.m	Determines damaged ply props by calling DamageMicro.m per ply	7
DamageMicro.m	Determines damaged effective composite properties	7
Functions/Margins/		
CalcLamMargins.m	Divides laminate problems into thermal preload and mechanical parts for ply level margin calculations	4
CalcMicroMargins.m	Divides micromechanics-based RUC problems into thermal preload and mechanical parts for micro scale margin calculations	4

(*continued on next page*)

Table 1.2 (*continued*)

Path/Name	Description	Chapter Introduced
FCriteria.m	Calculates margins of safety per failure theory	4
MicroMargins.m	Sets up variables for calling FCriteria.m for micro scale margins	4
Functions/Micromechanics/		
GetRUC.m	Obtains details of the RUC for GMC and HFGMC	5
GMC.m	Calculates mechanical and thermal strain concentration tensors and effective properties using the GMC micromechanics theory	5
HFGMC.m	Calculates mechanical and thermal strain concentration tensors and effective properties using the HFGMC micromechanics theory	6
LamMicroFields.m	Calculates microscale stress and strain fields within laminate plies	3
MicroFields.m	Calculates microscale (constituent level) stress and strain fields	3
MOC.m	Calculates mechanical and thermal strain concentration tensors and effective properties using the MOC micromechanics theory	3
MT.m	Calculates mechanical and thermal strain concentration tensors and effective properties using the MT micromechanics theory	3
Reuss.m	Calculates mechanical and thermal strain concentration tensors and effective properties using the Reuss micromechanics theory	3
RUCHex.m	Defines RUCs with hexagonal fiber packing	6
RUCRand.m	Defines RUCs with random fiber packing	6
RunMicro.m	Executes micromechanics theory specified in GetEffProps.m	3
Voigt.m	Calculates mechanical and thermal strain concentration tensors and effective properties using the Voigt micromechanics theory	3
Functions/Utilities/	Folder containing various utility functions	2
Functions/WriteResults/	Folder containing multiple output writing functions	2

the micromechanics model parameters are set in the function `GetEffProps.m`, whereas the constituent material properties are defined in the function `GetConstitProps.m`. For multiscale lamination theory problems, the plies are assigned materials in `LamProblemDef.m` that have micromechanics definitions (specified in `GetEffProps.m`). That is, for example, if material #100 has micromechanics model parameters defined in `GetEffProps.m`, this material can be assigned to plies in `LamProblemDef.m` to execute a multiscale lamination theory analysis.

In addition, the two scripts `MicroSimDamage.m` and `LamSimDamage.m` are used to execute micromechanics-based progressive damage simulations on unidirectional composite materials and composite laminates, respectively, in Chapter 7. These scripts are extended versions of `MicroAnalysis.m` and `LamAnalysis.m` and therefore similarly rely on `GetConstitProps.m`, `GetEffProps.m`, `MicroProblemDef.m`, and `LamProblemDef.m` to set up the problems. Finally, `RandomRUCHist.m` is a specialty script used in Chapter 6 to generate histograms of the effective properties predicted by HFGMC when using RUCs with random fiber packing arrangements.

1.5   Description of the book layout

Throughout this book, a basic knowledge of solid mechanics is assumed. Consequently, there is no chapter on the basics of solid and structural mechanics (e.g., introducing the concepts of stress and strain). Each chapter presents theory, gives a description of relevant MATLAB code, presents illustrative example problems, and (aside from Chapter 7) provides exercises. The exercises are designed to enhance understanding of the chapter's main concepts and to provide practical experience in the use of micromechanics-based composite analysis.

Chapter 2 presents classical lamination theory, wherein the plies are treated using only effective anisotropic properties (rather than using micromechanics to calculate their effective properties). Example problems address different types of laminates and classes of composite materials (PMCs, MMCs, and CMCs). Chapter 3 deals with the fundamentals of micromechanics, presenting the theoretical underpinnings of the subject in general, as well as details of the Voigt, Reuss, Mori-Tanaka (MT), and MOC analytical micromechanics theories. Emphasis is placed on the derivations of the mechanical and thermal strain concentration tensors for each method, which can be used to determine the effective properties of, and local fields within, composites. It is demonstrated that the strain concentration tensor is the core of any micromechanics theory. Example problems focus on comparing the methods in terms of their strain concentration tensors, predicted mechanical and thermal properties, and predicted local (constituent level) fields.

Chapter 4 presents a framework to predict the failure initiation in composites through the use of margins of safety. It involves determining the composite or laminate level loading at which a failure criterion is first met somewhere in the composite or laminate. Four classic failure criteria (max stress, max strain, Tsai–Hill, and Tsai–Wu) are presented, and their margin of safety equations are derived for mechanical, thermal, and thermomechanical loading. Applications demonstrate the utility of this approach for determining margins of safety for unidirectional and laminated composites, determination of allowable loads, as well as easy generation of multiaxial failure envelopes. Micromechanics enables identification of the underlying failure mechanisms, which vary with different combinations of multiaxial thermomechanical loading.

The GMC, which is obtained through geometric generalization of the MOC to an arbitrary number of subvolumes, is presented in Chapter 5. These additional subvolumes enable the theory to represent composites with more than two constituents (e.g., interface materials) and better discern the local field gradients in the composite (compared to MT and MOC). Thus, CMCs, which contain an explicit fiber-matrix interface material have been modeled using GMC in terms of effective properties, local fields, and failure envelopes.

Chapter 6 develops the HFGMC micromechanics theory, which relies on a second-order displacement field to provide higher-fidelity local field predictions in the composite constituents. This is in contrast to GMC whose first-order displacement field approximation results in constant stress and strain subcells. HFGMC's second-order approximation enables normal-shear field coupling, which drastically improves the theory's predictive capabilities. This improvement comes at a price in terms of computational efficiency while also introducing subcell grid refinement sensitivity, which is absent in the other theories.

Chapter 7 revisits the composite failure initiation discussed in Chapter 4, but now adds the important ability to capture damage progression. That is, during incremental load application, damage can initiate and then progress throughout the composite or laminate. The same failure theories introduced in Chapter 4 are used, in conjunction with the local fields, to assess local failure. This local failure is treated with the so-called subvolume elimination method, wherein, when failure occurs, the subcell (or material in the MT method) stiffness components are set to near zero. An array of PMC and CMC example problems are presented that highlight the various factors influencing progressive damage of unidirectional and laminated composites. The impact of HFGMC's more accurate local fields on the progressive damage predictions is demonstrated through comparison to MT and GMC.

1.6   Suggestions of how to use the book

This book is intended for two audiences: students and design/analysis professionals. It is also intended to be practical in the sense that the objective is to solve real micromechanics-based problems. Therefore an extensive MATLAB code has been written and provided to readers of the book, and the book should be thought of as supporting the application of the MATLAB code. Sufficiently detailed theory is presented to enable one to understand the underlying assumptions and limitations and thus to competently solve micromechanics-based composite problems. Comments in the code link directly to equation numbers in the book, and critical MATLAB code is presented adjacent to the theory in the relevant chapters. Examples and exercises are provided that were designed to be instructive and to reinforce the most important concepts related to composites and micromechanics. As such, most of the exercises involve use of the MATLAB code to solve practical composite problems and enable users to build intuition and confidence in working with composite materials and laminates.

For students, this book and the accompanying MATLAB code form the basis of an advanced undergraduate or graduate course on composite micromechanics. The students should have an understanding of basic solid mechanics and engineering materials (i.e., the concepts of displacement, stress, and strain, Hooke's Law, and thermoelastic material properties). It is not necessary for the student to have prior knowledge of classical lamination theory as this is covered in Chapter 2. However, Chapter 2 can be treated quickly and as more of a review in a course where the students have already covered lamination theory in a previous course. Chapter 3 covers the core of micromechanics. The fundamental tenets are presented, and the Voigt, Reuss, MT, and MOC theories are developed in a consistent manner, all based on each theory's mechanical and thermal strain concentration tensors. Then, Chapter 4 introduces the concept of composite failure based on the local stress and strain fields within the composite constituents, thus demonstrating the importance and utility of having access to these local fields. Therefore a basic course on micromechanics should cover Chapters 2–4. Chapters 5 and 6, which present the more advanced GMC and HFGMC micromechanics theories, can include less (or even very little) theoretical emphasis and focus on application of the MATLAB code. For the more advanced topic of progressive damage, see Chapter 7.

Design/analysis professionals are strongly advised to obtain and exercise the MAT-LAB code provided with the book. This is because far better appreciation, understanding, and insight can be gained by solving real micromechanics-based problems compared to reading and rederiving theory and solving analytical problems. Readers should familiarize themselves with the concept of strain concentration tensors and the example problems in Chapter 3, and see Chapter 4 for a practical application/discussion of composite margins of safety, including a consistent treatment of thermal strains. Chapter 7 shows the application of MT, GMC, and HFGMC to the technologically relevant topic of composite progressive damage and ultimate failure. Most importantly, the MATLAB code provided can be applied to any composite material system and also can be readily extended to address a wide array of industry-relevant composite problems.

References

Aboudi, J., Arnold, S.M., Bednarcyk, B.A., 2013. Micromechanics of Composite Materials: A Generalized Multiscale Analysis Approach. Elsevier, Oxford, United Kingdom.

Barbero, E.J., 2011. Introduction to Composite Materials Design. Second ed. CRC Press, New York, NY, United States.

Benveniste, Y., 1987. A new approach to the application of Mori-Tanaka's theory in composite materials. Mech. Mater. 6, 147–157.

Boyer, R.R., Cotton, J.D., Mohaghegh, M., Schafrik, R.E., 2015. Materials considerations for aerospace applications. MRS Bulletin, 40, 1055–1065.

Carlsson, L.A., Gillespie, J.W., 1990. Delaware Composites Design Encyclopedia. Six volume set. Technomic Publishing Co., Inc., Lancaster, PA, United States.

Christensen, R.M., 1979. Mechanics of Composite Materials. John Wiley & Sons, Inc., New York, NY, United States.

Daniel, I.M., Ishai, O., 1994. Engineering Mechanics of Composite Materials. Oxford University Press.

Dunn, D.G., 2010. The Effect of Fiber Volume Fraction in Hipercomp SiC-SiC Composites. Ph.D. Thesis. Alfred University, Alfred, NY, United States.

Engineering Village (2020): https://www.engineeringvillage.com/search/quick.url, Accessed May 8, 2020.

Fish, J., 2014. Practical Multiscaling. John Wiley & Sons, Chichester, United Kingdom.

JEC Composites (2011): Overview of the Worldwide Composites Industry: 2010–2015. JEC Composites Strategic Study.

Jones, R.M., 1975. Mechanics of Composite Materials. Hemisphere Publishing Corp., New York, NY, United States.

Granta Design Limited, 2012. CES Selector 2012. Version 11.9.9 http://www.grantadesign.com/products/ces/.

Harris, C.E., Starnes, J.H., Shuart, M.J., 2001. An Assessment of the State-of-the-Art in the Design and Manufacturing of Large Composite Structures for Aerospace Vehicles NASA/TM—2001-210844.

Herakovich, C.T., 1998. Mechanics of Fibrous Composites. John Wiley & Sons, Inc., New York, NY, United States.

Herakovich, C.T., 2012. Mechanics of composites: a historical review. Mechancis Research Communications 41, 1–20.

Hyer, M.W., 1998. Stress Analysis of Fiber-Reinforced Composite Materials. WCB McGraw-Hill, New York, NY, United States.

Kanoute, P., Boso, D.P., Chaboche, J.L., Schreer, B.A., 2009. Multiscale methods for composites: a review. Arch. Comput. Methods Eng. 16, 31–75.

MATLAB, 2018. version R2018. The MathWorks Inc, Natick, Massachusetts.

Miracle, D.B., Donaldson, S.L., 2001. ASM Handbook. Vol. 21 Composites. ASM International, Materials Park, OH, United States.

TextileToday (2020): https://www.textiletoday.com.bd/high-performance-synthetic-composites-manufacturing-recent-developments-and-applications/. Accessed April 18, 2020.

Zweben, C., Kelly, A., 2000. Comprehensive Composite Materials. Six volume set. Elsevier, New York, NY, United States.

Lamination theory using macromechanics

2

Chapter outline

2.1   Introduction

Macromechanics assumes that a material can be treated as a homogeneous continuum, with uniform properties, over a given length scale. This macroscale approach treats a composite as an anisotropic material, with its own experimentally measurable properties. Thus, the intrinsic, history dependent, interactive effects of the constituents (e.g., fiber, interface, and matrix) are embedded in the measured experimental properties.

Of course, composite materials have a microstructure and are, by definition, heterogeneous. Therefore, the question becomes, when is it appropriate to idealize the composite as homogeneous? Referring to Fig. 2.1, D is the minimum length associated with the homogeneous material volume. L is the minimum characteristic length of the problem/structure. d is the maximum length associated with microstructural features (e.g., fiber or repeating unit cell) present within the material volume. Clearly, in order for the assumption of homogeneity to be valid, d/D must be small. This is to say that, at the length scale you are observing, the material appears to be homogeneous, with

Practical Micromechanics of Composite Materials. DOI: 10.1016/C2019-0-04193-3

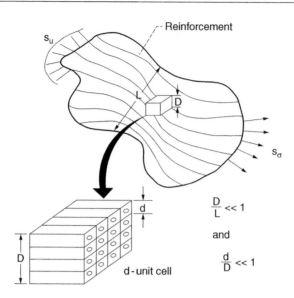

Figure 2.1 Relations between the scales of an RVE, RUC, and structure; S_u = displacement field, S_σ = stress field.

no visible microstructure. For example, if one magnifies a digital image, the pixels become visible. However, if the observation length is sufficiently large, no pixels are distinguishable. Similarly, in a composite material, if d/D is small, the material will "appear" to be homogeneous, and the constituents will be indistinguishable. Therefore, the properties of the material would represent a mixture of the constituent properties. Furthermore, assumed within continuum mechanics is the concept that stress, strain, and material properties are applicable to a material point. However, if microstructure is present but the composite is being idealized as homogenous, then there must be a volume associated with that point, and D is the length associate with this volume. The question is then, when is it appropriate to treat that point as a volume? Although this can never be truly satisfied, as D becomes small compared to L (the minimum characteristic length of the problem/structure), that is to say D/L ≪ 1, the approximation of the homogeneous material volume as a point will become better.

These concepts provide guidance on the applicability of both macromechanics and the associated experimental measurements for composites. For example, macromechanics applies well to composites with small-diameter fibers, in which there are typically hundreds of fibers through the thickness of a ply (typical of PMCs). Here d, the microstructural feature length, will be taken equal to the fiber diameter ($\sim$7 μm), and it will be assumed that a single ply thickness ($\sim$140 μm) is the minimum structural length L. Therefore, if D = 35 μm, then d/D = 0.2 and D/L = 0.25, and the conditions are reasonably satisfied. In contrast, in the case of a titanium matrix composite with large diameter SiC fibers (d $\sim$ 140 μm), the ply thickness is typically $\sim$237 μm as these composites contain only one fiber through the thickness of a ply. If d/D is maintained at 0.2, then D = 700 μm, and assuming a single ply (L = 237 μm) D/L = 3. Clearly this violates the assumptions associated with treating the titanium matrix composite ply as a homogeneous material. However, if we desired a similar D/L ratio as in the PMC case (0.25), then $L = 2800 \, \mu$m which would suggest at least a 12 ply thick laminate is needed. This agrees with previous results found by Aboudi et al. (2013) in Fig. 11.11, p. 816 wherein at least 8 plies were required for the response of a heterogenous titanium matrix composite to approach that of an equivalent homogenous material.

These concepts are also applicable to the experiments used to obtain the composite properties. For example, composite test specimens are typically quite small (e.g., thin tensile coupons with 25 mm by 12.5 mm gage sections), therefore if the composite has large microstructural features, care must be taken. For example in a 2D woven composite ply, the dimensions of the weave pattern can be on the order of cm. It should be noted that failure/strength properties (measurements and predictions) are much more sensitive to violation of these guidelines compared to elastic properties. However, by considering the statistical nature of these material properties, the impact of violating these guidelines can be minimized.

2.2 Linear thermoelastic behavior

Hooke's law provides the relationship between stresses and strains at a point within a material in the linear elastic regime. Herein, the thermal strains at a point within a material, induced by a temperature change (ΔT) from a reference temperature, are also considered. The one-dimensional form of Hooke's law is,

$$\sigma = \mathrm{E} \, (\varepsilon - \alpha \, \Delta T) \tag{2.1}$$

where σ is the stress, ε is the strain, E is the proportionality constant, known as Young's modulus, and α is the coefficient of thermal expansion (CTE). The Young's modulus and CTE are engineering properties that can be obtained from material coupon tests. The shear modulus (G) and the Poisson's ratio (ν) are additional engineering properties that can also be obtained from coupon tests.

If Eq. (2.1) is solved for the strain, the strain form of the one-dimensional Hooke's law can be written as,

$$\varepsilon = \frac{1}{\mathrm{E}} \sigma + \alpha \, \Delta T \tag{2.2}$$

and the proportionality constant is simply one divided by (i.e., the inverse of) the Young's modulus. The inverse of the Young's modulus is referred to as the compliance of the material.

In practice, materials experience multiaxial states of stress and strain. Thus, the one-dimensional equations above must be extended to account for all components of stress and strain, as described below. The discussion is limited to three technologically important types of materials: isotropic (with two independent engineering properties), transversely isotropic (with five independent engineering properties), and orthotropic (with nine independent engineering properties). The physical distinction among these three types of materials is the number of symmetry planes exhibited by the material's internal microstructure.

2.2.1 Isotropic Hooke's law

Isotropic materials exhibit the same behavior in all directions. Hooke's law for a thermoelastic, isotropic material in terms of stress, is given by,

$$\sigma_{ij} = C_{ijkl}\left(\varepsilon_{kl} - \alpha_{ij}\Delta T\right) \quad i,j,k,l = 1,2,3 \tag{2.3}$$

or, in terms of strain,

$$\varepsilon_{ij} = S_{ijkl}\sigma_{kl} + \alpha_{ij}\Delta T \quad i,j,k,l = 1,2,3 \tag{2.4}$$

where σ_{ij} and ε_{ij} are the multiaxial stress and strain components, respectively, C_{ijkl} are the elastic stiffness tensor components, S_{ijkl} are the elastic compliance tensor components, α_{ij} are the coefficients of thermal expansion, ΔT is the temperature change from a reference temperature, and the thermal strain is given by $\varepsilon_{ij}^{th} = \alpha_{ij}\Delta T$. Clearly the stiffness tensor (**C**) is analogous to the Young's modulus (E) in the one-dimensional case, whereas, the compliance tensor (**S**) is analogous to the inverse of the Young's modulus ($1/E$). Note that here, and in the remainder of this book, indicial and matrix notation will be used interchangeably.

Only a single coefficient of thermal expansion is needed for isotropic materials,

$$\begin{aligned}\alpha_{ij} &= \alpha \quad \text{for } i = j\\ \alpha_{ij} &= 0 \quad \text{for } i \neq j\end{aligned} \tag{2.5}$$

In matrix form, the stress form of isotropic Hooke's law can be written as,

$$\begin{bmatrix}\sigma_{11}\\\sigma_{22}\\\sigma_{33}\\\sigma_{23}\\\sigma_{13}\\\sigma_{12}\end{bmatrix} = \begin{bmatrix}C_{11}&C_{12}&C_{12}&0&0&0\\C_{12}&C_{11}&C_{12}&0&0&0\\C_{12}&C_{12}&C_{11}&0&0&0\\0&0&0&\frac{(C_{11}-C_{12})}{2}&0&0\\0&0&0&0&\frac{(C_{11}-C_{12})}{2}&0\\0&0&0&0&0&\frac{(C_{11}-C_{12})}{2}\end{bmatrix}\left(\begin{bmatrix}\varepsilon_{11}\\\varepsilon_{22}\\\varepsilon_{33}\\\gamma_{23}\\\gamma_{13}\\\gamma_{12}\end{bmatrix} - \begin{bmatrix}\alpha\Delta T\\\alpha\Delta T\\\alpha\Delta T\\0\\0\\0\end{bmatrix}\right) \tag{2.6}$$

where the components of the stiffness matrix, **C**,

$$C_{11} = \frac{E(1-v)}{(1+v)(1-2v)}, \quad C_{12} = \frac{Ev}{(1+v)(1-2v)} \tag{2.7}$$

are written in terms of the two independent material engineering constants for the isotropic material, E and v. Note that γ_{ij}, $i \neq j$, are the engineering shear strain components and are related to the tensorial shear strain components by $\gamma_{ij} = 2\varepsilon_{ij}$,

$i \neq j$. Likewise, the strain form of isotropic Hooke's law can be written as,

$$
\begin{bmatrix} \varepsilon_{11} \\ \varepsilon_{22} \\ \varepsilon_{33} \\ \gamma_{23} \\ \gamma_{13} \\ \gamma_{12} \end{bmatrix} = \begin{bmatrix} S_{11} & S_{12} & S_{12} & 0 & 0 & 0 \\ S_{12} & S_{11} & S_{12} & 0 & 0 & 0 \\ S_{12} & S_{12} & S_{11} & 0 & 0 & 0 \\ 0 & 0 & 0 & 2(S_{11}-S_{12}) & 0 & 0 \\ 0 & 0 & 0 & 0 & 2(S_{11}-S_{12}) & 0 \\ 0 & 0 & 0 & 0 & 0 & 2(S_{11}-S_{12}) \end{bmatrix}
$$

$$
\times \begin{bmatrix} \sigma_{11} \\ \sigma_{22} \\ \sigma_{33} \\ \sigma_{23} \\ \sigma_{13} \\ \sigma_{12} \end{bmatrix} + \begin{bmatrix} \alpha\Delta T \\ \alpha\Delta T \\ \alpha\Delta T \\ 0 \\ 0 \\ 0 \end{bmatrix} \tag{2.8}
$$

where the compliance matrix (**S**) components are given by,

$$
S_{11} = \frac{1}{E}, \quad S_{12} = -\frac{v}{E} \tag{2.9}
$$

2.2.2 Transversely isotropic Hooke's law

The stress form of Hooke's law for a transversely isotropic material, wherein the thermomechanical properties are symmetric about the axis (x_1-axis) that is normal to the plane of isotropy, x_2-x_3, is given by,

$$
\begin{bmatrix} \sigma_{11} \\ \sigma_{22} \\ \sigma_{33} \\ \sigma_{23} \\ \sigma_{13} \\ \sigma_{12} \end{bmatrix} = \begin{bmatrix} C_{11} & C_{12} & C_{12} & 0 & 0 & 0 \\ C_{12} & C_{22} & C_{23} & 0 & 0 & 0 \\ C_{12} & C_{23} & C_{22} & 0 & 0 & 0 \\ 0 & 0 & 0 & \frac{(C_{22}-C_{23})}{2} & 0 & 0 \\ 0 & 0 & 0 & 0 & C_{66} & 0 \\ 0 & 0 & 0 & 0 & 0 & C_{66} \end{bmatrix}
$$

$$
\times \left(\begin{bmatrix} \varepsilon_{11} \\ \varepsilon_{22} \\ \varepsilon_{33} \\ \gamma_{23} \\ \gamma_{13} \\ \gamma_{12} \end{bmatrix} - \begin{bmatrix} \alpha_1\Delta T \\ \alpha_2\Delta T \\ \alpha_2\Delta T \\ 0 \\ 0 \\ 0 \end{bmatrix} \right) \tag{2.10}
$$

where the stiffness matrix components can be expressed in terms of five independent elastic constants,

$$
E_1, E_2, v_{12}, v_{23}, G_{12} \tag{2.11}
$$

where E_1 and v_{12} are the axial Young's modulus and Poisson's ratio, respectively; E_2 and v_{23} are the transverse Young's modulus and Poisson's ratio, respectively; G_{12} is the axial shear modulus.

In addition, there are two CTEs in the axial (α_1) and transverse (α_2) directions. The stiffness matrix components are,

$$
C_{11} = E_1 + 4\kappa v_{12}{}^2, \quad C_{12} = 2\kappa v_{12} \tag{2.12}
$$

$$C_{22} = \kappa + \frac{0.5E_2}{(1 + \nu_{23})}, \quad C_{23} = \kappa - \frac{0.5E_1}{(1 + \nu_{23})} \tag{2.13}$$

$$C_{66} = G_{12} \tag{2.14}$$

with

$$\kappa = \frac{0.25E_1}{\left[0.5(1 - \nu_{23})\left(\frac{E_1}{E_2}\right) - \nu_{12}{}^2\right]} \tag{2.15}$$

The strain form of Hooke's law for a transversely isotropic material, with an x_2-x_3 plane of isotropy, is given by,

$$\begin{bmatrix} \varepsilon_{11} \\ \varepsilon_{22} \\ \varepsilon_{33} \\ \gamma_{23} \\ \gamma_{13} \\ \gamma_{12} \end{bmatrix} = \begin{bmatrix} S_{11} & S_{12} & S_{12} & 0 & 0 & 0 \\ S_{12} & S_{22} & S_{23} & 0 & 0 & 0 \\ S_{12} & S_{23} & S_{22} & 0 & 0 & 0 \\ 0 & 0 & 0 & 2(S_{22} - S_{23}) & 0 & 0 \\ 0 & 0 & 0 & 0 & S_{66} & 0 \\ 0 & 0 & 0 & 0 & 0 & S_{66} \end{bmatrix} \begin{bmatrix} \sigma_{11} \\ \sigma_{22} \\ \sigma_{33} \\ \sigma_{23} \\ \sigma_{13} \\ \sigma_{12} \end{bmatrix} + \begin{bmatrix} \alpha_1 \Delta T \\ \alpha_2 \Delta T \\ \alpha_2 \Delta T \\ 0 \\ 0 \\ 0 \end{bmatrix} \tag{2.16}$$

where

$$S_{11} = \frac{1}{E_1}, \quad S_{12} = -\frac{\nu_{12}}{E_1}, \quad S_{22} = \frac{1}{E_2}, \quad S_{23} = -\frac{\nu_{23}}{E_2} \tag{2.17}$$

$$2(S_{22} - S_{23}) = \frac{2(1 + \nu_{23})}{E_2} = \frac{1}{G_{23}}, \quad S_{66} = \frac{1}{G_{12}} \tag{2.18}$$

and G_{23} is the transverse shear modulus.

GetPlyProps.m, which is listed in MATLAB Function 2.1 and shown below, is used to specify the transversely isotropic mechanical properties shown in Eq. (2.11), along with the CTEs, for ply level materials. This is a function that should be used to introduce new ply level materials within a laminate analysis.

MATLAB Function 2.1 – GetPlyProps.m

This function defines effective ply material properties within a structure known as plyprops. This is where one can easily add additional materials; be they monolithic, effective composite, or obtained from some lower length scale method, for example micromechanics.

```
function [plyprops] = GetPlyProps(Geometry)

% ++++++++++++++++++++++++++++++++++++++++++++++++++++++++++++++++++++++++++++++
% Copyright 2020 United States Government as represented by the Administrator of the
% National Aeronautics and Space Administration. No copyright is claimed in the
% United States under Title 17, U.S. Code. All Other Rights Reserved. By downloading
% or using this software you agree to the terms of NOSA v1.3. This code was prepared
% by Drs. B.A. Bednarcyk and S.M. Arnold to complement the book "Practical
% Micromechanics of Composite Materials" during the course of their government work.
% ++++++++++++++++++++++++++++++++++++++++++++++++++++++++++++++++++++++++++++++
% Purpose: Provides material properties to Lamination Theory
% - Ply Properties can be defined as transversely isotropic directly or calculated
%   via micromechanics
% Input:
% - Geometry: Cell array containing laminate definition variables
% Output:
% - plyprops: Cell array containing ply material properties and allowables
% ++++++++++++++++++++++++++++++++++++++++++++++++++++++++++++++++++++++++++++++

NmatMax = 200;
plyprops = cell(1,NmatMax);

% -- Determine which mats are used in any problem
for NP = 1: length(Geometry)
    if isfield(Geometry{NP},'plymat')
        for I = 1: length(Geometry{NP}.plymat)
            mat = Geometry{NP}.plymat(I);

            plyprops{mat}.used = true;
        end
    end
end

% -- Monolithic Transversely Isotropic Materials
plyprops{1}.name = "IM7/8552";
plyprops{1}.E1 = 146.8e3;
plyprops{1}.E2 = 8.69e3;
plyprops{1}.G12 = 5.16e3;
plyprops{1}.v12 = 0.32;
plyprops{1}.a1 = -0.1e-6;
plyprops{1}.a2 = 31.e-6;

plyprops{2}.name = "Glass/Epoxy";
plyprops{2}.E1 = 43.5e3;
plyprops{2}.E2 = 11.5e3;
plyprops{2}.G12 = 3.45e3;
plyprops{2}.v12 = 0.27;
plyprops{2}.a1 = 6.84e-06;
plyprops{2}.a2 = 29.e-06;

plyprops{3}.name = "SCS-6/Ti-15-3";
plyprops{3}.E1 = 221.e3;
plyprops{3}.E2 = 145.e3;
plyprops{3}.G12 = 53.2e3;
plyprops{3}.v12 = 0.27;
plyprops{3}.a1 = 6.15e-06;
plyprops{3}.a2 = 7.9e-06;

plyprops{4}.name = "SiC/SiC";
plyprops{4}.E1 = 300.e3;
plyprops{4}.E2 = 152.4e3;
plyprops{4}.G12 = 69.5e3;
plyprops{4}.v12 = 0.17;
plyprops{4}.a1 = 4.88e-06;
plyprops{4}.a2 = 4.91e-06;

plyprops{5}.name = "Aluminum";
plyprops{5}.E1 = 70.e3;
plyprops{5}.E2 = 70.e3;
plyprops{5}.G12 = 26.923e3;
plyprops{5}.v12 = 0.3;
plyprops{5}.a1 = 23.e-06;
plyprops{5}.a2 = 23.e-06;

end
```

2.2.3 Orthotropic Hooke's law

Orthotropic thermoelastic materials have three mutually orthogonal planes of reflection symmetry, resulting in nine independent elastic constants. The stress form of orthotropic Hooke's law is given by,

$$
\begin{bmatrix} \sigma_{11} \\ \sigma_{22} \\ \sigma_{33} \\ \sigma_{23} \\ \sigma_{13} \\ \sigma_{12} \end{bmatrix} = \begin{bmatrix} C_{11} & C_{12} & C_{13} & 0 & 0 & 0 \\ C_{12} & C_{22} & C_{23} & 0 & 0 & 0 \\ C_{13} & C_{23} & C_{33} & 0 & 0 & 0 \\ 0 & 0 & 0 & C_{44} & 0 & 0 \\ 0 & 0 & 0 & 0 & C_{55} & 0 \\ 0 & 0 & 0 & 0 & 0 & C_{66} \end{bmatrix} \left(\begin{bmatrix} \varepsilon_{11} \\ \varepsilon_{22} \\ \varepsilon_{33} \\ \gamma_{23} \\ \gamma_{13} \\ \gamma_{12} \end{bmatrix} - \begin{bmatrix} \alpha_1 \Delta T \\ \alpha_2 \Delta T \\ \alpha_3 \Delta T \\ 0 \\ 0 \\ 0 \end{bmatrix} \right) \tag{2.19}
$$

where α_i are the three independent CTEs, and the components of the stiffness matrix are related to the engineering constants by,

$$
C_{11} = \frac{1 - \nu_{23}\nu_{32}}{E_2 E_3 \Delta} \tag{2.20}
$$

$$
C_{12} = \frac{\nu_{12} + \nu_{32}\nu_{13}}{E_1 E_3 \Delta} \tag{2.21}
$$

$$
C_{13} = \frac{\nu_{13} + \nu_{12}\nu_{23}}{E_1 E_2 \Delta} \tag{2.22}
$$

$$
C_{22} = \frac{1 - \nu_{13}\nu_{31}}{E_1 E_3 \Delta} \tag{2.23}
$$

$$
C_{23} = \frac{\nu_{23} + \nu_{21}\nu_{13}}{E_1 E_2 \Delta} \tag{2.24}
$$

$$
C_{33} = \frac{1 - \nu_{12}\nu_{21}}{E_1 E_2 \Delta} \tag{2.25}
$$

$$
C_{44} = G_{23} \tag{2.26}
$$

$$
C_{55} = G_{13} \tag{2.27}
$$

$$
C_{66} = G_{12} \tag{2.28}
$$

and,

$$
\Delta = \frac{1 - \nu_{12}\nu_{21} - \nu_{23}\nu_{32} - \nu_{31}\nu_{13} - 2\nu_{21}\nu_{32}\nu_{13}}{E_1 E_2 E_3} \tag{2.29}
$$

where E_1, E_2, and E_3 are the Young's moduli in the three orthogonal direction, v_{ij} are the Poisson's ratios, and G_{ij} are the shear moduli. Note that,

$$v_{32} = \frac{E_3 v_{23}}{E_2} \tag{2.30}$$

$$v_{31} = \frac{E_3 v_{13}}{E_1} \tag{2.31}$$

$$v_{21} = \frac{E_2 v_{12}}{E_1} \tag{2.32}$$

The strain form of orthotropic Hooke's law is given by,

$$
\begin{bmatrix} \varepsilon_{11} \\ \varepsilon_{22} \\ \varepsilon_{33} \\ \gamma_{23} \\ \gamma_{13} \\ \gamma_{12} \end{bmatrix} =
\begin{bmatrix}
S_{11} & S_{12} & S_{13} & 0 & 0 & 0 \\
S_{12} & S_{22} & S_{23} & 0 & 0 & 0 \\
S_{13} & S_{23} & S_{33} & 0 & 0 & 0 \\
0 & 0 & 0 & S_{44} & 0 & 0 \\
0 & 0 & 0 & 0 & S_{55} & 0 \\
0 & 0 & 0 & 0 & 0 & S_{66}
\end{bmatrix}
\begin{bmatrix} \sigma_{11} \\ \sigma_{22} \\ \sigma_{33} \\ \sigma_{23} \\ \sigma_{13} \\ \sigma_{12} \end{bmatrix} +
\begin{bmatrix} \alpha_1 \Delta T \\ \alpha_2 \Delta T \\ \alpha_3 \Delta T \\ 0 \\ 0 \\ 0 \end{bmatrix}
\tag{2.33}
$$

where,

$$S_{11} = \frac{1}{E_1}, \quad S_{12} = -\frac{v_{12}}{E_1}, \quad S_{13} = -\frac{v_{13}}{E_1} \tag{2.34}$$

$$S_{22} = \frac{1}{E_2}, \quad S_{23} = -\frac{v_{23}}{E_2}, \quad S_{33} = \frac{1}{E_3} \tag{2.35}$$

$$S_{44} = \frac{1}{G_{23}}, \quad S_{55} = \frac{1}{G_{13}}, \quad S_{66} = \frac{1}{G_{12}} \tag{2.36}$$

2.3   Lamination theory

Lamination theory, often referred to as Classical Lamination Theory (CLT), is a standard method for treating the mechanics of the most common form of composite materials, thin plates or shells (see for example the texts by Jones, 1975; Agarwal and Broutman, 1980; Daniel and Ishai, 1994, Herakovich, 1998; Hyer, 1998; Reddy, 2003). As in all plate and shell theories, the objective is to simplify what is truly a

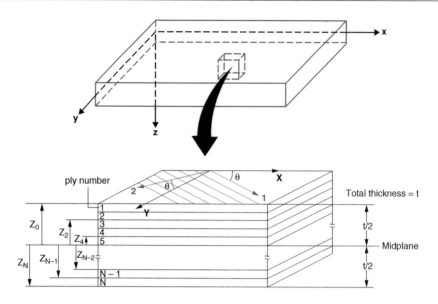

Figure 2.2 Lamination theory geometry, stacking sequence, and coordinate systems.

three-dimensional mechanics problem down to something tractable and easily handled using relatively straightforward mathematics. Lamination theory is the simplest of such theories, but it draws its greatest strengths from its simplicity. It enables a composite composed of multiple arbitrarily oriented layers, or plies (see Fig. 2.2), to be treated mechanically as a point within a plate or shell. It provides the relationship between the stress-like quantities and the strain-like quantities for a given point. Lamination theory accomplishes this based on the properties and geometry of the plies that make up the laminate at that point through homogenization. Further, the mechanics of each layer or ply can be determined through localization. That is, the stresses and strains in each ply, at every point, within a composite structure can be calculated from the plate/shell stress-like and strain-like quantities. Of course, given appropriate ply-level material stress/strain allowables, limits, or strengths, it is then possible to make a prediction of whether or not the composite will fail (or exceed the allowable limits). As demonstrated in Chapter 4, this capability makes lamination theory extremely useful. It can be used to design composite laminates to sustain structural loading, and indeed, this is a major use case for the theory world-wide. To take full advantage of lamination theory, its implementation as a computer code is necessary.

The basic assumptions of lamination theory (discussed in more detail below) are:

1. Each layer is a thin, homogeneous material with known effective properties
2. Each layer is in a state of plane stress
3. The laminate consists of perfectly bonded layers/plies
4. The laminate deforms according to the Kirchhoff-Love assumptions for bending and stretch- ing of thin plates, that is,

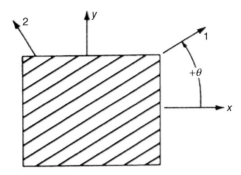

Figure 2.3 Ply local coordinate system, which is aligned with the fiber direction, and the ply orientation angle, θ.

  a. Normals to the midplane remain straight and normal to the deformed midplane after deformation
  b. Normals to the midplane do not change length

Additionally, throughout this text, it is assumed that the ply materials are linear thermoelastic and isotropic, transversely isotropic, or orthotropic. The laminate may be subjected to inplane force and moment resultants, midplane strains and curvatures, and a constant temperature change. The global laminate coordinate system (x, y, z), shown in Fig. 2.2, originates at the laminate midplane. The ply number is denoted by k, with $k = 1$ being the top (negativemost z-location) ply and $k = N$ being the bottom (positivemost z-location) ply, where N is the total number of plies in the laminate. The total thickness of the laminate is denoted by t, and the thickness of each ply is t_k. Each ply has its own local material coordinate system $(1, 2, 3)$, where 1 is always associated with the ply fiber direction, 2 is the in-plane direction transverse to the fibers, and 3 is the through-thickness transverse direction, always aligned with z. The ply orientation (θ_k) is defined by the angle between the fiber direction (1) and the global x-direction (see Fig. 2.2).

2.3.1   Mechanics of a ply

The plies that make up the laminate are also often referred to as layers or *laminae* (which is the plural form of *lamina*). The mechanical behavior of each ply is treated in its local $(1, 2, 3)$ coordinate system, with the 1-direction aligned with the ply-fiber direction, see Fig. 2.3. Because this local coordinate system is aligned with the fiber direction of the composite material of this ply, it is referred to as the *principal material coordinates* of the ply.

It is assumed that the ply-composite material is isotropic, transversely isotropic, or orthotropic, with the associated Hooke's law equations defined in Section 2.2 above. The assumption of *plane stress* in the ply requires that the three through-thickness stress components are zero,

$$\sigma_{13} = \sigma_{23} = \sigma_{33} = 0 \tag{2.37}$$

This fact enables the reduction of Hooke's law to the three in-plane stress and strain components as follows. Recall the strain form of orthotropic Hooke's law, Eq. (2.33), which, using Eq. (2.37), now becomes,

$$
\begin{bmatrix} \varepsilon_{11} \\ \varepsilon_{22} \\ \varepsilon_{33} \\ \gamma_{23} \\ \gamma_{13} \\ \gamma_{12} \end{bmatrix}
=
\begin{bmatrix}
S_{11} & S_{12} & S_{13} & 0 & 0 & 0 \\
S_{12} & S_{22} & S_{23} & 0 & 0 & 0 \\
S_{13} & S_{23} & S_{33} & 0 & 0 & 0 \\
0 & 0 & 0 & S_{44} & 0 & 0 \\
0 & 0 & 0 & 0 & S_{55} & 0 \\
0 & 0 & 0 & 0 & 0 & S_{66}
\end{bmatrix}
\begin{bmatrix} \sigma_{11} \\ \sigma_{22} \\ 0 \\ 0 \\ 0 \\ \sigma_{12} \end{bmatrix}
+
\begin{bmatrix} \alpha_1 \Delta T \\ \alpha_2 \Delta T \\ \alpha_3 \Delta T \\ 0 \\ 0 \\ 0 \end{bmatrix}
\tag{2.38}
$$

Using Eq. (2.38) to express ε_{33} in terms of σ_{11} and σ_{22} provides the implied through-thickness normal strain for the ply,

$$\varepsilon_{33} = S_{13}\sigma_{11} + S_{23}\sigma_{22} + \alpha_3 \Delta T \tag{2.39}$$

or, in terms of stiffness and strain components [from Eq. (2.19)],

$$
\begin{aligned}
\varepsilon_{33} = &-\frac{C_{13}}{C_{33}}(\varepsilon_{11} - \alpha_1 \Delta T) \\
&- \frac{C_{23}}{C_{33}}(\varepsilon_{22} - \alpha_2 \Delta T) + \alpha_3 \Delta T
\end{aligned}
\tag{2.40}
$$

Solving Eq. (2.38) for σ_{11} and σ_{22} in terms of ε_{11} and ε_{22} [using Eq. (2.40)] produces the *reduced* Hooke's law for the ply as,

$$
\begin{bmatrix} \sigma_{11} \\ \sigma_{22} \\ \tau_{12} \end{bmatrix}
=
\begin{bmatrix}
Q_{11} & Q_{12} & 0 \\
Q_{12} & Q_{22} & 0 \\
0 & 0 & Q_{66}
\end{bmatrix}
\begin{bmatrix} \varepsilon_{11} - \alpha_1 \Delta T \\ \varepsilon_{22} - \alpha_2 \Delta T \\ \gamma_{12} \end{bmatrix}
\tag{2.41}
$$

where Q_{ij} are the components of the *reduced stiffness matrix*, **Q**,

$$Q_{11} = \frac{S_{22}}{S_{11}S_{22} - S_{12}^2} = \frac{E_1}{1 - \nu_{12}\nu_{21}} \qquad\qquad Q_{12} = -\frac{S_{12}}{S_{11}S_{22} - S_{12}^2} = \frac{\nu_{12}E_2}{1 - \nu_{12}\nu_{21}}$$

$$\tag{2.42}$$

$$Q_{22} = \frac{S_{11}}{S_{11}S_{22} - S_{12}^2} = \frac{E_2}{1 - \nu_{12}\nu_{21}} \qquad\qquad Q_{66} = \frac{1}{S_{66}} = G_{12}$$

and E_1 and E_2 are the ply Young's moduli, ν_{12} and $\nu_{21} = \frac{E_2 \nu_{12}}{E_1}$ are the ply Poisson's ratios, G_{12} is the ply shear modulus, α_1 and α_2 are the ply CTEs, and ΔT is the change in temperature from a reference temperature. Note that, for consistency with common lamination theory nomenclature, the ply in-plane shear stress, σ_{12} is denoted as τ_{12}, although either can be used interchangeably.

2.3.2 Ply stress–strain coordinate transformations

To determine the laminate response from the ply-level equations, all of the ply-level equations must be in the same coordinate system rather than the individual local ply $(1, 2, 3)$ coordinate systems. Therefore, coordinate transformation equations are required. These equations are used to transform the stress, strain, CTE, and reduced stiffness components from Eq. (2.41) between the principle material $(1, 2, 3)$ coordinate system of each ply and the global laminate (x, y, z) coordinate system. The coordinate transformation is a specialized rotation by the angle θ in the 1-2 (and x-y) plane about the 3- (or z-) coordinate axis, as depicted in Fig. 2.3. Note that a positive angle is defined as counter clockwise rotation from the laminate x-axis to the ply 1-axis.

The transformations from laminate $(x$-$y)$ stresses and strains to the local ply stress and strains are given by,

$$\begin{bmatrix} \sigma_{11} \\ \sigma_{22} \\ \tau_{12} \end{bmatrix} = \mathbf{T}_1 \begin{bmatrix} \sigma_x \\ \sigma_y \\ \tau_{xy} \end{bmatrix} \qquad \begin{bmatrix} \varepsilon_{11} \\ \varepsilon_{22} \\ \gamma_{12} \end{bmatrix} = \mathbf{T}_2 \begin{bmatrix} \varepsilon_x \\ \varepsilon_y \\ \gamma_{xy} \end{bmatrix} \tag{2.43}$$

where,

$$\mathbf{T}_1 = \begin{bmatrix} \cos^2\theta & \sin^2\theta & 2\cos\theta \sin\theta \\ \sin^2\theta & \cos^2\theta & -2\cos\theta \sin\theta \\ -\cos\theta \sin\theta & \cos\theta \sin\theta & \cos^2\theta - \sin^2\theta \end{bmatrix} \tag{2.44}$$

$$\mathbf{T}_2 = \begin{bmatrix} \cos^2\theta & \sin^2\theta & \cos\theta \sin\theta \\ \sin^2\theta & \cos^2\theta & -\cos\theta \sin\theta \\ -2\cos\theta \sin\theta & 2\cos\theta \sin\theta & \cos^2\theta - \sin^2\theta \end{bmatrix} \tag{2.45}$$

Note that the MATLAB implementation of Eqs. (2.44) and (2.45) is given in MATLAB Function 2.2 – GetT.m.

In addition, the reduced stiffness matrix, $\mathbf{Q}$, in Eq. (2.41) must be transformed to the laminate $(x$-$y)$ coordinate system,

$$\bar{\mathbf{Q}} = \mathbf{T}_1^{-1} \mathbf{Q} \mathbf{T}_2 \tag{2.46}$$

where $\bar{\mathbf{Q}}$ is the transformed reduced stiffness matrix (see MATLAB Function 2.3 – Qbar.m), which is the stiffness of the ply in the laminate $(x$-$y)$ coordinate system. Therefore, Hooke's law for the ply in the laminate $(x$-$y)$ coordinate system is given by,

$$\begin{bmatrix} \sigma_x \\ \sigma_y \\ \tau_{xy} \end{bmatrix} = \begin{bmatrix} \bar{Q}_{11} & \bar{Q}_{12} & \bar{Q}_{16} \\ \bar{Q}_{12} & \bar{Q}_{22} & \bar{Q}_{26} \\ \bar{Q}_{16} & \bar{Q}_{26} & \bar{Q}_{66} \end{bmatrix} \begin{bmatrix} \varepsilon_x - \alpha_x \Delta T \\ \varepsilon_y - \alpha_y \Delta T \\ \gamma_{xy} - \alpha_{xy} \Delta T \end{bmatrix} \tag{2.47}$$

with ply CTEs given in laminate (x-y) coordinates as,

$$\begin{bmatrix} \alpha_x \\ \alpha_y \\ \alpha_{xy} \end{bmatrix} = \mathbf{T}_2^{-1} \begin{bmatrix} \alpha_1 \\ \alpha_2 \\ 0 \end{bmatrix} \tag{2.48}$$

As evident in Eq. (2.48), an orthotropic (or transversely isotropic) ply may have a nonzero shear CTE, α_{xy}, in laminate (x-y) coordinates, even though no shear CTE exists in the ply (1-2) coordinates. This shear CTE, α_{xy}, is an engineering shear quantity, meaning $\alpha_{xy}\Delta T$ is an engineering shear strain.

MATLAB Function 2.2 – GetT.m

This function calculates the two transformation matrices, $\mathbf{T}_1$ and $\mathbf{T}_2$, see Eqs. (2.44) and (2.45).

```
function [T1,T2] = GetT(A)
% ++++++++++++++++++++++++++++++++++++++++++++++++++++++++++++++++++++++++++
% Copyright 2020 United States Government as represented by the Administrator of the
% National Aeronautics and Space Administration. No copyright is claimed in the
% United States under Title 17, U.S. Code. All Other Rights Reserved. By downloading
% or using this software you agree to the terms of NOSA v1.3. This code was prepared
% by Drs. B.A. Bednarcyk and S.M. Arnold to complement the book "Practical
% Micromechanics of Composite Materials" during the course of their government work.
% ++++++++++++++++++++++++++++++++++++++++++++++++++++++++++++++++++++++++++
% Purpose: Calculate the two transformation matrices T1 and T2
% Input:
% - A: Ply orientation angle (deg) (theta - see Eqs. 2.44 and 2.45)
% Output:
% - T1 stress transformation matrix (Eq. 2.44)
% - T2 strain transformation matrix (Eq. 2.45)
% ++++++++++++++++++++++++++++++++++++++++++++++++++++++++++++++++++++++++++

m = cosd(A);
n = sind(A);

% -- T1 is transformation matrix for stress that maps from material to
%    global coordinate system (Eq. 2.44)
T1 = [m.^2, n.^2, 2*n.*m; ...
      n.^2, m.^2, -2*m.*n; ...
      -m.*n, m.*n, (m.^2)-(n.^2)];

% -- T2 is transformation matrix for strains that maps from material to
%    global coordinate system (Eq. 2.45)
T2 = [m.^2, n.^2, n.*m; ...
      n.^2, m.^2, -m.*n; ...
      -2*m.*n, 2*m.*n, (m.^2)-(n.^2)];

end
```

MATLAB Function 2.3 – `Qbar.m`

This function constructs the transformed reduced stiffness matrix, `Qbar`, matrices for a given ply, see Eqs. (2.46).

```
function Q_bar = Qbar(A, plyprops, matID)
% ++++++++++++++++++++++++++++++++++++++++++++++++++++++++++++++++++++++++++++
% Copyright 2020 United States Government as represented by the Administrator of the
% National Aeronautics and Space Administration. No copyright is claimed in the
% United States under Title 17, U.S. Code. All Other Rights Reserved. By downloading
% or using this software you agree to the terms of NOSA v1.3. This code was prepared
% by Drs. B.A. Bednarcyk and S.M. Arnold to complement the book "Practical
% Micromechanics of Composite Materials" during the course of their government work.
% ++++++++++++++++++++++++++++++++++++++++++++++++++++++++++++++++++++++++++++
% Purpose: Calculate the transformed reduced stiffness matrix
% Input:
% - A: Ply orientation angle theta (deg)
% - plyprops: Cell array containing ply material properties
% - matID: ply-level material id number
% Output:
% - Q_bar: transformed reduced stiffness matrix
% ++++++++++++++++++++++++++++++++++++++++++++++++++++++++++++++++++++++++++++

E1 = plyprops{matID}.E1;
E2 = plyprops{matID}.E2;
G12 = plyprops{matID}.G12;
v12 = plyprops{matID}.v12;

% -- Define plane-stress compliance matrix (Eqs. 2.17, 2.18)
S11 = 1/E1;
S12 = -v12/E1;
S22 = 1/E2;
S66 = 1/G12;

S = [S11 S12 0;S12 S22 0;0 0 S66];

% -- Define the reduced stiffness matrix in material coordinates (1,2,3)
Q = inv(S); % -- Equivalent to Eqs. 2.42

% -- Determine transformation matrices
[T1,T2] = GetT(A);

% -- Calculate transformed reduced stiffness matrix (Eq. 2.46)
Q_bar = inv(T1)*Q*T2;

end
```

2.3.3 Strain-displacement relations

According to the Kirchhoff-Love (Kirchhoff, 1850; Love, 1888) hypothesis for plates, a plane cross-section that is originally perpendicular to the midplane of the laminate (an x-y plane) remains planar and perpendicular to the midplane when the laminate is subjected to bending and extension. Furthermore, there is no change in length normal to the midplane. This implies not only that the out-of-plane shear strains are zero, but also that the through-thickness normal strain, ε_{33}, is zero. This is obviously inconsistent

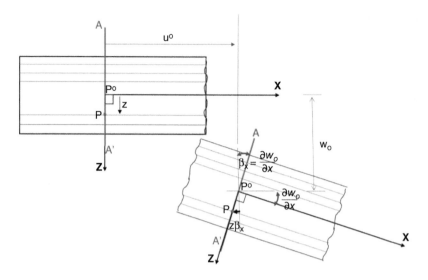

Figure 2.4 Kinematics of deformation of a laminate as viewed in the x-z plane.

with the nonzero ε_{33} implied by the plane stress assumption, see Eq. (2.39), but this inconsistency cannot be resolved within the context of the Kirchhoff-Love hypothesis. If the through-thickness strain, ε_{33}, is needed, it can be obtained from Eqs. (2.39) or (2.40).

As shown in Fig. 2.4, the x-direction rotation (slope) of the deformed midplane is β_x. If β_x is assumed to be small (that is $\partial w_0/\partial x \le 0.25$, which is approximately 15°), the tangent of β_x is approximately β_x (within a 2% error) so,

$$\tan \beta_x = \frac{\partial w_0}{\partial x} \approx \beta_x \qquad (2.49)$$

where w_0 is the midplane z-displacement. Similarly, for bending in the y-direction,

$$\tan \beta_y = \frac{\partial w_0}{\partial y} \approx \beta_y \qquad (2.50)$$

The in-plane displacement components in the x and y directions, u and v, can therefore be written as the sum of the midplane displacements, u_0 and v_0, and the displacements due to the rotations, β_x and β_y,

$$u = u_0 - z\frac{\partial w_0}{\partial x}, \quad v = v_0 - z\frac{\partial w_0}{\partial y} \qquad (2.51)$$

It can be seen in Fig. 2.1 that positive $\frac{\partial w_0}{\partial x}$ (and $\frac{\partial w_0}{\partial y}$) results in negative contributions to u (and v), as reflected in Eq. (2.51).

Using the standard strain-displacement relations,

$$\varepsilon_x = \frac{\partial u}{\partial x}, \qquad \varepsilon_y = \frac{\partial v}{\partial y}, \qquad \gamma_{xy} = \frac{\partial u}{\partial y} + \frac{\partial v}{\partial x} \tag{2.52}$$

the laminate in-plane strain components are written as

$$\begin{bmatrix} \varepsilon_x \\ \varepsilon_y \\ \gamma_{xy} \end{bmatrix} = \begin{bmatrix} \varepsilon_x^0 \\ \varepsilon_y^0 \\ \gamma_{xy}^0 \end{bmatrix} + z \begin{bmatrix} \kappa_x^0 \\ \kappa_y^0 \\ \kappa_{xy}^0 \end{bmatrix} \tag{2.53}$$

where

$$\begin{bmatrix} \varepsilon_x^0 \\ \varepsilon_y^0 \\ \gamma_{xy}^0 \end{bmatrix} = \begin{bmatrix} \dfrac{\partial u_0}{\partial x} \\[2ex] \dfrac{\partial v_0}{\partial y} \\[2ex] \dfrac{\partial u_0}{\partial y} + \dfrac{\partial v_0}{\partial x} \end{bmatrix} \tag{2.54}$$

$$\begin{bmatrix} \kappa_x^0 \\ \kappa_y^0 \\ \kappa_{xy}^0 \end{bmatrix} = \begin{bmatrix} -\dfrac{\partial^2 w_0}{\partial x^2} \\[2ex] -\dfrac{\partial^2 w_0}{\partial y^2} \\[2ex] -2\dfrac{\partial^2 w_0}{\partial x \partial y} \end{bmatrix} \tag{2.55}$$

ε_x^0, ε_y^0, and γ_{xy}^0, are the midplane strains, and κ_x^0, κ_y^0, and κ_{xy}^0, are the midplane curvatures (with units of 1/length).

Assuming that the midplane strains and curvatures do not vary with x and y (and are thus constant over the reference surface of a plate), but the midplane displacements, u_0, v_0 and w_0 are functions of x and y, Eqs. (2.54) and (2.55) can be integrated to determine the midplane displacements, following Hyer (1998):

$$u_0(x, y) = \varepsilon_x^0 x + C_1 y + C_2 \tag{2.56}$$

$$v_0(x, y) = \varepsilon_y^0 y + \left(\gamma_{xy}^0 - C_1 \right) x + C_3 \tag{2.57}$$

and

$$w_0(x, y) = -\frac{1}{2} \left(\kappa_x^0 x^2 + \kappa_y^0 y^2 + \kappa_{xy}^0 xy \right) + C_4 x + C_5 y + C_6 \tag{2.58}$$

where C_1 through C_6 are constants of integration that can be determined from the boundary conditions on the plate. For example, assuming the center of the plate is

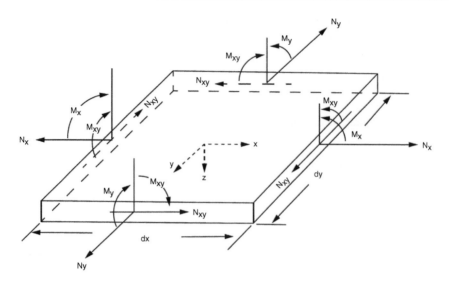

Figure 2.5 Generalized force and moments resultants used in CLT.

pinned in all directions $(u_0 = v_0 = w_0 = 0$ at $x = y = 0)$ provides $C_2 = C_3 = C_6 = 0$ and that the slopes $\partial w_0/\partial x = \partial w_0/\partial y = 0$ at the center as well, provides $C_4 = C_5 = 0$. Finally, arbitrarily suppressing rigid body rotation about the z-axis, results in $C_1 = 0$. Equations (2.56)–(2.58) are general expressions for the laminate plate displacements, provided that the midplane strains and curvatures are constant (independent of x and y) over the plate.

2.3.4   Laminate constitutive relations

Combining Eqs. (2.47) and (2.53) yields

$$
\begin{bmatrix} \sigma_x^k \\ \sigma_y^k \\ \tau_{xy}^k \end{bmatrix} = \begin{bmatrix} \bar{Q}_{11}^k & \bar{Q}_{12}^k & \bar{Q}_{16}^k \\ \bar{Q}_{12}^k & \bar{Q}_{22}^k & \bar{Q}_{26}^k \\ \bar{Q}_{16}^k & \bar{Q}_{26}^k & \bar{Q}_{66}^k \end{bmatrix} \left\{ \begin{bmatrix} \varepsilon_x^0 \\ \varepsilon_y^0 \\ \gamma_{xy}^0 \end{bmatrix} + z \begin{bmatrix} \kappa_x^0 \\ \kappa_y^0 \\ \kappa_{xy}^0 \end{bmatrix} - \begin{bmatrix} \alpha_x^k \\ \alpha_y^k \\ \alpha_{xy}^k \end{bmatrix} \Delta T \right\} \qquad (2.59)
$$

The laminate force resultants $(N_x, N_y,$ and $N_{xy})$, which are forces per unit length and moment resultants $(M_x, M_y,$ and $M_{xy})$, which are moments per unit length, described in Fig. 2.5, are obtained by integrating ply stresses over the laminate thickness t. Note that the force in the x-direction is $F_x = N_x\, dy$, the force in the y-direction is $F_y = N_y\, dx$, and the shear forces are $F_{xy} = N_{xy}\, dx$ and $F_{yx} = N_{yx}\, dy$. Similar expressions apply to the relations between moments and moment resultants.

The integration over the thickness, t, is performed in a piecewise manner across each layer as indicated below.

$$\begin{bmatrix} N_x \\ N_y \\ N_{xy} \end{bmatrix} = \int_{\frac{-t}{2}}^{\frac{t}{2}} \begin{bmatrix} \sigma_x \\ \sigma_y \\ \tau_{xy} \end{bmatrix} dz = \sum_{k=1}^{N} \int_{z_{k-1}}^{z_k} \begin{bmatrix} \sigma_x^k \\ \sigma_y^k \\ \tau_{xy}^k \end{bmatrix} dz \tag{2.60}$$

$$\begin{bmatrix} M_x \\ M_y \\ M_{xy} \end{bmatrix} = \int_{\frac{-t}{2}}^{\frac{t}{2}} \begin{bmatrix} \sigma_x \\ \sigma_y \\ \tau_{xy} \end{bmatrix} z\,dz = \sum_{k=1}^{N} \int_{z_{k-1}}^{z_k} \begin{bmatrix} \sigma_x^k \\ \sigma_y^k \\ \tau_{xy}^k \end{bmatrix} z\,dz \tag{2.61}$$

Substituting (2.59) into (2.60) and (2.61) yields,

$$\begin{bmatrix} N_x \\ N_y \\ N_{xy} \end{bmatrix} = \sum_{k=1}^{N} \begin{bmatrix} \bar{Q}_{11}^k & \bar{Q}_{12}^k & \bar{Q}_{16}^k \\ \bar{Q}_{12}^k & \bar{Q}_{22}^k & \bar{Q}_{26}^k \\ \bar{Q}_{16}^k & \bar{Q}_{26}^k & \bar{Q}_{66}^k \end{bmatrix}$$
$$\times \left\{ \int_{z_{k-1}}^{z_k} \begin{bmatrix} \varepsilon_x^0 \\ \varepsilon_y^0 \\ \gamma_{xy}^0 \end{bmatrix} dz + \int_{z_{k-1}}^{z_k} \begin{bmatrix} \kappa_x^0 \\ \kappa_y^0 \\ \kappa_{xy}^0 \end{bmatrix} z\,dz - \int_{z_{k-1}}^{z_k} \begin{bmatrix} \alpha_x^k \\ \alpha_y^k \\ \alpha_{xy}^k \end{bmatrix} \Delta T\,dz \right\} \tag{2.62}$$

$$\begin{bmatrix} M_x \\ M_y \\ M_{xy} \end{bmatrix} = \sum_{k=1}^{N} \begin{bmatrix} \bar{Q}_{11}^k & \bar{Q}_{12}^k & \bar{Q}_{16}^k \\ \bar{Q}_{12}^k & \bar{Q}_{22}^k & \bar{Q}_{26}^k \\ \bar{Q}_{16}^k & \bar{Q}_{26}^k & \bar{Q}_{66}^k \end{bmatrix}$$
$$\times \left\{ \int_{z_{k-1}}^{z_k} \begin{bmatrix} \varepsilon_x^0 \\ \varepsilon_y^0 \\ \gamma_{xy}^0 \end{bmatrix} z\,dz + \int_{z_{k-1}}^{z_k} \begin{bmatrix} \kappa_x^0 \\ \kappa_y^0 \\ \kappa_{xy}^0 \end{bmatrix} z^2\,dz - \int_{z_{k-1}}^{z_k} \begin{bmatrix} \alpha_x^k \\ \alpha_y^k \\ \alpha_{xy}^k \end{bmatrix} \Delta T z\,dz \right\} \tag{2.63}$$

Recognizing that the midplane strains, midplane curvatures, and ply CTEs are independent of z, the ply by ply integrals in Eqs. (2.62) and (2.63) can be performed, leading to the following definitions:

$$\begin{bmatrix} A_{11} & A_{12} & A_{16} \\ A_{12} & A_{22} & A_{26} \\ A_{16} & A_{26} & A_{66} \end{bmatrix} = \sum_{k=1}^{N} \begin{bmatrix} \bar{Q}_{11}^k & \bar{Q}_{12}^k & \bar{Q}_{16}^k \\ \bar{Q}_{12}^k & \bar{Q}_{22}^k & \bar{Q}_{26}^k \\ \bar{Q}_{16}^k & \bar{Q}_{26}^k & \bar{Q}_{66}^k \end{bmatrix} (z_k - z_{k-1}) \tag{2.64}$$

$$
\begin{bmatrix} B_{11} & B_{12} & B_{16} \\ B_{12} & B_{22} & B_{26} \\ B_{16} & B_{26} & B_{66} \end{bmatrix} = \frac{1}{2} \sum_{k=1}^{N} \begin{bmatrix} \bar{Q}_{11}^{k} & \bar{Q}_{12}^{k} & \bar{Q}_{16}^{k} \\ \bar{Q}_{12}^{k} & \bar{Q}_{22}^{k} & \bar{Q}_{26}^{k} \\ \bar{Q}_{16}^{k} & \bar{Q}_{26}^{k} & \bar{Q}_{66}^{k} \end{bmatrix} \left(z_k^2 - z_{k-1}^2 \right) \tag{2.65}
$$

$$
\begin{bmatrix} D_{11} & D_{12} & D_{16} \\ D_{12} & D_{22} & D_{26} \\ D_{16} & D_{26} & D_{66} \end{bmatrix} = \frac{1}{3} \sum_{k=1}^{N} \begin{bmatrix} \bar{Q}_{11}^{k} & \bar{Q}_{12}^{k} & \bar{Q}_{16}^{k} \\ \bar{Q}_{12}^{k} & \bar{Q}_{22}^{k} & \bar{Q}_{26}^{k} \\ \bar{Q}_{16}^{k} & \bar{Q}_{26}^{k} & \bar{Q}_{66}^{k} \end{bmatrix} \left(z_k^3 - z_{k-1}^3 \right) \tag{2.66}
$$

where the matrix $\mathbf{A}$ is called the extensional stiffness, $\mathbf{B}$ is called the coupling stiffness, and $\mathbf{D}$ is called the bending stiffness. Because of the symmetry of $\bar{\mathbf{Q}}$, each of these stiffness matrices must also be symmetric. Additionally,

$$
\begin{bmatrix} N_x^T \\ N_y^T \\ N_{xy}^T \end{bmatrix} = \sum_{k=1}^{N} \begin{bmatrix} \bar{Q}_{11}^{k} & \bar{Q}_{12}^{k} & \bar{Q}_{16}^{k} \\ \bar{Q}_{12}^{k} & \bar{Q}_{22}^{k} & \bar{Q}_{26}^{k} \\ \bar{Q}_{16}^{k} & \bar{Q}_{26}^{k} & \bar{Q}_{66}^{k} \end{bmatrix} \begin{bmatrix} \alpha_x^k \\ \alpha_y^k \\ \alpha_{xy}^k \end{bmatrix} \Delta T \left(z_k - z_{k-1} \right) \tag{2.67}
$$

$$
\begin{bmatrix} M_x^T \\ M_y^T \\ M_{xy}^T \end{bmatrix} = \frac{1}{2} \sum_{k=1}^{N} \begin{bmatrix} \bar{Q}_{11}^{k} & \bar{Q}_{12}^{k} & \bar{Q}_{16}^{k} \\ \bar{Q}_{12}^{k} & \bar{Q}_{22}^{k} & \bar{Q}_{26}^{k} \\ \bar{Q}_{16}^{k} & \bar{Q}_{26}^{k} & \bar{Q}_{66}^{k} \end{bmatrix} \begin{bmatrix} \alpha_x^k \\ \alpha_y^k \\ \alpha_{xy}^k \end{bmatrix} \Delta T \left(z_k^2 - z_{k-1}^2 \right) \tag{2.68}
$$

where $\mathbf{N}^T$ and $\boldsymbol{M}^T$ are the thermal force and moment resultants.

Using the above definitions, Eqs. (2.62) and (2.63) can be written as

$$
\begin{bmatrix} N_x + N_x^T \\ N_y + N_y^T \\ N_{xy} + N_{xy}^T \\ M_x + M_x^T \\ M_y + M_y^T \\ M_{xy} + M_{xy}^T \end{bmatrix} = \begin{bmatrix} A_{11} & A_{12} & A_{16} & B_{11} & B_{12} & B_{16} \\ A_{12} & A_{22} & A_{26} & B_{12} & B_{22} & B_{26} \\ A_{16} & A_{26} & A_{66} & B_{16} & B_{26} & B_{66} \\ B_{11} & B_{12} & B_{16} & D_{11} & D_{12} & D_{16} \\ B_{12} & B_{22} & B_{26} & D_{12} & D_{22} & D_{26} \\ B_{16} & B_{26} & B_{66} & D_{16} & D_{26} & D_{66} \end{bmatrix} \begin{bmatrix} \varepsilon_x^0 \\ \varepsilon_y^0 \\ \gamma_{xy}^0 \\ \kappa_x^0 \\ \kappa_y^0 \\ \kappa_{xy}^0 \end{bmatrix} \tag{2.69}
$$

where the matrix on the right-hand side is referred to as the **ABD** matrix. Given that each of **A**, **B**, and **D** are symmetric, the **ABD** matrix must also be symmetric. Eq. (2.69) can be written in compact form as,

$$
\begin{bmatrix} \mathbf{N} \\ \mathbf{M} \end{bmatrix} + \begin{bmatrix} \mathbf{N}^T \\ \mathbf{M}^T \end{bmatrix} = \begin{bmatrix} \mathbf{A} & \mathbf{B} \\ \mathbf{B} & \mathbf{D} \end{bmatrix} \begin{bmatrix} \varepsilon^0 \\ \kappa^0 \end{bmatrix} \tag{2.70}
$$

where $\mathbf{N}$, $\mathbf{M}$, $\mathbf{N}^T$, $\mathbf{M}^T$, $\boldsymbol{\varepsilon}^0$, and $\boldsymbol{\kappa}^0$ are rank three vectors whose components are readily identifiable from Eq. (2.69).

In practice, often times one knows the applied loading more often than midplane strains and curvatures, therefore the following is the more convenient form

$$
\begin{bmatrix}
\varepsilon_x^0 \\
\varepsilon_y^0 \\
\gamma_{xy}^0 \\
\kappa_x^0 \\
\kappa_y^0 \\
\kappa_{xy}^0
\end{bmatrix}
=
\begin{bmatrix}
a_{11} & a_{12} & a_{16} & b_{11} & b_{12} & b_{16} \\
a_{12} & a_{22} & a_{26} & b_{21} & b_{22} & b_{26} \\
a_{16} & a_{26} & a_{66} & b_{61} & b_{62} & b_{66} \\
b_{11} & b_{21} & b_{61} & d_{11} & d_{12} & d_{16} \\
b_{12} & b_{22} & b_{62} & d_{12} & d_{22} & d_{26} \\
b_{16} & b_{26} & b_{66} & d_{16} & d_{26} & d_{66}
\end{bmatrix}
\begin{bmatrix}
N_x + N_x^T \\
N_y + N_y^T \\
N_{xy} + N_{xy}^T \\
M_x + M_x^T \\
M_y + M_y^T \\
M_{xy} + M_{xy}^T
\end{bmatrix}
\tag{2.71}
$$

where lowercase **abd** terms are the components of the inverse of the **ABD** matrix, i.e., $\mathbf{abd} = [\mathbf{ABD}]^{-1}$. It should be noted that, due to the symmetry of the **ABD** matrix, its inverse, **abd**, must also be symmetric. Examining Eq. (2.71), it is clear that this requires **a** and **d** to be symmetric, but no such requirement exists for **b** as a result of the inversion of the entire **ABD** matrix. Therefore, the lower left of the **abd** matrix in Eq. (2.71) contains the transpose of **b**, and Eq. (2.71) can be written in compact form as,

$$
\begin{bmatrix}
\boldsymbol{\varepsilon}^0 \\
\boldsymbol{\kappa}^0
\end{bmatrix}
=
\begin{bmatrix}
\mathbf{a} & \mathbf{b} \\
\mathbf{b}^{\mathbf{T}} & \mathbf{d}
\end{bmatrix}
\left\{
\begin{bmatrix}
\mathbf{N} \\
\mathbf{M}
\end{bmatrix}
+
\begin{bmatrix}
\mathbf{N}^T \\
\mathbf{M}^T
\end{bmatrix}
\right\}
\tag{2.72}
$$

where $\mathbf{b}^{\mathbf{T}}$ is the transpose of the matrix **b**. Furthermore, the **a**, **b**, and **d** matrices can be written in terms of the inverses of the **A**, **B**, and **D** matrices as (Nettles, 1994),

$$
\begin{aligned}
\mathbf{a} &= \mathbf{A}^{-1} + \mathbf{A}^{-1}\,\mathbf{B}\,\mathbf{d}\,\mathbf{B}\,\mathbf{A}^{-1} \\
\mathbf{b} &= -\mathbf{A}^{-1}\,\mathbf{B}\,\mathbf{d} \\
\mathbf{d} &= \left(\mathbf{D} - \mathbf{B}\,\mathbf{A}^{-1}\,\mathbf{B}\right)^{-1}
\end{aligned}
\tag{2.73}
$$

These equations show that, in the special case of $\mathbf{B} = \mathbf{0}$ (which will be discussed later), $\mathbf{b} = \mathbf{0}$, $\mathbf{a} = \mathbf{A}^{-1}$, and $\mathbf{d} = \mathbf{D}^{-1}$, and, in this case, the effective laminate properties can be obtained from,

$$
E_x^* = \frac{1}{t\,a_{11}}, \qquad E_y^* = \frac{1}{t\,a_{22}}, \qquad v_{xy}^* = \frac{-a_{12}}{a_{11}}, \qquad G_{xy}^* = \frac{1}{t\,a_{66}}
\tag{2.74}
$$

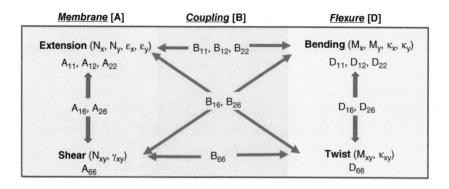

Figure 2.6 Laminate coupling relationships.

It should be noted that these equations are not valid if $\mathbf{B} \neq \mathbf{0}$ as, in this case, bending and extension are coupled, so the laminate does not behave like a material. Similarly, the <u>effective laminate CTEs</u> (again, valid only when $\mathbf{B} = \mathbf{0}$) are given by,

$$
\begin{bmatrix} \alpha_x^* \\ \alpha_y^* \\ \alpha_{xy}^* \end{bmatrix} = \frac{\mathbf{a}\,\mathbf{N}^T}{\Delta T}
\tag{2.75}
$$

where the effective shear CTE, α_{xy}^*, is an engineering shear quantity, meaning $\alpha_{xy}^* \Delta T$ is an engineering shear strain.

In order to determine the induced through-thickness normal strain in ply k that will result in a state of plane stress, Eq. (2.40) can be used. This equation can be written for ply k as,

$$
\varepsilon_{33}^k = -\frac{C_{13}^k}{C_{33}^k}\left(\varepsilon_{11}^k - \alpha_1^k \Delta T\right)
$$
$$
- \frac{C_{23}^k}{C_{33}^k}\left(\varepsilon_{22}^k - \alpha_2^k \Delta T\right) + \alpha_3^k \Delta T
\tag{2.76}
$$

The 6 by 6 matrix in Eq. (2.69) is referred to as the **ABD** matrix or the laminate stiffness matrix. This equation is the key to lamination theory. In general laminates, all terms of the **ABD** matrix are non-zero; however, in special, technologically relevant, types of laminates described in Section 2.6.1 some of these terms can be zero. Fig. 2.6 provides a visual illustrating the various laminate coupling relationships between extension, shear, bending, and twist. If the loading applied to the laminate is known, the laminate stiffness matrix is inverted, and the midplane strains and curvatures are calculated. Then using Eq. (2.53), the strains at every point in the laminate can be calculated, and from Eq. (2.59), the stresses can be calculated as well. Note that the **ABD** matrix is independent of the x and y dimensions of the laminate and is only influenced by the ply thicknesses, stacking arrangement (or sequence), and the specific material properties per ply.

MATLAB Function 2.4 – GetABD.m

This function is used to determine the **A**, **B** and **D** matrices associated with CLT, see Eqs. (2.64), (2.65), and (2.66), respectively.

```
function [Aij, Bij, Dij] = GetABD(Orient, plyprops, plymat, N, z)
% +++++++++++++++++++++++++++++++++++++++++++++++++++++++++++++++++++++++++++++++
% Copyright 2020 United States Government as represented by the Administrator of the
% National Aeronautics and Space Administration. No copyright is claimed in the
% United States under Title 17, U.S. Code. All Other Rights Reserved. By downloading
% or using this software you agree to the terms of NOSA v1.3. This code was prepared
% by Drs. B.A. Bednarcyk and S.M. Arnold to complement the book "Practical
% Micromechanics of Composite Materials" during the course of their government work.
% +++++++++++++++++++++++++++++++++++++++++++++++++++++++++++++++++++++++++++++++
% Purpose: Calculate the laminate ABD matrix
% Input:
% - Orient: Array of ply orientation angles theta (deg)
% - plyprops: Cell array containing ply material properties
% - plymat: Array of ply material id numbers
% - N: Total number of plies
% - z: Array of ply boundary z-coordinates
% Output:
% - Aij: Laminate extensional stiffness matrix
% - Bij: Laminate coupling stiffness matrix
% - Dij: Laminate bending stiffness matrix
% +++++++++++++++++++++++++++++++++++++++++++++++++++++++++++++++++++++++++++++++

% -- Preaallocate arrays
Aij = zeros(3,3);
Bij = zeros(3,3);
Dij = zeros(3,3);

Aij_vect = zeros(1,N);
Bij_vect = zeros(1,N);
Dij_vect = zeros(1,N);

% -- Calculate for A,B,D vectors by using for-loops to create summation
for i = 1:3
    for j = 1:3
        for k = 1:N   % -- k is current ply

            Qb = Qbar(Orient(k), plyprops, plymat(k));

            % -- .* and .^ are element by element operations
            Aij_vect(k) = Qb(i,j).*(z(k+1) - z(k));               % -- Eq. 2.64
            Bij_vect(k) = Qb(i,j).*((z(k+1)).^2 - (z(k)).^2);     % -- Eq. 2.65
            Dij_vect(k) = Qb(i,j).*((z(k+1)).^3 - (z(k)).^3);     % -- Eq. 2.66

        end

        Aij(i,j) = sum(Aij_vect);     % -- Eq. 2.64
        Bij(i,j) = sum(Bij_vect)/2;   % -- Eq. 2.65
        Dij(i,j) = sum(Dij_vect)/3;   % -- Eq. 2.66

    end
end
end
```

MATLAB Function 2.5 – GetThermLoadsNM.m

This function determines the thermal force and moment resultants associated with CLT, see Eqs. (2.67) and (2.68), respectively, as well as the ply-level CTEs in global x-y coordinates.

```
function [NMT, alphbar] = GetThermLoadsNM(Orient, plyprops, plymat, N, z)
% +++++++++++++++++++++++++++++++++++++++++++++++++++++++++++++++++++++++++++++
% Copyright 2020 United States Government as represented by the Administrator of the
% National Aeronautics and Space Administration. No copyright is claimed in the
% United States under Title 17, U.S. Code. All Other Rights Reserved. By downloading
% or using this software you agree to the terms of NOSA v1.3. This code was prepared
% by Drs. B.A. Bednarcyk and S.M. Arnold to complement the book "Practical
% Micromechanics of Composite Materials" during the course of their government work.
% +++++++++++++++++++++++++++++++++++++++++++++++++++++++++++++++++++++++++++++
% Purpose: Calculate the laminate thermal force and moment results, NT and MT **per
%          degree** - That is, these must be multiplied by DT to obtain actual force
%          and moment resultants in Eqs. 2.67 and 2.68
% Input:
% - Orient: Array of ply orientation angles theta (deg)
% - plyprops: Cell array containing ply material properties
% - plymat: Array of ply material id numbers
% - N: Total number of plies
% - z: Array of ply boundary z-coordinates
% Output:
% - NMT: Vector containing NT and MT (see Eq. 2.70) **per degree**
% - alphbar: Cell array containing ply CTEs in the global (x-y) coordinates
% +++++++++++++++++++++++++++++++++++++++++++++++++++++++++++++++++++++++++++++

% -- Preallocate alphbar
alphbar = cell(1,N);

% -- Obtain alphbar - cell array of ply CTEs in x-y coords
for k = 1:N
    a1 = plyprops{plymat(k)}.a1;
    a2 = plyprops{plymat(k)}.a2;
    alph =[a1; a2; 0]; % -- alph is the ply CTE vector, a12 = 0 in local
    [~,T2] = GetT(Orient(k));
    alphbar{k} = inv(T2) * alph; % -- Eq. 2.48
end

% -- Preallocate NT and MT
NT = zeros(3,1);
MT = zeros(3,1);

% -- Perform summation over plies
for k = 1:N
    Qb = Qbar(Orient(k), plyprops, plymat(k));
    NTply = Qb*alphbar{k}*(z(k+1) - z(k)); % -- Eq. 2.67 (divided by DT)
    NT = NT + NTply;
    MTply = Qb*alphbar{k}*(z(k+1)^2 - z(k)^2)/2; % -- Eq. 2.68 (divided by DT)
    MT = MT + MTply;
end

% -- Store NT and MT in single vector (NOTE: these are per degree)
NMT = [NT;MT]; % [Nx,Ny,Nxy,Mx,My,Mxy]

end
```

2.4 MATLAB implementation of classical lamination theory

The MATLAB code provided with this text is available from:

https://github.com/nasa/Practical-Micromechanics

In this section, the provided MATLAB (2018) code for solving laminate problems is discussed. A flow chart of the main problem driver script (AnalyzeLam.m) is shown in Fig. 2.7A, whereas the script itself is shown in MATLAB Script 2.6. Multiple laminate problems can be analyzed in a single execution of AnalyzeLam.m. The problem input is specified in MATLAB Function 2.7 – LamProblemDef.m and the MATLAB Function 2.1 – GetPlyProps.m. LamProblemDef.m is used to specify the problem name, ply angle orientations, ply materials, ply thicknesses, and laminate loading for all problems to be run. The LamProblemDef.m shown in MATLAB Function 2.7, includes two of the example problems discussed later in this chapter. The actual ply material properties are defined in GetPlyProps.m, wherein each ply material is assigned a unique identification number. As shown in MATLAB Function 2.7, the material i.d. numbers are then used to

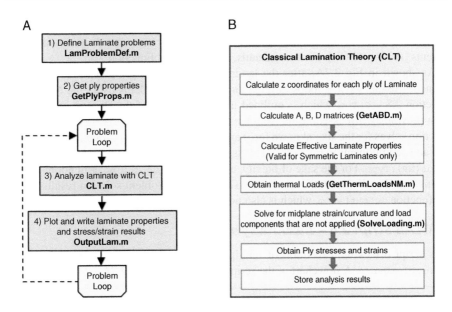

Figure 2.7 Flow charts describing steps involved in (A) solving a laminated composite problem in the MATLAB script `LamAnalysis.m` and (B) the MATLAB function `CLT.m`.

specify the material of each ply in the array `Geometry{NP}.plymat`. Likewise, the ply orientation angles and ply thicknesses are specified in the arrays `Geometry{NP}.Orient` and `Geometry{NP}.tply`, respectively. It is important that these three arrays all have the same length in a given problem, which is equal to the total number of plies in the laminate (an error is thrown by the code if this is not the case). Note that the problems defined in `LamProblemDef.m` must be numbered starting at 1 and be consecutive. Furthermore, the type of output (`OutInfo.Format = "txt"` or `"doc"`) and whether or not plots should be included in the output (`OutInfo.MakePlots = true` or false) are also specified in `LamProblemDef.m`. "`txt`" format generates ASCII text file output, whereas "`doc`" provides output in a Microsoft Word document. The "`doc`" output option is significantly slower than "`txt`" and only works on Windows operating systems.

Specification of applied laminate loads is also done in `LamProblemDef.m`. Any admissible combinations of force and moments resultants and midplane strains and curvatures, along with a temperature change, can be specified. That is, for example, either ε_x^0 or N_x can be specified, but not both. The same is true for all conjugate pairs of the force and moment resultants and midplane strains and curvatures. Therefore the loads input must specify the type and magnitude of the loading, for each of the six components. For example, to apply a mid-plane strain in the x-direction of 4 percent, $\varepsilon_x^0 = 0.04$, while specifying all other force and moment resultants to be zero, $N_y = N_{xy} = M_x = M_y = M_{xy} = 0$, one would specify;

```
Loads{NP}.Type = [EK, NM, NM, NM, NM, NM];
Loads{NP}.Value = [0.04, 0, 0, 0, 0, 0];
```

where EK indicates an applied strain or curvature and NM indicates a force or moment resultant, and the location in the `Loads{NP}.Type` vector corresponds to the six components [see Eq. (2.69)]. For more details, see MATLAB Function 2.7.

As shown in Fig. 2.7A, after the laminate problem and ply materials have been defined, `LamAnalysis.m` loops through and solves all problems using the MATLAB Function 2.8 – `CLT.m`, which is outlined in Fig. 2.7B. All of the steps in `CLT.m` have been previously discussed, aside from the MATLAB function `SolveLoading.m`. This function solves for the six unknown force and moment resultants or midplane strain and curvature components in Eq. (2.69). That is, the six components that are not specified in `Loads{NP}.Type`. For the sake of brevity, a listing of `SolveLoading.m` in not included in this text. Finally, step 4 in `LamAnalysis.m` outputs problem solution results using the function `OutputLam.m`, which is also not listed in this text.

MATLAB Script 2.6 – `LamAnalysis.m`

This script drives the execution of multiple CLT problems defined in the function `LamProblemDef.m`. A flowchart of the script is provided in Fig. 2.7A.

```
% +++++++++++++++++++++++++++++++++++++++++++++++++++++++++++++++++++++++++++
% Copyright 2020 United States Government as represented by the Administrator of the
% National Aeronautics and Space Administration. No copyright is claimed in the
% United States under Title 17, U.S. Code. All Other Rights Reserved. By downloading
% or using this software you agree to the terms of NOSA v1.3. This code was prepared
% by Drs. B.A. Bednarcyk and S.M. Arnold to complement the book "Practical
% Micromechanics of Composite Materials" during the course of their government work.
% +++++++++++++++++++++++++++++++++++++++++++++++++++++++++++++++++++++++++++
%
% Purpose: Driver script for laminate analysis problems using CLT. The problem input
%          is defined in the function LamProblemDef.m and GetPlyProps.m
%
% +++++++++++++++++++++++++++++++++++++++++++++++++++++++++++++++++++++++++++

% -- Clear memory and close files
clear;
close all;
fclose('all');
clc;

% -- Add needed function locations to the path
addpath('Functions/CLT');
addpath('Functions/Utilities');
addpath('Functions/WriteResults');
addpath('Functions/Micromechanics');
addpath('Functions/Margins');

%-------------------------------------------------------------------
% 1) Define Laminate Problems
%-------------------------------------------------------------------
[NProblems, OutInfo, Geometry, Loads] = LamProblemDef();

% -- Preallocate LamResults
LamResults = cell(1, NProblems);

%-------------------------------------------------------------------
% 2) Get ply properties
%-------------------------------------------------------------------
plyprops = GetPlyProps(Geometry);

% -- Loop through problems
for NP = 1: NProblems

    % -- Check for missing problem name
    if (ismissing(OutInfo.Name(NP)))
        OutInfo.Name(NP) = string(['Problem ', char(num2str(NP))]);
    end

    % -- Check for missing ply angles, thicknesses, materials
    if ~isfield(Geometry{NP},'Orient') || ~isfield(Geometry{NP},'tply') || ...
       ~isfield(Geometry{NP},'plymat')
        error(['Orient, tply, or plymat missing, Problem #', num2str(NP)]);
    end

    % -- Echo problem info to command window
    disp(['Problem #',num2str(NP),' - ', char(OutInfo.Name(NP))]);
    disp(['    Per ply angle orientations   [',num2str(Geometry{NP}.Orient), ']']);
    disp(['    Per ply thicknesses          [',num2str(Geometry{NP}.tply), ']']);
    disp(['    Per ply material assignments [',num2str(Geometry{NP}.plymat), ']']);

    % -- Check that the problem's ply materials, orientation, and thickness
    %    have consistent lengths
    if length(Geometry{NP}.tply) ~= length(Geometry{NP}.Orient) || ...
       length(Geometry{NP}.tply) ~= length(Geometry{NP}.plymat) || ...
       length(Geometry{NP}.Orient) ~= length(Geometry{NP}.plymat)
        error('Lengths of tply, Orient, and plymat are not consistent');
```

```
    end

    % -- Check that loads are specified for this problem
    if ~isfield(Loads{NP}, 'Type') || ~isfield(Loads{NP}, 'Value')
        error(strcat('Problem #', num2str(NP), ' Loads not properly defined'));
    end
    if ~isfield(Loads{NP}, 'DT')
        Loads{NP}.DT = 0;
    end

    % -- Check that the problem's ply materials have been defined
    for k = 1: length(Geometry{NP}.tply)
        if ~isfield(plyprops{Geometry{NP}.plymat(k)}, 'name')
            error(strcat('ply material #', num2str(Geometry{NP}.plymat(k)), ...
                        ' undefined ... check GetPlyProps'));
        end
    end

    % -- Check for slash character in OutInfo.Name
    k = strfind(OutInfo.Name(NP), '/');
    j = strfind(OutInfo.Name(NP), '\');
    if ~isempty(k) || ~isempty(j)
        error('Problem name contains a slash or backslash ... remove');
    end

    % -- Default temperature change (DT) to zero
    if ~isfield(Loads{NP}, 'DT')
        Loads{NP}.DT = 0;
    end

    %-----------------------------------------------------------------
    % 3) Analyze laminate with CLT per problem
    %-----------------------------------------------------------------
    [Geometry{NP}, LamResults{NP}] = CLT(plyprops, Geometry{NP}, Loads{NP});

    %-----------------------------------------------------------------
    % 4) Plot and write laminate properties and stress/strain results
    %-----------------------------------------------------------------
    [OutInfo] = OutputLam(OutInfo, NP, Geometry{NP}, Loads{NP}, plyprops, ...
                        LamResults{NP});

    % -- Display notification of file completion to command window
    disp(['  *** Problem ',char(num2str(NP)),' Completed ***'])
    disp(' ');
    close all;

end
```

MATLAB Function 2.7 – `LamProblemDef.m`

This function defines the problems to be analyzed, including the problem name, ply angle orientations, ply materials, ply thicknesses, and laminate loading. Certain output requests are also specified.

```matlab
function [NP, OutInfo, Geometry, Loads] = LamProblemDef()
% ++++++++++++++++++++++++++++++++++++++++++++++++++++++++++++++++++++++++++++
% Copyright 2020 United States Government as represented by the Administrator of the
% National Aeronautics and Space Administration. No copyright is claimed in the
% United States under Title 17, U.S. Code. All Other Rights Reserved. By downloading
% or using this software you agree to the terms of NOSA v1.3. This code was prepared
% by Drs. B.A. Bednarcyk and S.M. Arnold to complement the book "Practical
% Micromechanics of Composite Materials" during the course of their government work.
% ++++++++++++++++++++++++++++++++++++++++++++++++++++++++++++++++++++++++++++
%
% Purpose: Sets up and defines classical lamination theory (CLT) problems. Specifies
%          problem name, ply materials, ply thicknesses, ply orientations, loading,
%          and failure criteria.
% Input: None
% Output:
% - NP: Number of problems
% - OutInfo: Struct containing output information
% - Geometry: Cell array containing laminate definition variables per
%    problem
% - Loads: Cell array containing problem loading information per problem
% ++++++++++++++++++++++++++++++++++++++++++++++++++++++++++++++++++++++++++++

NProblemsMax = 200; % -- To preallocate cells
NP = 0; % -- Problem number counter
NM = 2; % -- Applied loading type identifier (force/moment resultants)
EK = 1; % -- Applied loading type identifier (midplane strains/curvatures)

% -- Preallocate cells
Geometry = cell(1,NProblemsMax);
Loads = cell(1,NProblemsMax);

% -- Type of output requested
OutInfo.Format = "txt";
%OutInfo.Format = "doc";

% -- Include or exclude plots from output
OutInfo.MakePlots = true;
%OutInfo.MakePlots = false;

%==============================================================
% -- Define all problems below
%==============================================================

    NP = NP + 1;
    OutInfo.Name(NP) = "Chapter 2 - Section 2.5.1 glass-epoxy [45]";
    Geometry{NP}.Orient = [45];
    Geometry{NP}.plymat = [2];
    Geometry{NP}.tply = [0.15];
    Loads{NP}.Type  = [NM, NM, NM, NM, NM, NM];
    Loads{NP}.Value = [ 1,  0,  0,  0,  0,  0];

    NP = NP + 1;
    OutInfo.Name(NP) = "Chapter 2 - Section 2.5.2 IM7-8552 [0,90,90,0] Nx";
    Geometry{NP}.Orient = [30,30];
    nplies = length(Geometry{NP}.Orient);
    Geometry{NP}.plymat(1:nplies) = 1;
    Geometry{NP}.tply(1:nplies) = 0.15;
    Loads{NP}.Type  = [NM, NM, NM, NM, NM, NM];
    Loads{NP}.Value = [ 1,  0,  0,  0,  0,  0];

%==============================================================
% -- End of problem definitions
%==============================================================

% -- Store applied loading type identifiers
for I = 1:NP
    Loads{I}.NM = NM;
    Loads{I}.EK = EK;
end
```

MATLAB Function 2.8 – `CLT.m`

This function constitutes the core of the classical lamination theory calculations.

```
function [Geometry, LamResults] = CLT(plyprops, Geometry, Loads, LamResults)
% ++++++++++++++++++++++++++++++++++++++++++++++++++++++++++++++++++++++++++++++++
% Copyright 2020 United States Government as represented by the Administrator of the
% National Aeronautics and Space Administration. No copyright is claimed in the
% United States under Title 17, U.S. Code. All Other Rights Reserved. By downloading
% or using this software you agree to the terms of NOSA v1.3. This code was prepared
% by Drs. B.A. Bednarcyk and S.M. Arnold to complement the book "Practical
% Micromechanics of Composite Materials" during the course of their government work.
% ++++++++++++++++++++++++++++++++++++++++++++++++++++++++++++++++++++++++++++++++
% Purpose: Perform classical lamination theory calculations
% - Calculate laminate ABD, effective props, thermal force and moment resultants,
%   unknown midplane strains/curvatures, and stresses/strains through the thickness
%   at top and bottom of each ply
% Input:
% - plyprops: Cell array containing ply material properties
% - Geometry: Struct containing laminate definition variables
% - Loads: Struct containing problem loading information
% - LamResults: Struct containing laminate analysis results
% Output:
% - Geometry: Updated struct containing laminate definition variables
% - LamResults: Updated struct containing laminate analysis results
% ++++++++++++++++++++++++++++++++++++++++++++++++++++++++++++++++++++++++++++++++

% -- Extract variables from structs for convenience
Orient = Geometry.Orient;
plymat = Geometry.plymat;
tl = Geometry.tply;
DT = Loads.DT;

% -- Check lengths of plymat, Orient, tl (must all be the same)
if isequal(length(plymat), length(Orient), length(tl))
    N = length(plymat);
    Geometry.N = N;
else
    error('Lengths of plymat, Orient, tl are not the same');
end

% -- Calculate z coordinate of each ply boundary
t = sum(tl);
tlx = cumsum(tl);
z = [-t/2, [-t/2+tlx]];

% -- Calculate ABD matrix and inverse
[Aij, Bij, Dij] = GetABD(Orient, plyprops, plymat, N, z);
ABD = [Aij, Bij; Bij, Dij];
ABDinv = inv(ABD);

% -- Calculate effective mechanical laminate properties (Eqs. 2.74)
%    (Valid for symmetric laminates Only, Bij = 0)
EX = 1/(ABDinv(1,1)*t);
EY = 1/(ABDinv(2,2)*t);
NuXY = -ABDinv(1,2)/ABDinv(1,1);
GXY =  1/(ABDinv(3,3)*t);

% -- Calculate thermal force and moment resultants **per degree** and
%    ply level CTEs transformed to global x-y coords (alphbar)
[NMT, alphbar] = GetThermLoadsNM(Orient, plyprops, plymat, N, z);

% -- Calculate laminate effective CTEs (Eq. 2.75)
%    (Valid for symmetric laminates Only, Bij = 0)
EKT = ABDinv*NMT; % -- Note, NMT is per degree at this point
AlphX = EKT(1);
AlphY = EKT(2);
AlphXY = EKT(3);
```

```
% -- Calculate actual thermal force and moment resultants
NMT = NMT*DT;

% -- Detemine unknown global NM and EK components based on specified loading
[NM, EK] = SolveLoading(6, ABD, -NMT, Loads);
Loads.NM = NM;

% -- Extract mid-plane strains (Eps_o) and Curvatures (Kappa)
Eps_o = EK(1:3);
Kappa = EK(4:6);

% ---------------------------------------------
% -- Localization of Laminate Stresses
% ---------------------------------------------
stress = [];
strain = [];
MCstress = [];
MCstrain = [];
LayerZ = zeros(1,N*2);

% -- Calculate stresses and strains at top and bottom points of each ply
for i = 0:N - 1

    LayerZ(2*i + 2) = z(i + 2);
    LayerZ(2*i + 1) = z(i + 1);

    % -- Eqs. 2.53, 2.59
    strainbot = Eps_o + z(i + 1)*(Kappa);
    stressbot = Qbar(Orient(i + 1), plyprops, ...
                     plymat(i + 1))*(strainbot - alphbar(i + 1)*DT);
    straintop = Eps_o + z(i + 2)*(Kappa);
    stresstop = Qbar(Orient(i + 1), plyprops, ...
                     plymat(i + 1))*(straintop - alphbar(i + 1)*DT);

    stress = [stress, stressbot, stresstop]; % -- recursively concat vectors
    strain = [strain, strainbot, straintop]; % -- recursively concat vectors

% -- Transform stress and strain to material coordinates (MC) per ply (Eqs. 2.43)
    Ang = Orient(i + 1);
    [T1, T2] = GetT(Ang);
    MCstrainbot = T2*strainbot;
    MCstressbot = T1*stressbot;
    MCstraintop = T2*straintop;
    MCstresstop = T1*stresstop;
    MCstress = [MCstress,MCstressbot,MCstresstop]; % -- recursively concat vectors
    MCstrain = [MCstrain,MCstrainbot,MCstraintop]; % -- recursively concat vectors

end

% -- Store geometry and results for return
Geometry.LayerZ = LayerZ;
LamResults.A = Aij;
LamResults.B = Bij;
LamResults.D = Dij;
LamResults.EX = EX;
LamResults.EY = EY;
LamResults.NuXY = NuXY;
LamResults.GXY = GXY;
LamResults.AlphX = AlphX;
LamResults.AlphY = AlphY;
LamResults.AlphXY = AlphXY;
LamResults.EK = EK;
LamResults.NM = NM;
LamResults.NMT = NMT;
LamResults.stress = stress;
LamResults.strain = strain;
```

```
LamResults.MCstress = MCstress;
LamResults.MCstrain = MCstrain;

end
```

2.5 Laminate example problems

In this section, example analysis results for unidirectional, cross-ply, and angle-ply composite laminates are presented. For examples that utilize the MATLAB code, the associated input data is included in `LamProblemDef.m`.

2.5.1 Unidirectional composites

Unidirectional laminates are those in which all fibers are aligned in a single direction, at an angle θ. These are the simplest of all laminates. Unidirectional [0] and [90] are typically used to obtain lamina or ply level properties since any angle other than [0] or [90] will exhibit extension-shear coupling when subjected to extension only or shear only, see Fig. 2.6. This extension-shear coupling, exhibited when A_{16} and $A_{26} \neq 0$, is one of the unique aspects of fiber reinforced composite behavior versus that of isotropic materials. Unidirectional laminates also exhibit no membrane–flexure coupling, i.e., $B_{ij} = 0$, nor bending-twist coupling since $D_{16} = D_{26} = 0$. The A_{16}, A_{26}, D_{16} and D_{26} components are zero because $\bar{Q}_{16} = \bar{Q}_{26} = 0$ when the fiber orientation angle is zero, i.e., [0], or ninety degrees, i.e., [90]. All other angles will produce nonzero values for $\bar{Q}_{16}$ and $\bar{Q}_{26}$, thus inducing extension-shear coupling. In such situations, subjecting a unidirectional test specimen to extension will produce both normal and shear stresses within the specimen.

2.5.1.1 Analytical results

Consider a unidirectional laminate with varying fiber angle subjected to a unit stress (1 MPa) in the global x-direction. The local stresses in the material coordinate system (1-2, aligned with fiber orientation angle) are given by Eq. (2.43) and plotted in Fig. 2.8. At angles of $0°$ and $90°$, the local shear stress is zero. At an angle of $45°$, the local shear stress is maximum at 0.5, while both the stress in the fiber direction, σ_{11}, and the transverse direction, σ_{22}, are 0.5 as well.

Explicit expressions for effective engineering constants, E_x, v_{xy}, E_y, v_{yx} ($= v_{xy}E_y/E_x$), and G_{xy} can be obtained from the local material coordinate properties, E_1, v_{12}, E_2, v_{21} ($= v_{12}E_2/E_1$), and G_{12}, as a function of angle θ, where $n = \sin(\theta)$ and $m = \cos(\theta)$, see Herakovich (1998),

$$E_x = \frac{E_1}{m^4 + \left(\frac{E_1}{G_{12}} - 2v_{12}\right)n^2m^2 + \frac{E_1}{E_2}n^4}$$

$$v_{xy} = \frac{v_{12}(n^4 + m^4) - \left(1 + \frac{E_1}{E_2} - \frac{E_1}{G_{12}}\right)n^2m^2}{m^4 + \left(\frac{E_1}{G_{12}} - 2v_{12}\right)n^2m^2 + \frac{E_1}{E_2}n^4}$$

$$E_y = \frac{E_2}{m^4 + \left(\frac{E_2}{G_{12}} - 2v_{21}\right)n^2m^2 + \frac{E_2}{E_1}n^4}$$

$$v_{yx} = \frac{v_{21}(n^4 + m^4) - \left(1 + \frac{E_2}{E_1} - \frac{E_2}{G_{12}}\right)n^2m^2}{m^4 + \left(\frac{E_2}{G_{12}} - 2v_{21}\right)n^2m^2 + \frac{E_2}{E_1}n^4}$$ (2.77)

$$G_{xy} = \frac{G_{12}}{n^4 + m^4 + 2\left(2\frac{G_{12}}{E_1}(1 + 2v_{12}) + 2\frac{G_{12}}{E_2} - 1\right)n^2m^2}$$

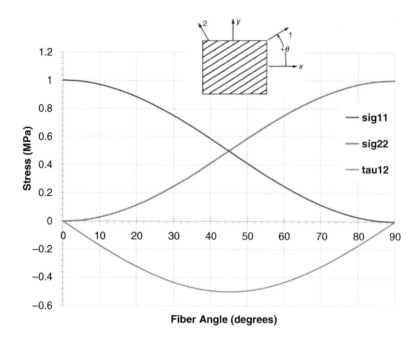

Figure 2.8 The local stresses in the material coordinate system (1-2), given an applied unit stress (1 MPa) in the global x-direction on a unidirectional composite. The fiber angle, θ, is that between the x- and 1-axes.

Table 2.1 Ply Material Properties, assembled from Ratcliffe et al. (2012), Herakovich (1998), Dunn (2010), and Arnold et al. (2016).

Mat	Material name	Density (g/cm³)	E_1 (GPa)	E_2 (GPa)	G_{12} (GPa)	ν_{12}	α_1 ($\times 10^{-6}/°C$)	α_2 ($\times 10^{-6}/°C$)
1	IM7/8552	1.58	146.8	8.69	5.16	0.32	-0.1	31
2	glass/epoxy	2.0	43.5	11.5	3.45	0.27	6.84	29
3	SCS-6/Ti-15-3	3.86	221	145	53.2	0.27	6.15	7.90
4	SiC/SiC	3.21	300	152.4	69.5	0.17	3.0	3.2
5	Al	2.7	70	70	26.923	0.3	23	23

Table 2.1 provides ply level material properties that are used throughout this book. The first four of these materials represent composite materials for the three general classes of composites: polymer matrix composites (PMCs), metal matrix composites (MMCs), and ceramic matrix composites (CMCs). The first is a common aerospace grade composite composed of IM7 carbon fibers and 8552 epoxy resin. The second is a generic glass fiber composite with an epoxy matrix, which is typically a lower performance and lower cost composite system compared to carbon/epoxy. The SCS-6

SiC fiber reinforcing a Ti-15-3 (titanium alloy) matrix is used for higher temperature applications, while SiC/SiC (SiC fibers reinforcing a SiC matrix) is used in even higher temperature applications, such as combustor liners. Aluminum properties are included for use in exercises at the end of the chapter.

Each of the effective unidirectional composite properties in Eq. (2.77) has been plotted versus fiber angle, θ, for the first four ply materials in Table 2.1. E_x and E_y are shown in Fig. 2.9, G_{xy} in Fig. 2.10, and ν_{xy} and ν_{yx} in Fig. 2.11. Fig. 2.9 illustrates how a slight change in fiber angle (say 5° or 10°) off of either [0] or [90] can cause significant changes in the effective Young's moduli, E_x and E_y. This effect can be greatly influenced by the material's degree of anisotropy, which can be characterized by the ratio E_1/E_2. For example, given the IM7/8552 composite material, wherein $E_1/E_2 = 17$, a 10° change in angle from 0° results in a 44% reduction of E_x. In contrast, for the glass/epoxy system, wherein $E_1/E_2 = 4.6$, such an angle change results in only a 23% reduction. Also, one sees that the influence of angle changes is greatly reduced when dealing with the metal matrix and ceramic matrix composites which have stiffness ratios $E_1/E_2 \leq 3$, see Fig. 2.9B.

Fig. 2.10 illustrates how the shear modulus, G_{xy}, is influenced by angle, and in particular, how it reaches a maximum value at $\pm 45°$, irrespective of material class (i.e., CMC, MMC, or PMC). Note that the glass/epoxy composite has the largest increase, 136.9%, over that at 0° or 90° followed by the IM7/8552 composite, 53.5%, then the SCS-6/Ti-15-3, 35.5%, and lastly the SiC/SiC composite, which exhibits only a 30.5% increase, over that at 0° or 90°. As shown in Eq. (2.77), this is driven by the ratios G_{12}/E_1 and G_{12}/E_2.

The effective Poisson's ratios, ν_{xy} and ν_{yx}, are plotted in Fig. 2.11 and exhibit significant variation with fiber angle. The solid lines represent ν_{xy} values, while the dashed lines represent ν_{yx} values. The angle at which the maximum in each of these Poisson's ratios occurs depends upon the material class. For example, in the case of the glass/epoxy composite, it is at $\theta = 30°$ and 60°, whereas, for the MMC, it occurs at 40° and 55°.

Lastly, the dependence of the effective composite coefficient of thermal expansion (CTE) on fiber angle in a unidirectional composite can be obtained from Eq. (2.48), which, in expanded form is,

$$
\begin{aligned}
\alpha_x &= \alpha_1 m^2 + \alpha_2 n^2 \\
\alpha_y &= \alpha_1 n^2 + \alpha_2 m^2 \\
\alpha_{xy} &= 2(\alpha_1 - \alpha_2)mn
\end{aligned}
\tag{2.78}
$$

Fig. 2.12 illustrates the variations in the effective CTEs as a function of fiber angle for a unidirectional PMC composed of IM7/8552 and a unidirectional MMC composed of SCS-6/Ti-15-3. Note how at [0] or [90] $\alpha_{xy} = 0$, yet at all other angles a shear thermal strain would be induced along with those in the x- and y-directions in response to an applied temperature change. Note also that the shear CTE reaches a maximum at $\pm 45°$

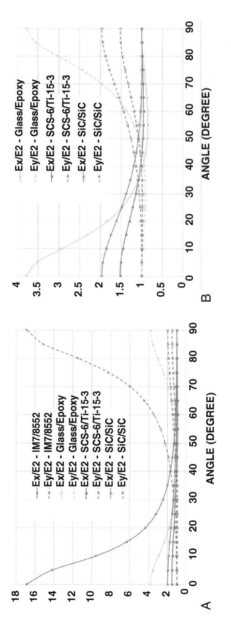

Figure 2.9 A) Variation of effective Young's modulus, E_x, and E_y, normalized with respect to E_2 in Table 2.1, B) same as A but excluding IM7/8552.

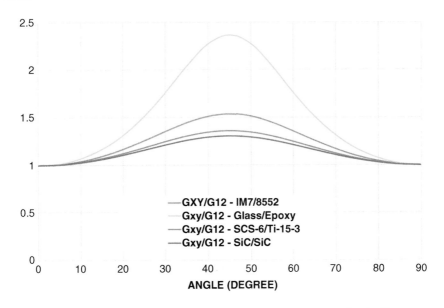

Figure 2.10 Variation of effective shear modulus, G_{xy}, with angle, normalized with respect to G_{12} in Table 2.1.

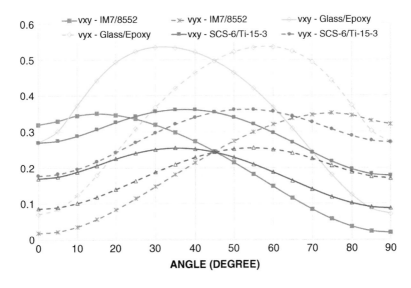

Figure 2.11 Variation of effective Poisson's ratios, ν_{xy} and ν_{yx}.

with the corresponding magnitude of $(\alpha_1 - \alpha_2)$. Clearly the influence of fiber angle in PMCs is much greater than in the case of the MMC, which is because of the much greater mismatch in properties.

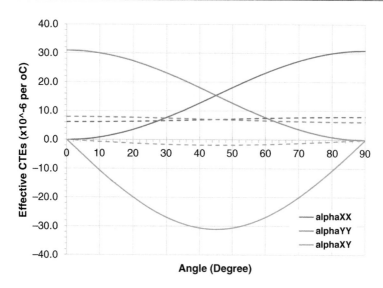

Figure 2.12 Variation of effective CTEs, α_x, α_y, and α_{xy}, as a function of fiber angle for IM7/8552 (solid lines) and SCS-6/Ti-15-3 (dashed lines) unidirectional composites.

2.5.1.2   MATLAB results

To solve unidirectional laminate problems using the MATLAB code, the specifics of the problem are defined using MATLAB Function 2.7 – LamProblemDef.m. The first problem defined in MATLAB Function 2.7 is:

```
NP = NP + 1;
OutInfo.Name(NP) = "Chapter 2 - Section 2.5.1 glass-epoxy [45]";
Geometry{NP}.Orient = [45];
Geometry{NP}.plymat = [2];
Geometry{NP}.tply = [0.15];
Loads{NP}.Type  = [NM, NM, NM, NM, NM, NM];
Loads{NP}.Value = [ 1,  0,  0,  0,  0,  0];
```

This defines a unidirectional glass/epoxy laminate (Geometry{NP}.plymat = [2]) with properties defined in MATLAB Function 2.1 GetPlyProps.m):

```
plyprops{2}.name = "Glass/Epoxy";
plyprops{2}.E1 = 43.5e3;
plyprops{2}.E2 = 11.5e3;
plyprops{2}.G12 = 3.45e3;
plyprops{2}.v12 = 0.27;
plyprops{2}.a1 = 6.84e-06;
plyprops{2}.a2 = 29.e-06;
```

The ply orientation of $45°$ (Geometry{NP}.Orient = [45]) relative to the global x-direction and a ply thickness of 0.15 mm (Geometry{NP}.tply = [0.15]) are specified. The MATLAB code also requires laminate level loading to be specified, as described in detail in Section 2.4. Here, a unit load, $N_x = 1$ MPa, has been applied, with all other force and moment components kept at zero, `Loads{NP}.Type` and `Loads{NP}.Value`. Finally, the options for setting the output format (`OutInfo.Format = "txt"`) and outputting plots (`OutInfo.MakePlots = true`) are specified near the top in MATLAB Function 2.7.

Once the problem has been defined in `LamProblemDef.m`, the MATLAB script `LamAnalysis.m` is executed. The output is written to a folder called "`Output`", as an ASCII text file: "`Composite Text Report - Chapter 2 - Section 2.5.1 glass-epoxy [45] {date and time}.txt`", where the actual date and time is included in the filename. The **ABD** and laminate effective properties from this output file are:

```
<==== ABD Matrices ====>
- A Matrix -
2857.9817 1822.9817 1223.5814
1822.9817 2857.9817 1223.5814
1223.5814 1223.5814 1865.5792

- B Matrix -
  0 0 0
  0 0 0
  0 0 0

- D Matrix -
5.3587 3.4181 2.2942
3.4181 5.3587 2.2942
2.2942 2.2942 3.498


<==== Laminate Effective Properties ====>
Ex Ey Nuxy Gxy Alpha_x Alpha_y Alpha_xy
──── ──── ─── ──── ──── ──── ───

10325.6102 10325.6102 0.49647 8172.6842 1.792e-05 1.792e-05 -2.216e-05
```

The normalized values $E_x/E_2 = 0.898$ and $G_{xy}/G_{12} = 2.37$ correspond to the results plotted in Fig. 2.9B and Fig. 2.10. Note that, on a Windows computer, if MS Word output is chosen (`OutInfo.Format = "doc"`), output will be written to a Word document called "`Composite Text Report - Chapter 2 - Section 2.5.1 glass-epoxy [45] {date and time}.doc`". Furthermore, for text-based output, plots will each be written to a separate file, whereas, for MS Word output, plots will be written to the same Word document as the other results. It should be noted that generation of MS Word output takes much longer than the text-based output.

2.5.2   Cross-ply laminates [0/90]$_s$

In this section the behavior of an IM7/8552 cross-ply laminate (a laminate in which all fibers are aligned with either the x- or y-axes) is presented, where the ply properties

Table 2.2 **ABD** matrices for two IM7/8552 cross-ply laminates with different ply stacking sequences.

	[0/90/90/0]			[90/0/0/90]		
A Matrix (N/mm)	46931.4841	1678.6555	0	46931.4841	1678.6555	0
	1678.6555	46931.4841	0	1678.6555	46931.4841	0
	0	0	3096	0	0	3096
B Matrix (N)	0	-1.4211e-14	0	-7.3896e-13	-1.4211e-14	0
	-1.4211e-14	-7.3896e-13	0	-1.4211e-14	0	0
	0	0	-2.8422e-14	0	0	-2.8422e-14
D Matrix (N·mm)	2345.8725	50.3597	0	470.0166	50.3597	0
	50.3597	470.0166	0	50.3597	2345.8725	0
	0	0	92.88	0	0	92.88

are given in Table 2.1. Specifically, two four ply laminate configurations, [0/90/90/0] and [90/0/0/90], with each ply thickness taken to be 0.15 mm (for a total laminate thickness of 0.6 mm) will be compared and discussed. The MATLAB definition of the [0/90/90/0] problem is given in MATLAB Function 2.7 – LamProblemDef.m:

```
NP = NP + 1;
OutInfo.Name(NP) = "Chapter 2 - Section 2.5.2 IM7-8552 [0,90,90,0] Nx";
Geometry{NP}.Orient = [30,30];
nplies = length(Geometry{NP}.Orient);
Geometry{NP}.plymat(1:nplies) = 1;
Geometry{NP}.tply(1:nplies) = 0.15;
Loads{NP}.Type  = [NM, NM, NM, NM, NM, NM];
Loads{NP}.Value = [ 1,  0,  0,  0,  0,  0];
```

The remaining example problems from this chapter are also provided in LamProblemDef.m. Therefore to execute any example one merely needs to uncomment the appropriate code.

The **ABD** matrices for the two cross-ply laminate configurations are shown in Table 2.2 where it is clear that each component of the **A** matrix is identical, the **B** matrix is essentially zero, and the components of the **D** matrix are identical except that D_{11} and D_{22} are switched depending upon the laminate specified. Consequently, one can classify these laminates as being symmetric (no extension-bending coupling, i.e., **B** = **0**). Higher D_{11} indicates higher bending stiffness in the x-direction, whereas higher D_{22} indicates higher bending stiffness in the y-direction. The effective laminate properties for both laminates are identical in both the x- and y-directions, i.e., the code output provides: $E_x = E_y = 78.119$ GPa, $\nu_{xy} = 0.035768$, $G_{xy} = 5.160$ GPa, $\alpha_x = \alpha_y = 2.1151 \times 10^{-6}$ /°C, and $\alpha_{xy} = 0$.

Next the ply stresses are plotted as a function of z in Fig. 2.13 assuming the application of a unit force resultant in the x-direction, $N_x = 1$ N/mm. Note that in each laminate, the [0] plies carry the majority of stress, i.e., 3.14 MPa, in the x-direction, while the [90] plies carry only 0.184 MPa. The stress in the y-direction is 0.053 MPa in the [0] plies while it is -0.053 MPa in the [90] plies, as these stresses must sum to zero since no loading is applied in the y-direction. No normal-shear coupling exists (since $A_{16} = A_{26} = 0$) and therefore $\sigma_{xy} = 0$ throughout the laminate. The mid-plane

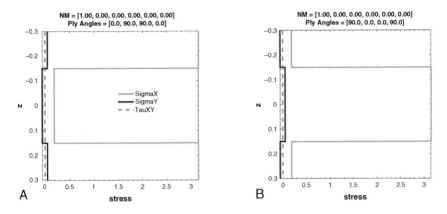

Figure 2.13 Ply level stresses (MPa) in IM7/8552 cross-ply laminates, (A) [0/90/90/0] and (B) [90/0/0/90], when subjected to a unit N_x force resultant, 1 N/mm.

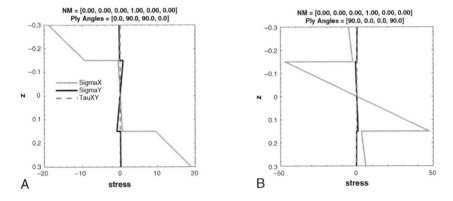

Figure 2.14 Ply level stresses (MPa) in IM7/8552 cross-ply laminates, (A) [0/90/90/0] and (B) [90/0/0/90], when subjected to a unit M_x moment resultant, 1 N·mm/mm.

micro ($\times 10^{-6}$) strains for both laminates are $\varepsilon_x = 21.335$ and $\varepsilon_y = -0.76311$, with no induced curvature since $\mathbf{B} = \mathbf{0}$.

Similarly, assuming the application of a *unit moment resultant* in the x-direction, $M_x = 1$ N·mm/mm, for both cross-ply laminates, the ply stresses are plotted in Fig. 2.14 as a function of z. In this case the placement of the [0] plies has a significant effect on both the overall magnitude and location of the maximum ply stress in the x-direction. Note that in the [0/90/90/0] laminate the maximum stress, i.e., 18.893 MPa, occurs at the top and bottom face, see Fig. 2.14A, while in the case of the [90/0/0/90] laminate it occurs at the interface between the first and second plies, and the interface of the third and fourth plies, with the magnitude being 47.225 MPa, see Fig. 2.14B. Also, linear variation of stress through each ply is clearly evident, as expected with bending. No bending-twist coupling exists (since $D_{16} = D_{26} = 0$) and therefore $\kappa_{xy} = 0$ throughout the laminate. The mid-plane strains for both laminates are zero, since $\mathbf{B} = \mathbf{0}$, and the curvature in the x-direction for the [90/0/0/90] laminate is, $\kappa_x = 0.00213$ mm^{-1}, whereas, for the [0/90/90/0] laminate, it is significantly smaller at $\kappa_x = 0.00042726$ mm^{-1}. Note that, for both laminates, the curvature in the y-direction is $\kappa_y = -4.5779 \times 10^{-5}$ mm^{-1}. Consequently, utilizing Eq. (2.58) we see that the vertical deflection of a plate, pinned at the center is

$$w_0(x, y) = -\frac{1}{2}\left(0.42726 \times 10^{-3}x^2 - 4.5779 \times 10^{-5}y^2\right)$$

for a [0/90/90/0] laminate

$$w_0(x, y) = -\frac{1}{2}\left(2.13 \times 10^{-3}x^2 - 4.5779 \times 10^{-5}y^2\right)$$

for a [90/0/0/90] laminate   (2.79)

Therefore, given a 250 mm (in the x-direction) by 125 mm (in the y-direction) composite plate, the [0/90/90/0] laminate gives a maximum deflection at ($x = 125$ mm, $y = 0$ mm) of −3.338 mm whereas for the [90/0/0/90] laminate, this deflection is −16.641 mm, which is almost a factor of five times more deflection. Clearly, in the case of bending, to minimize both deflection and stresses it is more effective to have the stiffer [0] plies on the outside of the laminated plate.

As a final example, consider a similar four ply, IM7/8552 balanced, cross-ply laminate, but now one that is *not symmetric*, i.e., [0/0/90/90]. Again, each ply thickness is taken to be 0.15 mm for a total laminate thickness of 0.6 mm, and the laminate will be first subjected to a unit force resultant in the x-direction and then to a unit moment resultant in the x-direction:

```
NP = NP + 1;
OutInfo.Name(NP) = "Chapter 2 - Section 2.5.2 IM7-8552 [0,0,90,90] Nx";
Geometry{NP}.Orient = [0,0,90,90];
nplies = length(Geometry{NP}.Orient);
Geometry{NP}.plymat(1:nplies) = 1;
Geometry{NP}.tply(1:nplies) = 0.15;
Loads{NP}.Type  = [NM, NM, NM, NM, NM, NM];
Loads{NP}.Value = [ 1,  0,  0,  0,  0,  0];

NP = NP + 1;
OutInfo.Name(NP) = "Chapter 2 - Section 2.5.2 IM7-8552 [0,0,90,90] Mx";
Geometry{NP}.Orient = [0,0,90,90];
nplies = length(Geometry{NP}.Orient);
Geometry{NP}.plymat(1:nplies) = 1;
Geometry{NP}.tply(1:nplies) = 0.15;
Loads{NP}.Type  = [NM, NM, NM, NM, NM, NM];
Loads{NP}.Value = [ 0,  0,  0,  1,  0,  0];
```

Table 2.3 illustrates the resulting **ABD** matrix. Comparing this laminate with the previous two shown in Table 2.2, it is clear that while the components of the **A** matrix are identical, the **B** matrix now has nonzero diagonal members, B_{11} and B_{22}, and the

Table 2.3 ABD matrix for an asymmetric cross-ply laminate.

	[0/0/90/90]		
A Matrix (N/mm)	46931.4841	1678.6555	0
	1678.6555	46931.4841	0
	0	0	3096
B Matrix (N)	-6252.8529	-1.4211e-14	0
	-1.4211e-14	6252.8529	0
	0	0	-2.8422e-14
D Matrix (N·mm)	1407.9445	50.3597	0
	50.3597	1407.9445	0
	0	0	92.88

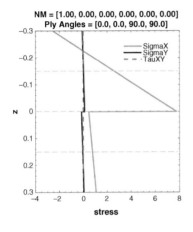

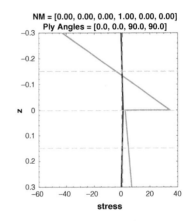

A   Applied unit force resultant, Nx

B   Applied unit moment resultant, Mx

Figure 2.15 Ply level stresses (MPa) in an IM7/8552 cross-ply [0/0/90/90] laminate when subjected to A) a unit N_x force resultant, 1 N/mm and B) a unit M_x moment resultant, 1 N·mm/mm.

components of the **D** matrix, aside from D_{11} and D_{22}, are identical. As a result of the nonzero **B**, one should expect extension-bending coupling to occur. In the case of an applied unit force resultant $N_x = 1$ N/mm, the resulting mid-plane micro ($\times 10^{-6}$) strains are $\varepsilon_x = 52.351$ and $\varepsilon_y = -1.8725$, but now with the addition of an induced curvature in the x-direction of, $\kappa_x = 2.325 \times 10^{-4}$ mm^{-1}. In the case of an applied unit moment $M_x = 1$ N·mm/mm, the resulting curvatures are $\kappa_x = 1.745 \times 10^{-3}$ mm^{-1} and $\kappa_y = -6.2417 \times 10^{-5}$ mm^{-1}, along with the additional induced mid-plane micro ($\times 10^{-6}$) strains of $\varepsilon_x = 232.5$ and $\varepsilon_y \approx 0$. Note that no normal-shear coupling exists (since $A_{16} = A_{26} = 0$) and therefore $\sigma_{xy} = 0$ throughout the laminate. The resulting stress profiles for each load state are shown in Fig. 2.15, with the linear distribution of stresses throughout the plies giving evidence to the presence of bending under both loading scenarios. As expected, the majority of the stress is carried in the [0] plies as loading is in the x-direction.

2.5.3   Local ply level stresses in material coordinates

To this point, effective properties and ply level stresses in the laminate (or global) coordinate system have been examined. However if one wants to examine the state of stress or strain in the material coordinate system (i.e., the axial (along the fiber) and transverse (perpendicular to the fiber) directions) all one need do is apply the transformations given in Eq. (2.43) to the laminate coordinate system (x-y) ply stresses or strains. These local, material coordinate stress and strain states are particularly important when examination of lower length scale (fiber or matrix) stresses or strains, or, of failure of the composite is desired (see Chapters 3 and 4). To illustrate the interaction between global and local fields, consider two laminates ([90/0/0/90] and [45/–45/ –45/45]) that, when loaded appropriately, give rise to identical local fields. For example, applying a uniaxial load of $N_x = 1$ N/mm to a four ply, 0.6 mm thick, [45/–45/–45/45] IM7/8552 laminate:

```
NP = NP + 1;
OutInfo.Name(NP) = "Chapter 2 - Section 2.5.3 IM7-8552 [45,-45,-45,45]";
Geometry{NP}.Orient = [45,-45,-45,45];
nplies = length(Geometry{NP}.Orient);
Geometry{NP}.plymat(1:nplies) = 1;
Geometry{NP}.tply(1:nplies) = 0.15;
Loads{NP}.Type  = [NM, NM, NM, NM, NM, NM];
Loads{NP}.Value = [ 1,  0,  0,  0,  0,  0];
```

This results in the following laminate midplane strains and curvatures and ply level stresses, in both global and local coordinate systems.

Laminate Midplane Strains and Curvatures [45/–45/–45/45]

Epsx0	Epsy0	Gamxy0	Kappax0	Kappay0	Kappaxy0
9.1035e-05	−7.0463e-05	−1.6538e-21	1.0571e-19	−9.1964e-20	−1.639e-20

Ply stresses through the thickness [45/–45/–45/45]

Z-location	SigmaX	SigmaY	TauXY		Sigma11	Sigma22	Tau12
−0.3	1.6667	8.0571e-16	0.71463		1.548	0.11871	−0.83333
−0.15	1.6667	5.4548e-16	0.71463		1.548	0.11871	−0.83333
−0.15	1.6667	4.8957e-16	−0.71463		1.548	0.11871	0.83333
0	1.6667	4.0014e-16	−0.71463		1.548	0.11871	0.83333
0	1.6667	4.0014e-16	−0.71463		1.548	0.11871	0.83333
0.15	1.6667	3.1071e-16	−0.71463		1.548	0.11871	0.83333
0.15	1.6667	2.5006e-17	0.71463		1.548	0.11871	−0.83333
0.3	1.6667	−2.3523e-16	0.71463		1.548	0.11871	−0.83333

Alternatively, an equivalent multiaxial load state ($N_x = 0.5$, N/mm $N_y = 0.5$ N/mm and $N_{xy} = 0.5$ N/mm) can be applied to a similar 0.6 mm, four ply [90/0/0/90] IM7/8552 laminate:

```
NP = NP + 1;
OutInfo.Name(NP) = "Chapter 2 - Section 2.5.3 IM7-8552 [90,0,0,90]";
Geometry{NP}.Orient = [90,0,0,90];
nplies = length(Geometry{NP}.Orient);
Geometry{NP}.plymat(1:nplies) = 1;
Geometry{NP}.tply(1:nplies) = 0.15;
Loads{NP}.Type  = [NM, NM, NM, NM, NM, NM];
Loads{NP}.Value = [0.5,0.5,0.5, 0,   0,   0,   0];
```

An identical local stress state (σ_{11}, σ_{22}, and σ_{12}) in each ply will be induced. This equivalent multiaxial state of stress was deduced by inspection of Fig. 2.8 at an angle of 45°, except here, the rotation (from the 1st ply at 90° to 45°) is negative and thus the shear must similarly be negative. The resulting midplane strains and curvatures, as well as ply stresses, are given below.

Laminate Midplane Strains and Curvatures [90/0/0/90]

Epsx0	Epsy0	Gamxy0	Kappax0	Kappay0	Kappaxy0
1.0286e-05	1.0286e-05	0.0001615	1.6514e-20	-2.922e-22	4.9419e-20

Ply stresses through the thickness [90/0/0/90]

Z-location	SigmaX	SigmaY	TauXY	Sigma11	Sigma22	Tau12
-0.3	0.11871	1.548	0.83333	1.548	0.11871	-0.83333
-0.15	0.11871	1.548	0.83333	1.548	0.11871	-0.83333
-0.15	1.548	0.11871	0.83333	1.548	0.11871	0.83333
0	1.548	0.11871	0.83333	1.548	0.11871	0.83333
0	1.548	0.11871	0.83333	1.548	0.11871	0.83333
0.15	1.548	0.11871	0.83333	1.548	0.11871	0.83333
0.15	0.11871	1.548	0.83333	1.548	0.11871	-0.83333
0.3	0.11871	1.548	0.83333	1.548	0.11871	-0.83333

Now however, the resulting midplane strains and laminate (*x-y* coordinate) ply stresses differ significantly between the two composite laminates. Such a result should be clear when one considers that both laminates' ply (or fiber) orientations are orthogonal to each other and thus one laminate is merely a −45-degree (clockwise) rotation from the other. This simple example clearly illustrates the often complex interaction between stacking sequence and global loading on the resulting local (material coordinate) ply level stress and strain states.

2.6 Laminate notation and classification

Laminates typically have multiple plies with varying orientations, and since the stacking sequence is an essential descriptor for laminates it is important that a standard,

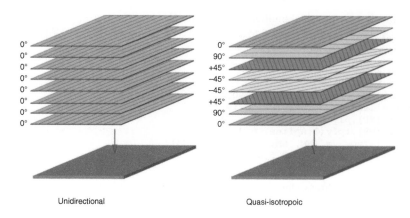

Unidirectional   Quasi-isotropoic

Figure 2.16 Example composite laminate layups.

shorthand, method of denoting laminates is defined. In this way, one does not have to list every angle explicitly for each ply when there are many plies. Furthermore, the stacking sequence is a simple way to classify laminates, for example unidirectional, cross-ply, and/or symmetric. Fig. 2.16 illustrates two eight-ply laminates, the first is unidirectional (all plies having their fibers oriented in the same direction theta, θ, in this case $0°$), denoted in short hand as $[0_8]$, while the second is a quasi-isotropic, symmetric laminate, denoted as $[0,90,\pm45]_s$. The subscript "s" stands for symmetric and indicates that the specified layup sequence is mirrored about the midplane of the laminate, see Fig. 2.2. Note the use of "s", declaring a laminate to be symmetric, also implies that thickness and material properties per ply are all the same. The subscript "T", can be used for clarity to indicate that the complete laminate layup has been specified. Other often used notation includes a numeric subscript, N, which means repeat N times; "U" which means unidirectional; "F" which means fabric, and "CL" which means cloth. Below are some example laminates specified using these notations:

$[0/-45/90/45]_s$	Contains 8 plies total; $[0/-45/90/45/45/90/-45/0]_T$
$[0/-45/90/45]_N$	Contains 4N plies total and is not symmetric
$[\pm45/0]_s$	Contains 6 plies; that is $[45/-45/0/0/-45/45]_T$
$[(0/45)_2/90]_s$	Contains 10 plies; that is $[0/45/0/45/90/90/45/0/45/0]_T$
$[0/45/-45/90]_{2s}$	Contains 16 plies total; $[0/45/-45/90/0/45/-45/90]_s$
$[0_2/45/-45/90_2]_s$	Contains 12 plies total; $[0/0/45/-45/90/90]_s$
$[0_U/45_{CL}]$	Contains 2 plies total; 1 uni @ $0°$ and 1 cloth @ $45°$

Although any type of laminate can be specified, special names (or classifications) have been given to certain types of laminates, e.g., symmetric, cross-ply, angle-ply, anti-symmetric, and balanced laminates, see Fig. 2.17. The physical interpretation of these laminates will be described in the next section. Here let it suffice to provide a definition of these classifications. As stated before, a *symmetric* laminate is one in which every ply (angle, thickness, and material ply properties) to one side of the

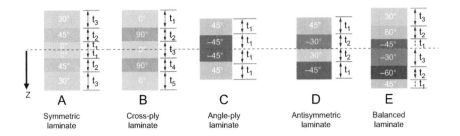

Figure 2.17 Typical classifications of laminates.

laminate midplane are mirrored about the midplane (present at the identical distance on the opposite side). A *cross-ply* laminate is one that has all plies oriented at either $0°$ or $90°$ (with at least one of each), e.g., Fig. 2.17B, which is also symmetric. An *angle-ply* laminate is one that has an equal number of equal thickness plies composed of the same material at a positive angle, $+\theta$, and negative angle, $-\theta$, e.g., Fig. 2.17C. *Antisymmetric* laminates have symmetrically located plies with mutually reversed orientations, see Fig. 2.17D, for example $[+\theta_1/-\theta_2/+\theta_2/-\theta_1]$, [0/90/0/90], or $[+\theta/-\theta]_T$. A laminate is said to be *balanced* if for every ply (of the same thickness and material) with a positive angle, $+\theta$, there exists a ply with a negative angle, $-\theta$, (where 0 and 90 plies are irrelevant) somewhere in the laminate, see Fig. 2.17E, e.g., [0/30/–45/–30/45/90] or $[0/90]_T$. A laminate does not have to be symmetric to be balanced.

Finally, a laminate is known as regular when the thickness of every ply is the same; however, if the ply thicknesses are not equal, then the laminate is called an irregular laminate and typically the ply thickness is denoted along with the ply angle. For example [0@t/90@2t/45@3t] would indicate that the first ply has thickness t while the second has twice the thickness of the first and the third has three times the thickness of the first ply.

2.6.1   Special laminates

In general, the **A**, **B**, and **D** matrices are fully populated; where **A**, the extensional stiffness, relates midplane normal and shear strains to normal and shear force resultants; **B**, the bending-extensional coupling stiffness relates bending curvatures to force resultants and midplane strains to moment resultants; **D** is the bending stiffness matrix and relates bending curvatures to moment resultants. Given specific stacking sequences, terms within these matrices become zero and, in fact, for certain composite layups, all entries within a given stiffness matrix are zero. For example, from Eq. (2.65), if plies are arranged with identically oriented plies at equal distance from the midplane (so it is symmetric about the midplane) then the matrix **B** is always zero and no bending-extensional coupling exists. All asymmetric (or non-symmetric) laminates have bending-extension coupling, that is, $B_{ij} \neq 0$. A number of other specialized laminate configurations, shown in Fig. 2.17, are described in Table 2.4

Table 2.4 Specialized laminates and the impact on the **ABD** matrix (for equal thickness plies of same material).

Type		A	B	D	Comment
Unidirectional	[0] or [90]	$A_{16}=A_{26}=0$	$B_{ij}=0$	$D_{16}=0$ $D_{26}=0$	Symmetric
	$[\theta]$		$B_{ij}=0$		Symmetric
Balanced		$A_{16}=A_{26}=0$			All off-axis (non-0, non-90) plies occur in pairs
Cross-Ply	Symmetric	$A_{16}=A_{26}=0$	$B_{ij}=0$	$D_{16}=0$ $D_{26}=0$	Balanced
	Asymmetric	$A_{16}=A_{26}=0$	$B_{12}=B_{16}=B_{26}=B_{66}=0$ $B_{22}=-B_{11}\neq0$	$D_{16}=0$ $D_{26}=0$	
Angle Ply	General	$A_{16}=A_{26}=0$			Balanced
	Symmetric	$A_{16}=A_{26}=0$	$B_{ij}=0$		
	Antisymmetric	$A_{16}=A_{26}=0$	$B_{11}=B_{12}=B_{22}=B_{66}=0$ $B_{16}\neq0$, $B_{26}\neq0$	$D_{16}=0$ $D_{26}=0$	
Quasi-Isotropic	$[\Delta\theta_1, \Delta\theta_2, ..., \Delta\theta_N]$ N fiber angles $\Delta\theta = \pi/N$ and $N \geq 3$	$A_{11}=A_{22}$, $A_{66}=\frac{1}{2}(A_{11}-A_{12})$ $A_{16}=A_{26}=0$			Any equal number of *all* N angles can be in laminate
	Symmetric	$A_{11}=A_{22}$, $A_{66}=\frac{1}{2}(A_{11}-A_{12})$ $A_{16}=A_{26}=0$	$B_{ij}=0$		

along with their impact on specific terms within the **A**, **B**, and **D** matrices. In the case of unidirectional or cross-ply laminates, bending-twisting coupling is eliminated. Extensional-shear coupling can be eliminated by specifying what is called a balanced laminate construction, i.e., for every $+\theta$ ply, there is a $-\theta$ ply present. These plies do not have to be adjacent to satisfy this requirement nor does the laminate have to be symmetric. Balanced laminates always have $A_{16} = A_{26} = 0$. The MATLAB code can be used to construct various examples for each of these types of laminates and, for example, to study the effects of small changes in ply orientations and ply thicknesses to see the sensitivity of such manufacturing defects on the **A**, **B**, and **D** matrices. For practical structures it is necessary to understand not only mechanical loading, but also thermal loading, so one should investigate the effects of temperature change, see the exercises below for some informative examples.

2.6.2   Design rules for laminates

Fiber reinforced composites in general have various advantages over conventional structural materials such as metals. These include but are not limited to their high stiffness to density (E/ρ) and high strength to density (σ/ρ) ratios, thermal stability, and fatigue and/or environmental resistance, which make them a desirable material for weight sensitive applications such as in aerospace components. More important is the ability to tailor the thermomechanical response of a composite laminate to a specific application. Yet it is precisely this often-complex, tailorable, anisotropic behavior that can make the design of composites significantly more difficult than with isotropic metals, since one has increased degrees of freedom coming from the specification of individual ply thicknesses, effective material properties, and orientation angles. Consequently, one needs to have a thorough understanding/appreciation of the unique coupling effects present in composites that are not typically present in isotropic materials (e.g., metals) before undertaking design with composites. The classical lamination theory, discussed previously, provides just such an understanding and allows one to achieve the "best designs" by capitalizing on the advantages and minimizing any disadvantages along the way. However, design rules are often instituted to account for effects that lamination theory does not include, e.g., delamination, stress concentrations, and manufacturing issues with respect to specific stacking sequences.

To assist the designer, a number of "lessons learned" or "best-practices" have been developed over time to minimize design and manufacturing risks when using composite materials (Kassapoglou, 2011; Composite World, 2002). For example, most fibers should be aligned in the direction of the principal loads or stresses, thus axially in the case of beams and longitudinally and circumferentially in the case of pressure vessels. Shells, even under pure axial loads, can buckle in both the axial and circumferential directions, therefore hoop (circumferential) fibers are needed for stability. During manufacturing, constant ply angles are difficult to maintain, especially on doubly-curved surfaces, therefore realistic variations should be accounted for in the analysis, and limiting the number of ply angles in a given laminate can be very beneficial. Non-symmetric laminates hold their shape only at cure temperature, that is they will warp (for open sections) or develop significant residual stresses (for closed sections) at other temperatures. Symmetric laminates avoid this problem. Quasi-isotropic laminates are isotropic under extensional loading only; strength and bending stiffness are not isotropic.

A major design consideration, especially for PMCs, is delamination, which is typically induced due to the development of interlaminar stresses (σ_{zz}, τ_{xz}, τ_{yz}). These stresses are neglected in CLT and are a result of three-dimensional effects typically induced at free-edges, notches, cut-outs, ply-drops or terminations, and other material discontinuities. These stresses can be calculated (but are beyond the scope of this book) and have been shown to decay rapidly inside the laminate [see Jones (1975), Hyer (1998), Herakovich (1998), and Mittelstedt and Becker (2007)]. Interlaminar stresses can be controlled/minimized through the choice of material, ply orientation, stacking sequence, ply thickness, and/or through tailoring near discontinuities (e.g., functionally grading the material, see Aboudi et al. (2013)). Similarly, suppression of interlaminar stresses near free edges can be achieved using through-thickness reinforcement, e.g., z-pinning, stitching, use of 3D woven materials, etc. Such techniques, however, can add significant manufacturing costs.

Finally, thermal effects can play an important role in any design. CLT clearly illustrates that the effective coefficient of thermal expansion depends upon stacking sequence, ply orientation, ply thickness, and material selection. Consequently, these coefficients can be tailored to give the laminate desired thermal expansions (e.g., close to zero). This can be very important to a designer when large thermal excursions or gradients exist. Remember, however, even though no external (laminate level) thermal strains may be induced, significant internal stresses may result and lead to premature failure if not accounted for. Examples of composite design rules are given in Table 2.5.

2.7   Summary

This chapter presents classical lamination theory, wherein the composite laminate plies are represented using effective anisotropic properties (rather than using micromechanics to calculate their effective properties). Towards this end, the chapter begins with a review of linear, thermoelastic, constitutive equations for isotopic, transversely isotropic, and orthotropic materials. Then the mechanics of a ply, along with the ply stress-strain coordinate transformations and strain-displacement relations, are introduced. This culminates in the development of the thermomechanical laminate constitutive relations, that is, the construction of the **ABD** matrix along with the mechanical and thermal force and moment resultant expressions. The associated MATLAB implementation of these classical laminate expressions, as well as other common functions and solution flow charts, are then provided. Next, a number of example problems addressing different types of laminates (unidirectional, cross-ply, and angle-ply) and classes of composite materials (PMCs, MMCs, and CMCs) are presented. Finally, a brief overview of laminate notation and classification and common design rules are provided.

Table 2.5 Typical "best design practices" for composite structures (Kassapoglou, 2011; Composite World, 2002).

Composite Practice	Reason Applied
Try to employ "Balance" and "Symmetry" conditions whenever possible	Minimize/eliminate value of "B-matrix", avoid bending, coupling, warping and twisting effects
Avoid stacking too many adjacent plies all at one angle	Minimize delamination and residual stresses. Minimize the size of through-thickness matrix cracks due to thermal or transverse loading during service. Avoid ply stacks > 0.6-0.8 mm, approx. 4-5 plies
Angle plies carry shear loads; $\pm45°$ plies are the most efficient.	In-plane shear is carried in tension and compression, as well as shear, in $\pm45°$ layers
Angle differences between plies should be minimized, e.g. $[0°/45°/90°/-45°]$ rather than $[0°/90°/45°/-45°]$	Minimize interlaminar shear stresses caused by shear coupling. Reduces free-edge stresses (which cause delamination). Avoids microcracking–particularly for cryogenic or wide temperature excursion applications.
Add $\pm45°$ plies with at least one pair on laminate surfaces	Provides better damage tolerance for many applications
Add fabric (woven) ply to inner or outer layer	Fabric ply can absorb more impact damage as well as minimize drilling "breakout" for holes
Use larger faction of $\pm$ plies in shear regions	Shear loads are best handled with additional +/– plies in a structure.
When designing with only $0°$, $90°$ and $\pm45°$ plies, all laminate designs should have a minimum of one layer in each direction	$0°$ layers for longitudinal loads, $90°$ layers for transverse loads, $\pm45°$ layers for shear loads. Protects against secondary load cases not considered in design
When designing with only $0°$, $90°$ and $\pm45°$ plies, at least 10% of the total number of plies should be oriented in each direction	
Maximum of 60% of any one ply angle in laminates	Protects against secondary load cases not considered in design
Minimize stress concentrations and point loads	Composites, in contrast to metals, can be elastic to failure, with little or no material nonlinearity to minimize sudden failures
Use a minimum of 8 plies for the skins Use a minimum of 4 plies for the stiffeners No more than four $0°$ plies can be used consecutively	Minimize manufacturing/imperfection sensitivity and maintain damage tolerance
Drop/add plies according to the 20/1 rule (drop 1 ply over a length of 20 ply thickness) Never drop/add exterior plies Drop/add plies keeping the most symmetry and balance	Minimize stress concentrations

2.8   Exercises

Exercise 2.1. Given a four-ply unidirectional laminate, $[0_4]$, 0.6 mm thick, composed of IM7/8552 (ply properties given in Table 2.1), determine the corresponding **A**, **B**, and **D** matrices assuming equally thick plies. Determine the effective laminate engineering properties. Calculate the mid-plane strains given an applied $N_x = 1$ N/mm and midplane curvatures given an applied $M_x = 1$ N·mm/mm. Calculate and plot the ply stresses in global coordinates for each loading case.

Exercise 2.2. Given the four-ply unidirectional laminate of Exercise 2.1, apply midplane strains, $\left[\varepsilon_x^0, \ \varepsilon_y^0, \ \gamma_{xy}^0 \right] = [1.1353 \times 10^{-5}, -3.6331 \times 10^{-6}, 0]$, and midplane curvatures, $\left[\kappa_x^0, \ \kappa_y^0, \ \kappa_{xy}^0 \right] = [3.7844 \times 10^{-4}, -1.211 \times 10^{-4}, 0]$/mm. Calculate the force and moment resultants.

Exercise 2.3. Given the four-ply unidirectional laminate of Exercise 2.1, vary the ply angle between $[0_4]$ and $[90_4]$ with an increment of 15 degrees. For the resulting seven unidirectional laminates, determine and plot versus angle the effective laminate engineering properties. Results should be consistent with Fig. 2.9–Fig. 2.12.

Exercise 2.4. Given the four-ply unidirectional laminates of Exercise 2.3, examine the **A**, **B**, and **D** matrices. Which laminates have extension-shear coupling? Which laminates have bending-twist coupling? Calculate and plot versus angle the mid-plane strains for each laminate given an applied $N_x = 1$ N/mm.

Exercise 2.5. Given the four-ply unidirectional laminates of Exercise 2.4 but now using the SCS-6/Ti-15-3 material system (ply properties given in Table 2.1), calculate and plot versus angle the mid-plane strains for each laminate given an applied $N_x = 1$ N/mm.

Exercise 2.6. A laminate with unit thickness is a special case. Consider a four-ply unidirectional laminate, $[0_4]$, 1 mm total thickness, composed of IM7/8552. What N_x should be applied to get a $\sigma_{xx} = 6$ MPa in all plies? What N_x should be applied to result in a $\sigma_{xx} = 10$ MPa in all plies? Given the laminate of Exercise 2.1 (0.6 mm total thickness) what N_x should be applied to result in $\sigma_{xx} = 10$ MPa in all plies?

Exercise 2.7. Consider a unidirectional laminate with thickness 2t, composed of IM7/8552 subjected to a load M_x resulting in a curvature κ_x^0. If the thickness is now t, what M_x would be required to give the same curvature κ_x^0? Hint the required moment scales with **D**.

Exercise 2.8. Consider a four-ply regular cross-ply laminate, 0.6 mm total thickness, composed of IM7/8552. Determine the laminate **A**, **B**, and **D** matrices for the following stacking sequences: [0/0/90/90], [0/90/90/0], [0/90/0/90], [90/0/0/90]. Classify these laminates and indicate (based on Table 2.4) which features in their **A**, **B**, or **D** matrices confirm the classifications.

Exercise 2.9. Consider a four-ply regular angle-ply laminate, 0.6 mm total thickness, composed of IM7/8552. Determine the laminate **A**, **B**, and **D** matrices for the following stacking sequences: [45/–45/–45/45], [30/–30/–30/30], [45/–45/45/–45], [45/–45/–30/30]. Classify these laminates and indicate (based on Table 2.4) which features in their **A**, **B**, or **D** matrices confirm the classifications.

Exercise 2.10. Consider an eight-ply regular laminate that is 1.2 mm thick and composed of IM7/8552. Determine the laminate **A**, **B**, and **D** matrices and the effective laminate properties for the following stacking sequence: $[-45/0/+45/90]_s$. Which components of the **A** are zero and which are equal? Perform the following calculation using the effective laminate properties, $E_x/2(1 + v_{xy})$, and compare to the laminate effective G_{xy} obtained. Apply $N_x = 1$ N/mm and calculate σ_x in the zero plies. Next apply $N_y = 1$ N/mm to the laminate and calculate the σ_y in the 90 plies. This laminate is an example of a quasi-isotropic.

Exercise 2.11. An important class of laminates is called quasi-isotropic because the extensional elastic response of such composites is isotropic in the sense that the **A** matrix has 2 independent components: $A_{11} = A_{22}$, $A_{16} = A_{26} = 0$, $A_{66} = \frac{1}{2}(A_{11} - A_{12})$, see Table 2.4. Quasi-isotropy results when the angle between a laminate's ply angles is:

$$\Delta\theta = \pi / N$$

where N must be ≥ 3 and the laminate can contain any equal number of *all* N ply angles. The plies must also be equal thickness and comprised of the same material. Consider four laminates with $\Delta\theta = 180° (= 0° = \pi/1)$, $90° (= \pi/2)$, $60° (= \pi/3)$, and $45°(= \pi/4)$ and show whether or not each laminate is quasi-isotropic. Does the ply sequence affect quasi-isotropy and why or why not? Provide an example layup of a *symmetric* quasi-isotropic laminate with minimum number of plies.

Exercise 2.12. Given a two-ply angle-ply laminate, [–30/+30], 1 mm thick, composed of IM7/8552, determine the corresponding **A**, **B**, and **D** matrices and calculate and plot the ply stresses in x-y coordinates given an applied load $N_x = 10$ N/mm. To determine the impact of potential manufacturing defects, repeat the exercise for variation in ply angle of 1°, that is, [–29/+31]. What is the percent change in maximum ply σ_x between the two laminates?

Exercise 2.13. Consider a $[\pm30/0]_s$ composite plate of dimensions 125 mm in the y-axis direction by 250 mm in the x-axis direction (see Fig. 2.5) whose ply thickness is 0.15 mm composed of IM7/8552. Calculate the required applied force resultants and actual forces along the edge of the plate to result in only a midplane strain, $\varepsilon_x^0 = 0.01$ with all other strains and curvatures remaining zero. How do the applied force resultants and forces change for a square 250 mm plate?

Exercise 2.14. Consider a $[\pm30/0]_s$ composite plate of dimensions 125 mm in the y-axis direction by 250 mm in the x-axis direction (see Fig. 2.5) whose ply thickness is 0.15 mm composed of IM7/8552. Calculate the required moment resultants to have only a twisting curvature, $\kappa_{xy}^0 = 0.00222$ mm^{-1} with all other strains and curvatures remaining zero. Calculate the induced midplane bending deflection (w_0) at the corner of the plate, assuming that no deflection takes place at the center of the plate and that the midplane strains and curvatures are constant. How do the moment resultants and deflection change for a square 250 mm plate?

Exercise 2.15. How thick would an aluminum plate (with aluminum properties given in Table 2.1) have to be to have the same midplane bending deflection (w_0) at the corner as the rectangular composite plate in Exercise 2.14, if it were the same length and width and loaded with the same moment resultants ($\mathbf{M} = [2.6481, 0.9584, 3.8664$ N·mm/mm])? How much heavier would the aluminum plate be than the rectangular plate in Exercise 2.14.

Exercise 2.16. Consider the square plate described in Exercise 2.14 but now with a layup of $[\pm 45/0]_s$. Calculate the deflection at the corner, if it were the same length and width and loaded with the same moment resultants ($\mathbf{M} = [2.6481, 0.9584, 3.8664]$ N·mm/mm). How does this deflection compare to that in Exercise 2.14? How do the curvatures compare?

Exercise 2.17. Consider a regular cross-ply symmetric laminate, $[0/90]_{Ns}$, determine the minimum total number of plies, required to keep $\varepsilon_x^0 \leq 0.01$ for all composite material systems in Table 2.1 when subjected to $N_x = 1800$ N/mm, with all other force and moment results zero. For PMC materials use a ply thickness of 0.15 mm, and for MMC and CMC materials use a ply thickness of 0.25 mm. Specify final layup and total thickness for each system. If minimizing mass were the only design consideration for this laminate, which composite material would be selected? Hint: the layups must be in multiples of four plies.

Exercise 2.18. Consider a four-ply unidirectional laminate, $[0_4]$, 0.6 mm thick, composed of IM7/8552, and apply only a thermal load, $\Delta T = 100$ °C. Calculate the midplane strains and compare them to the effective coefficients of thermal expansion multiplied by ΔT. Plot the ply stresses in x-y coordinates; why are they zero?

Exercise 2.19. Consider a regular cross-ply laminate, $[90/0/0/90]$, 0.6 mm thick, composed of IM7/8552, and apply only a thermal load, $\Delta T = 100$ °C. Calculate midplane strains and curvatures and plot the ply level stresses in x-y coordinates. Why are the ply stresses non-zero despite the fact that no mechanical loading has been applied? Calculate the average σ_x and σ_y in the laminate. Change the lay-up to $[90/90/0/0]$ and calculate mid-plane strains and curvatures and plot the ply level stresses in x-y coordinates. Why is there bending? Are the average stresses still zero? Why or why not?

Exercise 2.20. Consider the asymmetric regular cross-ply laminate, $[90/90/0/0]$, of Exercise 2.19. What force and moment resultants should be applied to ensure that the mid-plane strains and curvatures are zero? How do these applied loads relate to the thermal force and moment resultants? Plot the ply level stresses in x-y coordinates.

Exercise 2.21. Consider a regular quasi-isotropic laminate, $[-45/0/45/90]_s$, composed of IM7/8552 with a ply thickness of 0.15 mm. Apply $N_x = 300$ N/mm with all other force and moment resultants zero and no thermal loading. Calculate and plot the ply level stresses. Next apply $N_x = 300$ N/mm with all other force and moment resultants zero in combination with $\Delta T = 150$ °C. Calculate and plot the ply level stresses in x-y coordinates. Compare the axial stress in ply coordinates (σ_{11}) in the [0] and [90] plies between the two load cases. Did the magnitudes of these stresses increase or decrease and by what percentage.

Exercise 2.22. Consider four regular quasi-isotropic laminates, $[-45/0/45/90]_s$ with a ply thickness of 0.15 composed of IM7/8552, glass/epoxy, SCS-6/Ti-15-3, and SiC/SiC, and apply $\Delta T = 150$ °C with no mechanical loading. Calculate and plot the ply level stresses in x-y coordinates. Compare the maximum ply σ_x, along with the ply level properties, among the four composite materials. What aspect of the ply properties is the main driver of the trend in the induced maximum ply σ_x?

References

Aboudi, J., Arnold, S.M., Bednarcyk, B.A., 2013. Micromechanics of Composite Materials: A Generalized Multiscale Analysis Approach. Elsevier, Oxford, United Kingdom.

Agarwal, B.D., Broutman, L.J., 1980. Analysis and Performance of Fiber Composites. John Wiley & Sons, Inc., New York, NY, Unite States.

Arnold, S.M., Mital, S., Murthy, P.L., Bednarcyk, B.A., 2016. Multiscale modeling of random microstructures in SiC/SiC ceramic matrix composites within MAC/GMC framework. In: American Society for Composites 31st Annual Technical Conference, Williamsburg, VA, United States, September 19-22.

Composite World (2002): https://www.compositesworld.com/articles/laminate-design-rules, Accessed April 18, 2020.

Daniel, I.M., Ishai, O., 1994. Engineering Mechanics of Composite Materials. Oxford University Press.

Dunn, D.G., 2010. The Effect of Fiber Volume Fraction in Hipercomp SiC-SiC Composites. Ph.D. Thesis, Alfred University Alfred, NY, United States.

Jones, R.M., 1975. Mechanics of Composite Materials. Hemisphere Publishing Corp., New York, NY.

Herakovich, C.T., 1998. Mechanics of Fibrous Composites. John Wiley & Sons, Inc., New York, NY.

Hyer, M.W., 1998. Stress Analysis of Fiber-Reinforced Composite Materials. WCB McGraw-Hill, New York, NY, United States (Updated version published in 2009).

Kassapoglou, C., 2011. Design and Analysis of Composite Structures: With Applications to Aerospace. AIAA.

Kirchhoff, G., 1850. Über das Gleichgewicht und die Bewegung einerelastischen Scheibe. Journal für Reins und Angewandte Mathematik 40, 51–88.

Love, A.E.H., 1888. On the small free vibrations and deformations of elastic shells. Philos. Trans. Roy. Soc. (London), Vol. série A, N° 17, pp. 491–549.

MATLAB, 2018. version R2018. The MathWorks Inc, Natick, Massachusetts, United States.

Mittelstedt, C., Becker, W., 2007. Free-edge Effects in Composite Laminates. Appl. Mech. Rev. 60 (5), 217–245.

Nettles, A., 1994. Basic Mechanics of Laminated Composite Plates. NASA RP-1351, Oct. 1994.

Ratcliffe, J.G., Czabaj, M.W., O'Brien, T.K., 2012. Characterizing Delamination Migration in Carbon/Epoxy Tape Laminates. In: 2012 American Society for Composites 27th Technical Conference 15, US-Japan Conference on Composite Materials; 1-3 October 2012.

Reddy, J.N., 2003. Mechanics of Laminated Composite Plates and Shells: Theory and Analysis. CRC Press, Boca Raton, FL.

Closed form micromechanics

3

Chapter outline

Practical Micromechanics of Composite Materials. DOI: 10.1016/C2019-0-04193-3

3.1   Introduction

In Chapter 2 the fundamental concepts for analyzing a composite using macromechanics and lamination theory were outlined. At the core of macromechanics is the
assumption that a material volume element, and therefore each ply, need only be
described by experimentally measurable engineering material properties (typically
isotropic, transversely isotropic, or orthotropic). Then, by using classical lamination
theory (CLT), complex composite layup sequences, involving different ply thicknesses
and ply angles, can be combined to obtain the overall laminate stiffness matrix
(**ABD**) relating midplane strains and curvatures to generalized force and moment
resultants. From the laminate compliance matrix (inverse of the **ABD, abd**) one can
obtain effective laminate engineering properties. Fig. 3.1 illustrates both the ply and
laminate level length scales (which were discussed in detail in Chapter 2), however, an
additional lower length scale, associated with the constituent materials, is now shown.
Micromechanics enables this connection of the constituent length scale to that of the
homogenized ply by calculating the effective properties/response for each ply based on
the constituent properties and microstructural features. Consequently, in this chapter,
Chapter 3, the fundamental concepts of micromechanics are presented, along with four
classical micromechanics theories: the Voigt approximation, the Reuss approximation,
the Mori-Tanaka (MT) Method and the Method of Cells (MOC). It is shown that, at
the heart of all of these theories, is the ability to obtain strain concentration tensors,
which relate the local strains throughout the composite material constituents to the
global or average strains acting on a composite material. Once these concentration
tensors are known, the effective, homogenized, mechanical properties and effective
coefficients of thermal expansion for the composite can be readily calculated. These
composite material responses can represent the plies within a laminate, which can
then be analyzed using CLT as described in Chapter 2. Thus, a key characteristic
of a micromechanical theory is its ability to predict strain (or stress) concentration
tensors. Further, the accuracy of any micromechanics theory is largely determined
by the accuracy of its strain concentration tensors. As stated, the strain concentration
tensors provide the local strain fields within the composite constituents, and from these
local strains, the local stresses can be calculated as well.

3.1.1   Macromechanics versus micromechanics

The distinguishing feature between micromechanical and macromechanical approaches to composite analysis is the lowest considered length scale. *Micromechanics*
begins at the micro scale (i.e., constituent level) and macromechanics begins at the
homogenized material level within the global scale (i.e., ply level). Each approach
has unique needs in terms of experimental characterization and verification. In the
micro scale approach, models for the constituents are employed, in conjunction with
a homogenization procedure, to predict the composite behavior. The macro scale
approach treats the composite (or ply) as an effective, anisotropic material, with its
own experimentally measurable properties. Consequently, the material itself (via a

Table 3.2 Fundamental concepts and equations for the micromechanics of heterogeneous materials, where $i, j, k, \ell = 1, 2, 3$.

Average stress	$\bar{\sigma}_{ij} = \frac{1}{V} \int_V \sigma_{ij} \, dV$	(3.1)
Average strain (for perfect bonding)	$\bar{\varepsilon}_{ij} = \frac{1}{V} \int_V \varepsilon_{ij} \, dV$	(3.2)
Homogeneous displacement boundary conditions	$u_i(S) = \varepsilon_{ij}^0 x_j$	(3.3)
Homogeneous traction boundary conditions	$t_i(S) = \sigma_{ij}^0 n_j$	(3.4)
Average strain theorem	$\bar{\varepsilon}_{ij} = \varepsilon_{ij}^0$	(3.5)
Average stress theorem	$\bar{\sigma}_{ij} = \sigma_{ij}^0$	(3.6)
Effective composite constitutive equations	$\bar{\sigma}_{ij} = C_{ijk\ell}^* \bar{\varepsilon}_{k\ell}$	(3.7)
	or $\bar{\varepsilon}_{ij} = S_{ijk\ell}^* \bar{\sigma}_{k\ell}$	

reproduce properties of the material as a whole. Alternatively, for scales below the RVE one must consider the microstructure of the material explicitly.

3.2.2 Fundamental concepts and related equations

There are several concepts fundamental to the micromechanical analysis of heterogeneous materials that are presented in this section. Rather than provide the full derivations, the resulting equations are presented in tabular form (see Table 3.2), along with some discussion to place the concepts and equations in context. The reader is referred (for example) to Hill (1963), Christensen (1979), and Aboudi et al. (2013) for more details on these concepts.

Under an imposed macroscopically homogeneous stress or deformation field on an RVE, the *average stress and strain tensors* are respectively defined by Eqs. (3.1) and (3.2), where V is the volume of the RVE. Thus, given the local stress or strain fields (no overbar) throughout the RVE (which may contain multiple materials), the average stresses and strains (with overbar) can be determined.

Homogeneous boundary conditions applied on the surface of a homogeneous body will produce a homogeneous field. Either displacements (u_i) or tractions (t_i) can be imposed at the boundary (S), as given in Eqs. (3.3) and (3.4), respectively. In Eq. (3.3), ε_{ij}^0 are constant strains applied at the surface, S, and x_j are the Cartesian coordinates that describe the location of a point in the body. In Eq. (3.4), σ_{ij}^0 are constant stresses at the surface and **n** is the unit outward normal vector to S. These homogeneous boundary conditions enable the loading on a composite to be defined simply by specifying either ε_{ij}^0 or σ_{ij}^0. This concept will be used extensively throughout this chapter as the classic closed-form micromechanics theories are introduced.

To calculate the average strains in a composite material, it would seem that one must solve the elasticity problem of the RVE subjected to the homogeneous displacement boundary conditions, Eq. (3.3). However, the *average strain theorem* proves that the average strains $\bar{\varepsilon}_{ij}$ are identical to the constant strains ε_{ij}^0 applied on S, Eq. (3.5). Likewise, the *average stress theorem* proves that, for homogeneous traction boundary conditions on S, see Eq. (3.4), the average stress field in the composite, $\bar{\sigma}_{ij}$, is identical

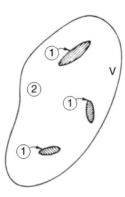

Figure 3.2 A schematic representation of a two-phase heterogeneous material with volume V.

to the constant stress, σ_{ij}^0, applied at S, Eq. (3.6). The average stress and strain theorems are obvious in the case of a homogeneous material, but less so for a composite material composed of multiple phases. For a composite, the internal stresses and strains may be distributed in a very complex way, but under homogeneous displacement boundary conditions, the average strain will always equal the constant strains ε_{ij}^0 applied on S.

The **effective composite constitutive equations** provide the relationship between the average composite stress and average composite strain. In the linearly elastic case, the average stress and strain are related by the effective stiffness tensor, $\mathbf{C^*}$, or the effective compliance tensor, $\mathbf{S^*}$, as shown in Eq. (3.7), where $\mathbf{S^*} = \mathbf{C^{*-1}}$. Clearly, if the composite's $\mathbf{C^*}$ can be obtained, the average composite elastic behavior can be treated in the exact same way as a homogeneous material. For example, the $\mathbf{C^*}$ could be used to represent a ply within a laminate. Furthermore, the composite effective engineering properties (e.g., effective Young's moduli) can be obtained from $\mathbf{C^*}$. Calculating $\mathbf{C^*}$ and the effective engineering properties of a composite is a primary purpose of micromechanics.

3.3 Strain and stress concentration tensors for two-phase composites

Here the concept of strain and stress concentration tensors, and their relationship to effective material properties, will be established. Consider a heterogeneous material composed of two phases, denoted by 1 and 2, as shown in Fig. 3.2. Average stress and strain values will be distinguished by an overbar, where the average of any quantity is defined by the integral of the quantity over a specified region divided by the volume of the region (see Eqs. (3.1) and (3.2)). Thus, $\bar{\sigma}$ and $\bar{\varepsilon}$ are the averages over the entire volume, whereas, for example, $\bar{\sigma}^{(1)}$ and $\bar{\varepsilon}^{(1)}$ are the averages over region (phase) 1. Consequently, the average stress and strain for a perfectly-bonded two-phase composite are,

$$\bar{\sigma} = v_1\bar{\sigma}^{(1)} + v_2\bar{\sigma}^{(2)} \tag{3.8}$$

$$\bar{\varepsilon} = v_1 \bar{\varepsilon}^{(1)} + v_2 \bar{\varepsilon}^{(2)} \tag{3.9}$$

where v_1 and v_2 (with $v_1 + v_2 = 1$) are the fractional concentrations by volume of the phases. The relations between stress and strain tensors at any point in the phases, assuming linear elasticity, can be written as,

$$\sigma^{(1)} = \mathbf{C}^{(1)}\varepsilon^{(1)}, \qquad \sigma^{(2)} = \mathbf{C}^{(2)}\varepsilon^{(2)} \tag{3.10}$$

with inverses,

$$\varepsilon^{(1)} = \mathbf{S}^{(1)}\sigma^{(1)}, \qquad \varepsilon^{(2)} = \mathbf{S}^{(2)}\sigma^{(2)} \tag{3.11}$$

Here, as in the previous chapter, $\mathbf{C}$ and $\mathbf{S}$ refer to the material stiffness and compliance tensors, respectively, which are uniform within each phase. Volume averaging Eqs. (3.10) and (3.11) (resulting in overbars on the stress and strain) and substituting into Eqs. (3.8) and (3.9) results in,

$$\bar{\sigma} = v_1 \mathbf{C}^{(1)}\bar{\varepsilon}^{(1)} + v_2 \mathbf{C}^{(2)}\bar{\varepsilon}^{(2)} \tag{3.12}$$

$$\bar{\varepsilon} = v_1 \mathbf{S}^{(1)}\bar{\sigma}^{(1)} + v_2 \mathbf{S}^{(2)}\bar{\sigma}^{(2)} \tag{3.13}$$

There is a unique dependence of the average strains in the phases upon the overall strain in the composite. Let this be written as,

$$\bar{\varepsilon}^{(1)} = \mathbf{A}_1\bar{\varepsilon}, \qquad \bar{\varepsilon}^{(2)} = \mathbf{A}_2\bar{\varepsilon} \tag{3.14}$$

where $\mathbf{A}_1$ and $\mathbf{A}_2$ are called the *strain concentration tensors*. It must be noted that these concentration tensors relate the *mechanical strains*, and Eq. (3.14) must be extended to account for thermal strains (see Section 3.9.2). It can be shown by substituting Eq. (3.14) into Eq. (3.9) that,

$$v_1 \mathbf{A}_1 + v_2 \mathbf{A}_2 = \mathbf{I} \tag{3.15}$$

where $\mathbf{I}$ is the identity matrix. In deriving the above it is also necessary to use the fact that $v_1 + v_2 = 1$.

Combining Eq. (3.14) with Eq. (3.12) yields

$$\bar{\sigma} = \mathbf{C}^*\bar{\varepsilon} \tag{3.16}$$

where

$$\mathbf{C}^* = v_1 \mathbf{C}^{(1)}\mathbf{A}_1 + v_2 \mathbf{C}^{(2)}\mathbf{A}_2 \tag{3.17}$$

and $\mathbf{C}^*$ is the *effective stiffness* of the composite. Equivalently, if

$$\bar{\sigma}^{(1)} = \mathbf{B}_1\bar{\sigma}, \qquad \bar{\sigma}^{(2)} = \mathbf{B}_2\bar{\sigma}, \tag{3.18}$$

with $v_1 \mathbf{B}_1 + v_2 \mathbf{B}_2 = \mathbf{I}$, and, $\mathbf{B}_1$ and $\mathbf{B}_2$ are called the *stress concentration tensors*, then

$$\bar{\varepsilon} = \mathbf{S}^* \bar{\sigma} \tag{3.19}$$

where

$$\mathbf{S}^* = v_1 \mathbf{S}^{(1)} \mathbf{B}_1 + v_2 \mathbf{S}^{(2)} \mathbf{B}_2 \tag{3.20}$$

which is the *effective compliance* of the composite.

The relationship between the stress and strain concentration tensors can be established for a given phase k, where here $k = 1, 2$. The phase constitutive equation is given by,

$$\bar{\sigma}^{(k)} = \mathbf{C}^{(k)} \bar{\varepsilon}^{(k)} \tag{3.21}$$

Substituting using the phase strain and stress concentration tensors,

$$\mathbf{B}_k \bar{\sigma} = \mathbf{C}^{(k)} \mathbf{A}_k \bar{\varepsilon} \tag{3.22}$$

and then using Eq. (3.16),

$$\mathbf{B}_k \mathbf{C}^* \bar{\varepsilon} = \mathbf{C}^{(k)} \mathbf{A}_k \bar{\varepsilon} \tag{3.23}$$

Therefore,

$$\mathbf{B}_k = \mathbf{C}^{(k)} \mathbf{A}_k \mathbf{C}^{*-1} \tag{3.24}$$

The strain and stress concentration concepts can be generalized such that an RVE can be discretized into any number of regions, and, provided the stiffness in each region is constant, a unique strain and stress concentration tensor exists for each region. These concentration tensors relate the average fields in the regions to the average fields of the composite, and the generalizations of Eqs. (3.17) and (3.20) are applicable.

It should be emphasized that *the strain/stress concentration tensors are the key to micromechanics*. As shown, given either the strain concentration tensors or stress concentration tensors for a composite, one can immediately calculate the composite effective properties as well as all local fields throughout the composite. This constitutes the complete solution of the micromechanics problem. Consequently, the development of each micromechanics theory below comes down to establishing the theory's strain or stress concentration tensors.

3.4   The Voigt approximation

The simplest model for the evaluation of the average elastic moduli of composite materials was introduced by Voigt (1887) for the estimation of the average properties of polycrystals. This approximation assumes that the strain *throughout* the composite is uniform (e.g., $\bar{\boldsymbol{\varepsilon}}^{(1)} = \bar{\boldsymbol{\varepsilon}}^{(2)} = \bar{\boldsymbol{\varepsilon}}$), and it then follows from Eq. (3.14) that, for a two-phase material,

$$\mathbf{A}_1 = \mathbf{A}_2 = \mathbf{I} \tag{3.25}$$

yielding from Eq. (3.17),

$$\mathbf{C}^* = v_1 \mathbf{C}^{(1)} + v_2 \mathbf{C}^{(2)} \tag{3.26}$$

Equation (3.26) provides the effective stiffness tensor of the composite in terms of the volume averaged stiffnesses of the individual phases. It is equivalent to the rule of mixtures applied to the constituent stiffness tensors. Note that, due to the linear nature of Eq. (3.26), if both constituents are isotropic, the composite effective stiffness tensor will be isotropic as well. The Voigt approximation provides excellent results in the fiber direction of a continuously reinforced composite, since both the fiber and matrix are subject to the same applied strain due to compatibility requirements. The MATLAB implementation of the Voigt approximation is given in MATLAB Function 3.1.

3.5   The Reuss approximation

The assumption of Reuss (1929) is that the stress *throughout* the composite (in all phases) is uniform and equal to the average stress (e.g., $\bar{\boldsymbol{\sigma}}^{(1)} = \bar{\boldsymbol{\sigma}}^{(2)} = \bar{\boldsymbol{\sigma}}$). This implies from Eq. (3.18) that, for a two-phase composite,

$$\mathbf{B}_1 = \mathbf{B}_2 = \mathbf{I} \tag{3.27}$$

and from Eq. (3.20) the effective compliance is given by,

$$\mathbf{S}^* = v_1 \mathbf{S}^{(1)} + v_2 \mathbf{S}^{(2)} \tag{3.28}$$

This is equivalent to the rule of mixtures applied to the constituent compliance tensors. Note that, due to the linear nature of Eq. (3.28), if both constituents are isotropic, the composite effective compliance tensor will be isotropic as well. It should also be noted that neither the Voigt nor Reuss assumption is correct: the implied Voigt stresses are such that the tractions at phase boundaries would not be continuous, while the implied Reuss strains are such that inclusions and matrix could not remain bonded. However, they do provide upper and lower bounds for the effective stiffness tensor components of the composite. The MATLAB implementation of the Reuss approximation is given in MATLAB Function 3.2.

MATLAB Function 3.1 – *Voigt.m*

This function performs micromechanics calculations based on the Voigt approximation.

```
function [Cstar, CTEs, Af_V, Am_V, ATf_V, ATm_V] = Voigt(Fconstit, Mconstit, Vf)
% +++++++++++++++++++++++++++++++++++++++++++++++++++++++++++++++++++++++++++++++++++
% Copyright 2020 United States Government as represented by the Administrator of the
% National Aeronautics and Space Administration. No copyright is claimed in the
% United States under Title 17, U.S. Code. All Other Rights Reserved. By downloading
% or using this software you agree to the terms of NOSA v1.3. This code was prepared
% by Drs. B.A. Bednarcyk and S.M. Arnold to complement the book "Practical
% Micromechanics of Composite Materials" during the course of their government work.
% +++++++++++++++++++++++++++++++++++++++++++++++++++++++++++++++++++++++++++++++++++
% Purpose: Calculate the mechanical and thermal strain concentration tensors for the
%          Voigt micomechanics theory
% Input:
% - Fconstit: Struct containing fiber material properties
% - Mconstit: Struct containing matrix material properties
% - Vf: Fiber volume fraction
% Output:
% - Cstar: Effective stiffness matrix
% - CTEs: Effective CTEs
% - Af_V: Voigt fiber strain concentration tensor
% - Am_V: Voigt matrix strain concentration tensor
% - ATf_V: Voigt fiber thermal strain concentration tensor
% - ATm_V: Voigt matrix thermal strain concentration tensor
% +++++++++++++++++++++++++++++++++++++++++++++++++++++++++++++++++++++++++++++++++++

% -- Get fiber stiffness matrix
Cf = GetCFromProps(Fconstit);

ALPHAf=[Fconstit.aL; Fconstit.aT; Fconstit.aT; 0; 0; 0];   % -- fiber CTE vector

% -- Get matrix stiffness matrix
Cm = GetCFromProps(Mconstit);

ALPHAm=[Mconstit.aL; Mconstit.aL; Mconstit.aL; 0; 0; 0];   % -- matrix CTE vector

% -- Voigt strain concentration tensors (Eq. 3.25)
Af_V = eye(6);
Am_V = eye(6);

% -- Voigt effective stiffness matrix (Eq. 3.26)
Cstar = Vf*Cf*Af_V + (1 - Vf)*Cm*Am_V;


% ------------ THERMAL --------------

% -- Voigt thermal strain concentration tensors (Eq. 3.153)
ATf_V = zeros(6,1);
ATm_V = zeros(6,1);

% -- Effective compliance matrix
SS=inv(Cstar);

% -- Stress concentration tensors (Eq. 3.24)
BHf=Cf*Af_V*SS;
BHm=Cm*Am_V*SS;

% -- Transposes of stress concentration tensors
BHTf=transpose(BHf);
BHTm=transpose(BHm);

% -- Effective CTEs (Eq. 3.138)
CTEs = Vf*BHTf*ALPHAf+(1 - Vf)*BHTm*ALPHAm;

end
```

MATLAB Function 3.2 – `Reuss.m`

This function performs micromechanics calculations based on the Reuss approximation.

```matlab
function [Cstar, CTEs, Af_R, Am_R, ATf_R, ATm_R] = Reuss(Fconstit,Mconstit,Vf)
% ++++++++++++++++++++++++++++++++++++++++++++++++++++++++++++++++++++++++++++++++
% Copyright 2020 United States Government as represented by the Administrator of the
% National Aeronautics and Space Administration. No copyright is claimed in the
% United States under Title 17, U.S. Code. All Other Rights Reserved. By downloading
% or using this software you agree to the terms of NOSA v1.3. This code was prepared
% by Drs. B.A. Bednarcyk and S.M. Arnold to complement the book "Practical
% Micromechanics of Composite Materials" during the course of their government work.
% ++++++++++++++++++++++++++++++++++++++++++++++++++++++++++++++++++++++++++++++++
% Purpose: Calculate the effective stiffness, effective CTES, and the mechanical and
%          thermal strain concentration tensors for the Reuss micomechanics theory
% Input:
% - Fconstit: Struct containing fiber material properties
% - Mconstit: Struct containing matrix material properties
% - Vf: Fiber volume fraction
% Output:
% - Cstar: Effective stiffness matrix
% - CTEs: Effective CTEs
% - Af_R: Reuss fiber strain concentration tensor
% - Am_R: Reuss matrix strain concentration tensor
% - ATf_R: Reuss fiber thermal strain concentration tensor
% - ATm_R: Reuss matrix thermal strain concentration tensor
% ++++++++++++++++++++++++++++++++++++++++++++++++++++++++++++++++++++++++++++++++

% -- Get fiber stiffness and compliance matrices
Cf = GetCFromProps(Fconstit);
Sf = inv(Cf);

ALPHAf=[Fconstit.aL; Fconstit.aT; Fconstit.aT; 0; 0; 0];   % -- fiber CTE vector

% -- Get matrix stiffness matrix
Cm= GetCFromProps(Mconstit);
Sm = inv(Cm);

ALPHAm=[Mconstit.aL; Mconstit.aL; Mconstit.aL; 0; 0; 0];   % -- matrix CTE vector

% -- Reuss stress concentration tensors (Eq. 3.27)
Bf_R = eye(6);
Bm_R = eye(6);

% -- Reuss effective compliance matrix (Eq. 3.28)
Sstar = Vf*Sf*Bf_R + (1 - Vf)*Sm*Bm_R;

% -- Effecive stiffness matrix
Cstar = inv(Sstar);

% -- Reuss strain concentration tensors (Eq. 3.22)
Af_R = inv(Cf)*Bf_R*Cstar;
Am_R = inv(Cm)*Bm_R*Cstar;

% ------------ THERMAL --------------

% -- Transposes of stress concentration tensors
BHTf=transpose(Bf_R);
BHTm=transpose(Bm_R);

% -- Effective CTEs (Eq. 3.138)
CTEs = Vf*BHTf*ALPHAf+(1 - Vf)*BHTm*ALPHAm;

% -- Reuss thermal strain concentration tensors (Eqs. 3.147 & 3.148)
ATf_R = (eye(6) - Af_R)*inv(Cf - Cm)*(Cf*ALPHAf - Cm*ALPHAm);
ATm_R = (eye(6) - Am_R)*inv(Cf - Cm)*(Cf*ALPHAf - Cm*ALPHAm);

end
```

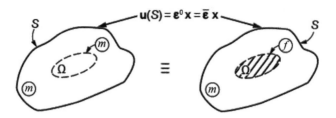

Figure 3.3 Eshelby equivalent inclusion method.

3.6 Eshelby equivalent inclusion method: the dilute case

Here the solution of an inclusion embedded in an infinite matrix phase, subjected to conditions of uniform deformation at large distances from the inclusion, known as the dilute case, will be examined. This dilute case solution forms the foundation of several classical micromechanics theories, most notably the Mori-Tanaka method, presented in the next section. In the specific case of an ellipsoidal inclusion, the Eshelby equivalent inclusion principle provides a means for the determination of the elastic fields in the inclusion. Consider an infinite homogeneous matrix, whose elastic stiffness is $\mathbf{C}^{(m)}$, containing an ellipsoidal domain Ω composed of the same matrix material (see Fig. 3.3), having an *eigenstrain* $\boldsymbol{\varepsilon}^*$. According to Eshelby (1957, 1959), an eigenstrain is a stress-free strain (e.g., thermal strain). Since the natural change of shape of the domain Ω is restrained by its surroundings, a state of internal stress is obtained. The total strain $\boldsymbol{\varepsilon}$ in Ω is given by the sum of the eigenstrain and the resulting elastic strain. The latter elastic strain is related to the stress by Hooke's law for the matrix material, that is,

$$\sigma = \mathbf{C}^{(m)}\left(\boldsymbol{\varepsilon} - \boldsymbol{\varepsilon}^*\right) \text{ in } \Omega \tag{3.29}$$

Eshelby (1957) has shown that if $\boldsymbol{\varepsilon}^*$ is uniform inside the ellipsoidal domain Ω, then the total strain $\boldsymbol{\varepsilon}$ is uniform as well, and the following linear transformation is valid

$$\boldsymbol{\varepsilon} = \mathbf{P}\boldsymbol{\varepsilon}^* \text{ in } \Omega \tag{3.30}$$

where $\mathbf{P}$ is called the Eshelby tensor. Note the components of the Eshelby tensor are a function of the inclusion shape and material properties of the body. Now consider that this infinite matrix phase containing the ellipsoidal domain Ω with eigenstrain $\boldsymbol{\varepsilon}^*$ is subjected to an applied homogeneous displacement boundary condition, Eq. (3.3), at infinity. Because Ω contains the same matrix material, the stress within Ω becomes

$$\sigma = \mathbf{C}^{(m)}\left(\boldsymbol{\varepsilon}^0 + \boldsymbol{\varepsilon} - \boldsymbol{\varepsilon}^*\right) \text{ in } \Omega \tag{3.31}$$

where $\sigma^0 = \mathbf{C}^{(m)}\boldsymbol{\varepsilon}^0$. Now, consider replacing the ellipsoidal domain Ω, whose stiffness tensor is $\mathbf{C}^{(m)}$, with an "equivalent" inclusion material whose stiffness is $\mathbf{C}^{(f)}$, see Fig. 3.3, and that is subjected to the same homogeneous displacement boundary conditions and the same total strain $\boldsymbol{\varepsilon}$. The stress in the domain Ω is then,

$$\sigma^{(f)} = \mathbf{C}^{(f)}(\boldsymbol{\varepsilon}^0 + \boldsymbol{\varepsilon}) \text{ in } \Omega \tag{3.32}$$

The purpose of the eigenstrain is to achieve equivalency of the stresses in the domain Ω when occupied by either the matrix or the equivalent fiber such that $\sigma^{(f)} = \sigma$. Setting Eqs. (3.31) and (3.32) equal gives,

$$\mathbf{C}^{(m)}(\boldsymbol{\varepsilon}^0 + \boldsymbol{\varepsilon} - \boldsymbol{\varepsilon}^*) = \mathbf{C}^{(f)}(\boldsymbol{\varepsilon}^0 + \boldsymbol{\varepsilon}) \tag{3.33}$$

Eq. (3.32) implies that $(\boldsymbol{\varepsilon}^0 + \boldsymbol{\varepsilon}) = \bar{\boldsymbol{\varepsilon}}^{(f)}$, and then if follows from Eq. (3.33) that

$$\mathbf{C}^{(m)}\boldsymbol{\varepsilon}^* = (\mathbf{C}^{(m)} - \mathbf{C}^{(f)}) \bar{\boldsymbol{\varepsilon}}^{(f)} \tag{3.34}$$

From Eqs. (3.32) and (3.30) the strain in the inclusion is given by:

$$\bar{\boldsymbol{\varepsilon}}^{(f)} = \boldsymbol{\varepsilon}^0 + \boldsymbol{\varepsilon} = \boldsymbol{\varepsilon}^0 + \mathbf{P}\boldsymbol{\varepsilon}^* \tag{3.35}$$

which can be rearranged as,

$$\boldsymbol{\varepsilon}^* = \mathbf{P}^{-1}(\bar{\boldsymbol{\varepsilon}}^{(f)} - \boldsymbol{\varepsilon}^0) \tag{3.36}$$

Then substituting in Eq. (3.36) into Eq. (3.34), one obtains

$$\mathbf{C}^{(m)}\mathbf{P}^{-1}(\bar{\boldsymbol{\varepsilon}}^{(f)} - \boldsymbol{\varepsilon}^0) = (\mathbf{C}^{(m)} - \mathbf{C}^{(f)}) \bar{\boldsymbol{\varepsilon}}^{(f)} \tag{3.37}$$

or,

$$[\mathbf{C}^{(m)}\mathbf{P}^{-1} - (\mathbf{C}^{(m)} - \mathbf{C}^{(f)})] \bar{\boldsymbol{\varepsilon}}^{(f)} = \mathbf{C}^{(m)}\mathbf{P}^{-1}\boldsymbol{\varepsilon}^0 \tag{3.38}$$

By multiplying both sides of Eq. (3.38) by $\mathbf{P}\mathbf{C}^{(m)-1}$, the following expression is obtained

$$[\mathbf{I} - \mathbf{P}\mathbf{C}^{(m)-1}(\mathbf{C}^{(m)} - \mathbf{C}^{(f)})] \bar{\boldsymbol{\varepsilon}}^{(f)} = \boldsymbol{\varepsilon}^0 \tag{3.39}$$

Hence comparing Eq. (3.39) to Eq. (3.14), and using the average strain theorem, Eq. (3.5), it is clear that the strain concentration tensor, $\mathbf{A}_f$, of an ellipsoidal inclusion within an infinite matrix can be identified as follows:

$$\mathbf{A}_f^{dilute}\bar{\boldsymbol{\varepsilon}}^{(f)} = \varepsilon_0 \text{ where } \mathbf{A}_f^{dilute} = [\mathbf{I} - \mathbf{P}\mathbf{C}]^{-1} \tag{3.40}$$

where,

$$\mathbf{C} = \mathbf{C}^{(m)-1}\left(\mathbf{C}^{(m)} - \mathbf{C}^{(f)}\right) \tag{3.41}$$

From this solution for the ellipsoidal inclusion, many special cases of the Eshelby tensor, such as sphere, elliptic cylinder, spheroid, crack, etc., have been derived by Mura (1982) in tensorial form. For instance in the case of a spherical inclusion, the Eshelby tensor, $\mathbf{P}$, nonzero terms in contracted notation are:

$$
\begin{aligned}
P_{11} &= P_{22} = P_{33} = (7 - 5v_m)/[15(1 - v_m)] \\
P_{12} &= P_{23} = P_{31} = P_{13} = P_{21} = P_{32} = (5v_m - 1)/[15(1 - v_m)] \\
P_{44} &= P_{55} = P_{66} = 2(4 - 5v_m)/[15(1 - v_m)]
\end{aligned} \tag{3.42}
$$

whereas, for a cylindrical inclusion that is infinitely long in the 1-direction, the nonzero components of $\mathbf{P}$ become:

$$
\begin{aligned}
P_{22} &= P_{33} = \left[\tfrac{3}{4} + \tfrac{1}{2}(1 - 2v_m)\right]/[2(1 - v_m)] \\
P_{12} &= P_{13} = P_{21} = P_{31} = v_m/[2(1 - v_m)] \\
P_{23} &= P_{32} = \left[\tfrac{1}{4} - \tfrac{1}{2}(1 - 2v_m)\right]/[2(1 - v_m)] \\
P_{44} &= 2\left[\tfrac{1}{4} + \tfrac{1}{2}(1 - 2v_m)\right]/[2(1 - v_m)] \\
P_{55} &= P_{66} = 2/4
\end{aligned} \tag{3.43}
$$

Note that, in both cases, the Poisson ratio, v_m, is associated with the matrix. Also, the factor of 2 in the contracted notation $\mathbf{P}$ shear components above comes from combining the two appropriate tensorial shear terms given by Mura (1982), i.e., $P_{44} = P_{2323} + P_{2332}$, $P_{55} = P_{1313} + P_{1331}$, $P_{66} = P_{1212} + P_{1221}$.

3.7 The Mori-Tanaka (MT) method

The work of Tanaka and Mori (1972) and Mori and Tanaka (1973), originally concerned with calculating the average internal stress in a matrix of a material containing precipitates using eigenstrains, was the starting point of a series of papers on the determination of the effective behavior of composites, (e.g., Taya and Arsenault (1989)) and has become one of the more popular methods due to its simplicity in implementation. Benveniste (1987) reformulated this method by elucidating the nature of the approximation involved in the theory, which enhanced its utilization considerably (see Fig. 1.7). This reference includes many of the related works on the subject.

Consider the problem of a two phase composite, containing multiple inclusions, wherein $\boldsymbol{\varepsilon}^0$ and $\boldsymbol{\sigma}^0$ are the externally applied homogeneous boundary conditions at the boundary, S, see Fig. 3.4. Section 3.3 provides expressions for the effective stiffness and compliance in terms of the strain concentration tensors. Using

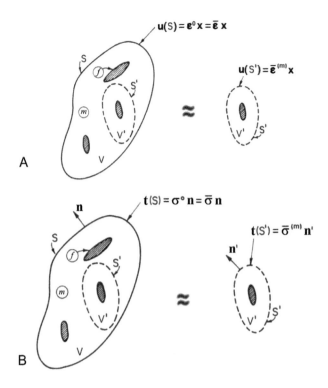

Figure 3.4 A schematic representation of the Mori-Tanaka method for (A) displacement boundary conditions, and, (B) traction boundary conditions. The entire multiphased material (left) is replaced by a single inclusion embedded in the matrix (right).

Eq. (3.15) wherein phase 1 is the fiber (or inclusion) and phase 2 is the matrix,

$$\mathbf{A}_m = \frac{1}{v_m}\left(\mathbf{I} - v_f\mathbf{A}_f\right) \tag{3.44}$$

and substituting into (3.17) gives,

$$\mathbf{C}^* = v_f\mathbf{C}^{(f)}\mathbf{A}_f + \mathbf{C}^{(m)}\left(\mathbf{I} - v_f\mathbf{A}_f\right) \tag{3.45}$$

which can be written as,

$$\mathbf{C}^* = \mathbf{C}^{(m)} + v_f\left(\mathbf{C}^{(f)} - \mathbf{C}^{(m)}\right)\mathbf{A}_f \tag{3.46}$$

It can similarly be shown that,

$$\mathbf{S}^* = \mathbf{S}^{(m)} + v_f\left(\mathbf{S}^{(f)} - \mathbf{S}^{(m)}\right)\mathbf{B}_f \tag{3.47}$$

where the concentration tensors $\mathbf{A}_f$ and $\mathbf{B}_f$, from Eqs. (3.14) and (3.18), are such that the average strain and stress in the fiber are given by

$$\bar{\varepsilon}^{(f)} = \mathbf{A}_f \varepsilon^0, \quad \bar{\sigma}^{(f)} = \mathbf{B}_f \sigma^0, \tag{3.48}$$

The dilute solution assumes that the average strain (or stress) in the matrix is approximately equal to the externally applied field (be it strain or stress). Furthermore it is known, due to the average strain or stress theorem, that the average strain (or stress) in a body is exactly equal to the applied strain (or stress) field. That is,

$$\bar{\varepsilon}^{(m)} \approx \varepsilon^0 = \bar{\varepsilon} \quad \text{(Dilute)} \tag{3.49}$$

$$\bar{\sigma}^{(m)} \approx \sigma^0 = \bar{\sigma} \quad \text{(Dilute)} \tag{3.50}$$

In the general case where the volume concentration is not small, the above relations are *not valid*, as the magnitude of the average strain in the matrix and inclusions is influenced by their interaction. Therefore the strain on the surface S' in Fig. 3.4 varies from point to point. However, the Mori-Tanaka method assumes that the strain on the surface S' is constant and equal to the average strain in the matrix $\bar{\varepsilon}^{(m)}$. This assumption implies that the average strain $\bar{\varepsilon}^{(f)}$ in the interacting inclusions can be approximated by that of a single inclusion embedded in an infinite matrix subjected to the uniform *average matrix* strain $\bar{\varepsilon}^{(m)}$. This is illustrated in Fig. 3.4A, which shows that the problem to be solved, according to the Mori-Tanaka method, is that of a single inclusion in a certain large volume V', which is enclosed by a surface S', and subjected to the boundary conditions

$$\mathbf{u}(S') = \bar{\varepsilon}^{(m)}\mathbf{x} \tag{3.51}$$

Consequently,

$$\bar{\varepsilon}^{(f)} = \mathbf{T}\bar{\varepsilon}^{(m)} \tag{3.52}$$

where the strain concentration tensor $\mathbf{T}$ is determined from the solution of a single particle embedded in an infinite matrix (i.e., the dilute solution see Eq. (3.40)) subjected to boundary conditions (3.51). Combining Eqs. (3.49) and (3.9), wherein 1 is replaced by f (fiber or inclusion) and 2 by m (matrix), yields,

$$\bar{\varepsilon} = v_m\bar{\varepsilon}^{(m)} + v_f\bar{\varepsilon}^{(f)} = \varepsilon^0 \tag{3.53}$$

Substituting Eq. (3.52) into Eq. (3.53), and rearranging yields

$$\bar{\varepsilon}^{(m)} = \left(v_m \mathbf{I} + v_f \mathbf{T}\right)^{-1} \varepsilon^0 \tag{3.54}$$

Then substituting this result back into Eq. (3.52) and associating the result with the first of Eq. (3.48) one obtains,

$$\bar{\varepsilon}^{(f)} = \mathbf{T}\left(v_m \mathbf{I} + v_f \mathbf{T}\right)^{-1} \varepsilon^0 \tag{3.55}$$

Comparing Eq. (3.55) to Eq. (3.48), it is clear that the strain concentration tensor in terms of $\mathbf{T}$ for the Mori-Tanaka method is given by,

$$\mathbf{A}_f = \mathbf{T}\left(v_m \mathbf{I} + v_f \mathbf{T}\right)^{-1} \tag{3.56}$$

Given Eq. (3.52) and the Mori-Tanaka assumption that the average matrix strain is analogous to the applied strain from the dilute solution, one can associate the tensor $\mathbf{T}$ with the Eshelby dilute solution strain concentration tensor, $\mathbf{T} = \mathbf{A}_f^{\text{dilute}}$. Consequently, the strain concentration tensor [Eq. (3.56)] of the Mori-Tanaka method (denoted herein by $\mathbf{A}_f^{MT}$) can be expressed in terms of $\mathbf{A}_f^{\text{dilute}}$ in the form,

$$\mathbf{A}_f^{MT} = \mathbf{A}_f^{\text{dilute}}\left[v_m \mathbf{I} + v_f \mathbf{A}_f^{\text{dilute}}\right]^{-1} \tag{3.57}$$

where

$$\mathbf{A}_f^{\text{dilute}} = [\mathbf{I} - \mathbf{PC}]^{-1} \tag{3.58}$$

where $\mathbf{C}$ was defined in Eq. (3.41). It must be noted that the Mori-Tanaka method, like other classical methods [see Aboudi et al. (2013)], lacks coupling between its normal and shear fields. That is, if one applies a far-field strain, ε^0, consisting only of normal components, only normal strains in the constituents will result with zero shear strain components. This can be concluded from the fact that only average fields in each constituent are considered. In fact, examining Eq. (3.58), one can see that any contribution to the normal-shear coupling terms in $\mathbf{A}_f^{\text{dilute}}$ [e.g., $(\mathbf{A}_f^{\text{dilute}})_{1112}$, $(\mathbf{A}_f^{\text{dilute}})_{2213}$, etc.] must come from the Eshelby tensor, $\mathbf{P}$, as the stiffness tensors for orthotropic constituents do not contain such terms. However, the Eshelby tensors for spherical and cylindrical inclusions, Eqs. (3.42) and (3.43), do not include such terms either. In these two widely-considered cases, the form of the strain concentration equation, $\bar{\varepsilon}^{(f)} = \mathbf{A}_f^{MT} \varepsilon^0$, in the case of normal far-field loading is,

$$
\begin{bmatrix} \bar{\varepsilon}_{11}^{(f)} \\ \bar{\varepsilon}_{22}^{(f)} \\ \bar{\varepsilon}_{33}^{(f)} \\ \bar{\varepsilon}_{23}^{(f)} \\ \bar{\varepsilon}_{13}^{(f)} \\ \bar{\varepsilon}_{12}^{(f)} \end{bmatrix} = \begin{bmatrix} \bullet & \bullet & \bullet & 0 & 0 & 0 \\ \bullet & \bullet & \bullet & 0 & 0 & 0 \\ \bullet & \bullet & \bullet & 0 & 0 & 0 \\ 0 & 0 & 0 & \bullet & 0 & 0 \\ 0 & 0 & 0 & 0 & \bullet & 0 \\ 0 & 0 & 0 & 0 & 0 & \bullet \end{bmatrix} \begin{bmatrix} \varepsilon_{11}^{0} \\ \varepsilon_{22}^{0} \\ \varepsilon_{33}^{0} \\ 0 \\ 0 \\ 0 \end{bmatrix}
\tag{3.59}
$$

Hence, the local shear strain terms due to normal loading are zero. The same is true for the matrix constituent. This feature of micromechanics theories has been termed "lack of shear coupling". Other limitations of the Mori-Tanaka method are discussed by Ferrari (1991).

Finally, the effective stiffness or compliance of the composite can easily be obtained now that the concentration tensor for the inclusion (fiber) is known. This is achieved by substituting Eq. (3.57) into Eq. (3.46), to reveal the effective stiffness of the composite

$$
\mathbf{C}^{*} = \mathbf{C}^{(m)} + v_{f} \left(\mathbf{C}^{(f)} - \mathbf{C}^{(m)} \right) \mathbf{A}_{f}^{\text{dilute}} \left(v_{m} \mathbf{I} + v_{f} \mathbf{A}_{f}^{\text{dilute}} \right)^{-1}
\tag{3.60}
$$

Clearly if $v_{f} \to 0$ then $\mathbf{C}^{*} \to \mathbf{C}^{(m)}$, while if $v_{f} \to 1$ then $\mathbf{C}^{*} \to \mathbf{C}^{(f)}$, thus illustrating that the assumptions inherent in MT have enabled an extension of the dilute solution of Eshelby ($v_{f} \ll 1$) to those involving interacting inclusions.

Following a similar procedure, Fig. 3.4B, but now with traction homogeneous boundary conditions and with $\bar{\sigma}^{(f)} = \mathbf{B}_{f}^{\text{dilute}} \bar{\sigma}^{(m)}$, yields the MT method's effective compliance:

$$
\mathbf{S}^{*} = \mathbf{S}^{(m)} + v_{f} \left(\mathbf{S}^{(f)} - \mathbf{S}^{(m)} \right) \mathbf{B}_{f}^{\text{dilute}} \left(v_{m} \mathbf{I} + v_{f} \mathbf{B}_{f}^{\text{dilute}} \right)^{-1}
\tag{3.61}
$$

where $\mathbf{B}_{f}^{\text{dilute}}$ denotes the stress concentration tensor of an isolated inclusion and is related to $\mathbf{A}_{f}^{\text{dilute}}$ by

$$
\mathbf{B}_{f}^{\text{dilute}} = \mathbf{C}^{(f)} \mathbf{A}_{f}^{\text{dilute}} \mathbf{S}^{(m)}
\tag{3.62}
$$

It was shown by Benveniste (1987) that the results of Eqs. (3.60) and (3.61) are consistent in the sense that $\mathbf{S}^{*} = (\mathbf{C}^{*})^{-1}$. It should also be noted that the MT theory has been extended to admit multiple concentric coatings about the inclusion, each with constant stress and strain fields (see Barral et al., 2020).

The MATLAB implementation of the MT method is shown in MATLAB Function 3.3.

MATLAB Function 3.3 – `MT.m`

This function performs micromechanics calculations for the Mori-Tanaka (MT) micromechanics theory.

```matlab
function [Cstar, CTEs, Af_MT, Am_MT, ATf_MT, ATm_MT] = MT(Fconstit, Mconstit, Vf)
% ++++++++++++++++++++++++++++++++++++++++++++++++++++++++++++++++++++++++++++++++
% Copyright 2020 United States Government as represented by the Administrator of the
% National Aeronautics and Space Administration. No copyright is claimed in the
% United States under Title 17, U.S. Code. All Other Rights Reserved. By downloading
% or using this software you agree to the terms of NOSA v1.3. This code was prepared
% by Drs. B.A. Bednarcyk and S.M. Arnold to complement the book "Practical
% Micromechanics of Composite Materials" during the course of their government work.
% ++++++++++++++++++++++++++++++++++++++++++++++++++++++++++++++++++++++++++++++++

% Purpose: Calculate the effective stiffness, effective CTES, and the mechanical and
%          thermal strain concentration tensors for the Mori-Tanaka (MT)
%          micromechanics theory
% Input:
% - Fconstit: Struct containing fiber material properties
% - Mconstit: Struct containing matrix material properties
% - Vf: Fiber volume fraction
% Output:
% - Cstar: Effective stiffness matrix
% - CTEs: Effective CTEs
% - Af_MT: MT fiber strain concentration tensor
% - Am_MT: MT matrix strain concentration tensor
% - ATf_MT: MT fiber thermal strain concentration tensor
% - ATm_MT: MT matrix thermal strain concentration tensor
% ++++++++++++++++++++++++++++++++++++++++++++++++++++++++++++++++++++++++++++++++

% -- Prevent divide by zero for Vf = 1
if Vf > 0.99999999
    Vf = 0.99999999;
end

% -- Get fiber stiffness matrix
Cf = GetCFromProps(Fconstit);

ALPHAf=[Fconstit.aL; Fconstit.aT; Fconstit.aT; 0; 0; 0];  % -- Fiber CTE vector

% -- Get matrix stiffness matrix
Cm= GetCFromProps(Mconstit);
Num = Mconstit.vL;

ALPHAm=[Mconstit.aL; Mconstit.aL; Mconstit.aL; 0; 0; 0];  % -- Matrix CTE vector

% -- Eshelby Tensor for cylindrical fiber (Eq. 3.43)
[P] = EshelbyCyl(Num);

% -- Eshelby Tensor for spherical inclusion (Eq. 3.42)
%[P] = EshelbySphere(Num);

C = (Cm^-1)*(Cm - Cf);  % -- Eq. 3.41

% -- Dilute fiber strain concentration tensor (Eq. 3.40)
Af_dilute = inv(eye(6) - P*C);

% -- MT fiber strain concentration tensor (Eq. 3.57)
Af_MT = Af_dilute*inv((1-Vf)*eye(6) + Vf*Af_dilute);

% -- MT effective stiffness matrix (Eq. 3.46)
Cstar = Cm + Vf*(Cf - Cm)*Af_MT;

% -- MT matrix strain concentration tensor (Eq. 3.44)
Am_MT = (eye(6)- Vf*Af_MT)/(1 - Vf);

% ------------ THERMAL --------------
%  -- Effective compliance matrix
SS=inv(Cstar);

% -- Stress concentration tensors (Eq. 3.24)
BHf=Cf*Af_MT*SS;
BHm=Cm*Am_MT*SS;

% -- Transposes of stress concentration tensors
BHTf=transpose(BHf);
BHTm=transpose(BHm);
```

```
% -- Effective CTEs (Eq. 3.138)
CTEs = Vf*BHTf*ALPHAf+(1 - Vf)*BHTm*ALPHAm;

% -- MT thermal strain concentration tensors (Eqs. 3.147 & 3.148)
ATf_MT = (eye(6) - Af_MT)*inv(Cf - Cm)*(Cf*ALPHAf - Cm*ALPHAm);
ATm_MT = (eye(6) - Am_MT)*inv(Cf - Cm)*(Cf*ALPHAf - Cm*ALPHAm);

end

%--------------------------------------------------------------
%--------------------------------------------------------------

function [P] = EshelbyCyl(Num)
    % -- Eq. 3.43
    P(2,1) = Num/(2.*(1. - Num));
    P(3,1) = Num/(2.*(1. - Num));
    P(2,2) = (1./(2.*(1. - Num))) * (3./4. + (1. - 2.*Num)/2.);
    P(3,3) = P(2,2);
    P(2,3) = (1./(2.*(1. - Num))) * (1./4. - (1. - 2.*Num)/2.);
    P(3,2) = P(2,3);
    % -- Engineering shear strains - multiply tensorial P by 2
    P(4,4) = (2./(2.*(1. - Num))) * (1./4. + (1. - 2.*Num)/2.);
    P(5,5) = 2./4.;
    P(6,6) = 2./4.;
end

%--------------------------------------------------------------
%--------------------------------------------------------------
```

3.8   The method of cells for continuously fiber-reinforced materials (doubly periodic)

Here the method of cells (MOC) for the micromechanical analysis of composite materials is presented. Like the previous methods, the present approach is approximate. It relies on the fundamental assumption that the two-phased composite has a periodic microstructure in which the reinforcing material (e.g., fibers) is arranged in a periodic manner. This assumption allows the analysis of a single repeating unit cell (RUC), or fundamental building block, rather than the whole composite with its many fibers. The analysis of the single RUC consists of the imposition of displacement and traction continuity conditions at the interfaces within the unit cell as well as at the interfaces between neighboring unit cells, in conjunction with the equilibrium conditions. For elastic composites, in which both phases are linear elastic materials, the final micromechanics analysis leads to determination of the strain concentration tensors and effective stiffness tensor. The main advantage of the MOC, however, over previous methods is its applicability (generalization) to composites whose constituents experience nonlinear behavior (e.g., damage, inelasticity), see Aboudi et al. (2013). This nonlinear generalization of other methods is not always possible.

In Fig. 3.5 a continuously reinforced composite material is idealized as a doubly-periodic array of fibers embedded in a matrix. Double-periodicity means that the geometric pattern is *repeated infinitely* in two directions (the x_2- and x_3- directions in Fig. 3.5). The MOC repeating unit cell (RUC), and its four subvolumes, or "subcells," are identified, whose centroids (indicated by the red dots in Fig. 3.5) represent one fiber point and three matrix points, respectively. The rectangular graphical representation of the subcells indicates the region of influence of its centroid, *not the actual fiber or matrix shape*. It should be emphasized that due to the MOC enforcement of continuity,

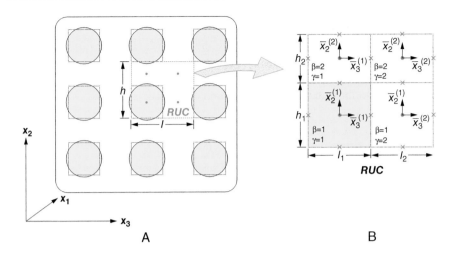

Figure 3.5 (A) A composite with a doubly periodic array of fibers, which are continuous in the x_1-direction. (B) A repeating unit cell (RUC) with four subcells $\beta,\gamma = 1,2$.

no influence of corners (i.e., stress risers) is present. Therefore, although graphically the fiber is usually depicted as square, it really has no shape and is more appropriately consider to be pseudocircular. The rectangular subcells are used to depict the region of influence of each subcell centroid. The fact that three matrix subcells are present enables the MOC to capture some variation of the matrix fields. In contrast, mean field approaches such as the Mori-Tanaka (MT) method discern only the average matrix field variables (i.e., a single value applicable to the entire matrix).

3.8.1 Geometry and basic relations

In the MOC, the cross sectional area of the fiber is $h_1 l_1$, and h_2 and l_2 represents its spacing in the matrix, see Fig. 3.5. As a result of this doubly-periodic fiber arrangement, it is sufficient to analyze an RUC as shown in Fig. 3.5B. This RUC contains four subcells denoted by $\beta,\gamma = 1,2$. Let four local coordinate systems $(x_1, \bar{x}_2^{(\beta)}, \bar{x}_3^{(\gamma)})$ be introduced, all of which have origins that are located at the centroid of each subcell.

Since the average behavior of the composite is sought, as will be shown, a first-order theory in which the displacement in the subcell is expanded linearly in terms of the distances from the center of the subcell (i.e., in terms of $\bar{x}_2^{(\beta)}$ and $\bar{x}_3^{(\gamma)}$) can be used. Note that the presented framework is not limited to a first-order displacement field. Later in this book, a second-order theory is used in the high-fidelity generalized method of cells (Chapter 6). For the MOC, the following first-order displacement expansion in each subcell is considered:

$$u_i^{(\beta\gamma)} = w_i^{(\beta\gamma)}(\mathbf{x}) + \bar{x}_2^{(\beta)}\varphi_i^{(\beta\gamma)} + \bar{x}_3^{(\gamma)}\psi_i^{(\beta\gamma)} \quad i = 1, 2, 3 \tag{3.63}$$

where $w_i^{(\beta\gamma)}(\mathbf{x})$ are the displacement components of the centroid of the subcell and the microvariables, $\varphi_i^{(\beta\gamma)}$ and $\psi_i^{(\beta\gamma)}$, characterize the linear dependence of the displacements on the local subcell coordinates $\bar{x}_2^{(\beta)}$, $\bar{x}_3^{(\gamma)}$. In Eq. (3.63) and in the sequel, repeated β or γ indices do not imply summation. The microvariables can be related to the subcell strain components through the standard strain-displacement relations:

$$\varepsilon_{ij}^{(\beta\gamma)} = \frac{1}{2}\left[\partial_j u_i^{(\beta\gamma)} + \partial_i u_j^{(\beta\gamma)}\right] \qquad i, j = 1, 2, 3 \tag{3.64}$$

where $\partial_1 = \partial/\partial x_1$, $\partial_2 = \partial/\partial \bar{x}_2^{(\beta)}$ and $\partial_3 = \partial/\partial \bar{x}_3^{(\gamma)}$.

3.8.2 Interfacial displacement continuity conditions

Continuity of interfacial displacements between subcells, interior to the unit cell, requires that

$$u_i^{(1\gamma)}\Big|_{\bar{x}_2^{(1)}=\frac{h_1}{2}} = u_i^{(2\gamma)}\Big|_{\bar{x}_2^{(2)}=\frac{-h_2}{2}} \tag{3.65}$$

and,

$$u_i^{(\beta 1)}\Big|_{\bar{x}_3^{(1)}=\frac{l_1}{2}} = u_i^{(\beta 2)}\Big|_{\bar{x}_3^{(2)}=\frac{-l_2}{2}} \tag{3.66}$$

These continuity conditions are applied in an average sense; the integrals of the displacement components along the boundary are required to be continuous. Thus,

$$\int_{\frac{-l_\gamma}{2}}^{\frac{l_\gamma}{2}} \left(u_i^{(1\gamma)}\Big|_{\bar{x}_2^{(1)}=\frac{h_1}{2}} - u_i^{(2\gamma)}\Big|_{\bar{x}_2^{(2)}=\frac{-h_2}{2}}\right) d\bar{x}_3^{(\gamma)} = 0 \tag{3.67}$$

and,

$$\int_{\frac{-h_\beta}{2}}^{\frac{h_\beta}{2}} \left(u_i^{(\beta 1)}\Big|_{\bar{x}_3^{(1)}=\frac{l_1}{2}} - u_i^{(\beta 2)}\Big|_{\bar{x}_3^{(2)}=\frac{-l_2}{2}}\right) d\bar{x}_2^{(\beta)} = 0 \tag{3.68}$$

Substituting for $u_i^{(\beta\gamma)}$ given in Eq. (3.63), inserting it into Eqs. (3.67) and (3.68), and performing the integration yields,

$$w_i^{(1\gamma)} + \frac{h_1}{2}\phi_i^{(1\gamma)} = w_i^{(2\gamma)} - \frac{h_2}{2}\phi_i^{(2\gamma)} \tag{3.69}$$

and,

$$w_i^{(\beta 1)} + \frac{l_1}{2}\psi_i^{(\beta 1)} = w_i^{(\beta 2)} - \frac{l_2}{2}\psi_i^{(\beta 2)} \tag{3.70}$$

Continuity between adjacent RUCs is also required due to periodicity. Considering first the x_2-direction yields,

$$u_i^{(2\gamma)}\Big|_{\bar{x}_2^{(2)}=\frac{h_2}{2}} = \left[u_i^{(1\gamma)}\Big|_{\bar{x}_2^{(1)}=\frac{-h_1}{2}}\right]_{\text{above}} \tag{3.71}$$

and,

$$u_i^{(1\gamma)}\Big|_{\bar{x}_2^{(1)}=\frac{-h_1}{2}} = \left[u_i^{(2\gamma)}\Big|_{\bar{x}_2^{(2)}=\frac{h_2}{2}}\right]_{\text{below}} \tag{3.72}$$

where "above" and "below" refer to the adjacent RUCs. Applying these conditions in an average sense, substituting with Eq. (3.63), and integrating yields,

$$w_i^{(2\gamma)} + \frac{h_2}{2}\phi_i^{(2\gamma)} = \left[w_i^{(1\gamma)} - \frac{h_1}{2}\phi_i^{(1\gamma)}\right]_{\text{above}} \tag{3.73}$$

and,

$$\left[w_i^{(2\gamma)} + \frac{h_2}{2}\phi_i^{(2\gamma)}\right]_{\text{below}} = w_i^{(1\gamma)} - \frac{h_1}{2}\phi_i^{(1\gamma)} \tag{3.74}$$

The quantities from the RUCs above and below are represented using a Taylor series expansion of the form

$$f(x_0 + \Delta x) = f(x_0) + \frac{\partial f}{\partial x}\Big|_{x_0}\Delta x + \frac{\partial^2 f}{\partial x^2}\Big|_{x_0}(\Delta x)^2 + \dots \tag{3.75}$$

On the right-hand side of Eq. (3.73), $w_i^{(1\gamma)}$ is a function of $\mathbf{x}$, while $\varphi_i^{(1\gamma)}h_1/2$ is a constant. From Eq. (3.75), we have,

$$\left[w_i^{(1\gamma)}\right]_{\text{above}} = w_i^{(1\gamma)} + (h_1 + h_2)\frac{\partial w_i^{(1\gamma)}}{\partial x_2} + (h_1 + h_2)^2\frac{\partial^2 w_i^{(1\gamma)}}{\partial x_2^2} + \dots \tag{3.76}$$

Retaining only terms up to the first order in the subcell dimension h_β in Eq. (3.76), and substituting into Eq. (3.73) yields,

$$w_i^{(2\gamma)} + \frac{h_2}{2}\phi_i^{(2\gamma)} = w_i^{(1\gamma)} + (h_1 + h_2)\frac{\partial w_i^{(1\gamma)}}{\partial x_2} - \frac{h_1}{2}\phi_i^{(1\gamma)} \tag{3.77}$$

Similarly, Eq. (3.74) becomes,

$$w_i^{(2\gamma)} - (h_1 + h_2)\frac{\partial w_i^{(2\gamma)}}{\partial x_2} + \frac{h_2}{2}\phi_i^{(2\gamma)} = w_i^{(1\gamma)} - \frac{h_1}{2}\phi_i^{(1\gamma)} \tag{3.78}$$

Subtracting Eq. (3.78) from Eq. (3.77) yields,

$$\frac{\partial w_i^{(2\gamma)}}{\partial x_2} = \frac{\partial w_i^{(1\gamma)}}{\partial x_2} \tag{3.79}$$

and solving Eq. (3.69) for $w_i^{(2\gamma)}$ and substituting into Eq. (3.78) yields,

$$h_1\varphi_i^{(1\gamma)} + h_2\varphi_i^{(2\gamma)} = (h_1 + h_2)\frac{\partial w_i^{(2\gamma)}}{\partial x_2} \tag{3.80}$$

A similar application of displacement continuity between adjacent RUCs in the x_3-direction yields,

$$\frac{\partial w_i^{(\beta 1)}}{\partial x_3} = \frac{\partial w_i^{(\beta 2)}}{\partial x_3} \tag{3.81}$$

and,

$$l_1\psi_i^{(\beta 1)} + l_2\psi_i^{(\beta 2)} = (l_1 + l_2)\frac{\partial w_i^{(\beta 2)}}{\partial x_3} \tag{3.82}$$

Because the RUC is infinitely long in the x_1-direction, the strain components must be independent of x_1. This condition, in combination with Eqs. (3.63) and (3.64) yields,

$$\frac{\partial w_1^{(11)}}{\partial x_1} = \frac{\partial w_1^{(12)}}{\partial x_1} = \frac{\partial w_1^{(21)}}{\partial x_1} = \frac{\partial w_1^{(22)}}{\partial x_1} \tag{3.83}$$

Equations (3.79), (3.81), and (3.83) are satisfied by assuming that common gradients of the displacement functions w_i exist such that,

$$\frac{\partial w_i^{(\beta\gamma)}}{\partial x_j} = \frac{\partial w_i}{\partial x_j} \tag{3.84}$$

From Eqs. (3.63), (3.64), and (3.84) it is clear that the subcell total strain components are independent of $\bar{x}_2^{(\beta)}$ and $\bar{x}_3^{(\gamma)}$ and are thus constant within a given subcell. The fact that the subcell total strain components are constant within a given subcell requires the subcell stress components to be constant within a given subcell, provided the subcell

is deforming elastically. These arguments show that the average fields and pointwise
fields within the subcells are identical. That is,

$$\bar{\sigma}_{ij}^{(\beta\gamma)} = \sigma_{ij}^{(\beta\gamma)}, \qquad \bar{\varepsilon}_{ij}^{(\beta\gamma)} = \varepsilon_{ij}^{(\beta\gamma)} \tag{3.85}$$

This also implies that the equations of equilibrium are identically satisfied in each
subcell.

For a continuously reinforced composite (see Fig. 3.5), the strain field is independent
of x_1. Therefore, utilizing Eqs. (3.63) and (3.64), along with Eq. (3.84), the subcell
strain field is given by,

$$
\begin{aligned}
\varepsilon_{11}^{(\beta\gamma)} &= \frac{\partial}{\partial x_1} w_1, \\
\varepsilon_{22}^{(\beta\gamma)} &= \phi_2^{(\beta\gamma)}, \\
\varepsilon_{33}^{(\beta\gamma)} &= \psi_3^{(\beta\gamma)}, \\
2\varepsilon_{23}^{(\beta\gamma)} &= \phi_3^{(\beta\gamma)} + \psi_2^{(\beta\gamma)}, \\
2\varepsilon_{13}^{(\beta\gamma)} &= \psi_1^{(\beta\gamma)} + \frac{\partial}{\partial x_1} w_3, \\
2\varepsilon_{12}^{(\beta\gamma)} &= \phi_1^{(\beta\gamma)} + \frac{\partial}{\partial x_1} w_2
\end{aligned}
\tag{3.86}
$$

Note that the shear strains in these equations are tensorial quantities, and the engineering shear strains, $\gamma_{ij} = 2\varepsilon_{ij}$, $i \neq j$. The volume average strain components (assuming
perfect bonding) in the composite are given by

$$\bar{\varepsilon}_{ij} = \frac{1}{V} \sum_{\beta,\gamma=1}^{2} h_\beta l_\gamma \varepsilon_{ij}^{(\beta\gamma)} \tag{3.87}$$

where $V = (h_1 + h_2)(l_1 + l_2)$.

It can be shown from Eqs. (3.80), (3.82), (3.84), (3.86), and (3.87) (see Chapter 3
Appendix for details) that,

$$\bar{\varepsilon}_{ij} = \frac{1}{2}\left(\frac{\partial w_i}{\partial x_j} + \frac{\partial w_j}{\partial x_i} \right) \tag{3.88}$$

Equation (3.88) relates the gradients of the subcell centroid displacements to average
RUC strains. It will be used to relate the subcell strains to the average RUC strains.

The subcell constitutive equations are used to relate the subcell stresses and strains. For orthotropic materials, these equations are given by

$$
\begin{bmatrix}
\sigma_{11}^{(\beta\gamma)} \\
\sigma_{22}^{(\beta\gamma)} \\
\sigma_{33}^{(\beta\gamma)} \\
\sigma_{23}^{(\beta\gamma)} \\
\sigma_{13}^{(\beta\gamma)} \\
\sigma_{12}^{(\beta\gamma)}
\end{bmatrix}
=
\begin{bmatrix}
C_{11}^{(\beta\gamma)} & C_{12}^{(\beta\gamma)} & C_{13}^{(\beta\gamma)} & 0 & 0 & 0 \\
C_{12}^{(\beta\gamma)} & C_{22}^{(\beta\gamma)} & C_{23}^{(\beta\gamma)} & 0 & 0 & 0 \\
C_{13}^{(\beta\gamma)} & C_{23}^{(\beta\gamma)} & C_{33}^{(\beta\gamma)} & 0 & 0 & 0 \\
0 & 0 & 0 & C_{44}^{(\beta\gamma)} & 0 & 0 \\
0 & 0 & 0 & 0 & C_{55}^{(\beta\gamma)} & 0 \\
0 & 0 & 0 & 0 & 0 & C_{66}^{(\beta\gamma)}
\end{bmatrix}
\begin{bmatrix}
\varepsilon_{11}^{(\beta\gamma)} \\
\varepsilon_{22}^{(\beta\gamma)} \\
\varepsilon_{33}^{(\beta\gamma)} \\
2\varepsilon_{23}^{(\beta\gamma)} \\
2\varepsilon_{13}^{(\beta\gamma)} \\
2\varepsilon_{12}^{(\beta\gamma)}
\end{bmatrix}
$$

$$(3.89)$$

It should be noted that, for transversely isotropic constituents with 2-3 isotropy, $C_{33}^{(\beta\gamma)} = C_{22}^{(\beta\gamma)}$, $C_{13}^{(\beta\gamma)} = C_{12}^{(\beta\gamma)}$, $C_{55}^{(\beta\gamma)} = C_{66}^{(\beta\gamma)}$, and $C_{44}^{(\beta\gamma)} = \frac{1}{2}(C_{22}^{(\beta\gamma)} - C_{23}^{(\beta\gamma)})$.

The volume average stress components in the composite are given by

$$
\bar{\sigma}_{ij} = \frac{1}{V} \sum_{\beta,\gamma=1}^{2} h_\beta l_\gamma \sigma_{ij}^{(\beta\gamma)}
$$

$$(3.90)$$

Using Eq. (3.86), the subcell constitutive Eq. (3.89) can be written as,

$$
\begin{aligned}
\sigma_{11}^{(\beta\gamma)} &= C_{11}^{(\beta\gamma)} \bar{\varepsilon}_{11} + C_{12}^{(\beta\gamma)} \phi_2^{(\beta\gamma)} + C_{13}^{(\beta\gamma)} \psi_3^{(\beta\gamma)} \\
\sigma_{22}^{(\beta\gamma)} &= C_{12}^{(\beta\gamma)} \bar{\varepsilon}_{11} + C_{22}^{(\beta\gamma)} \phi_2^{(\beta\gamma)} + C_{23}^{(\beta\gamma)} \psi_3^{(\beta\gamma)} \\
\sigma_{33}^{(\beta\gamma)} &= C_{13}^{(\beta\gamma)} \bar{\varepsilon}_{11} + C_{23}^{(\beta\gamma)} \phi_2^{(\beta\gamma)} + C_{33}^{(\beta\gamma)} \psi_3^{(\beta\gamma)} \\
\sigma_{23}^{(\beta\gamma)} &= C_{44}^{(\beta\gamma)} \left(\psi_2^{(\beta\gamma)} + \phi_3^{(\beta\gamma)} \right) \\
\sigma_{13}^{(\beta\gamma)} &= C_{55}^{(\beta\gamma)} \left(\psi_1^{(\beta\gamma)} + \frac{\partial w_3}{\partial x_1} \right) \\
\sigma_{12}^{(\beta\gamma)} &= C_{66}^{(\beta\gamma)} \left(\phi_1^{(\beta\gamma)} + \frac{\partial w_2}{\partial x_1} \right)
\end{aligned}
$$

$$(3.91)$$

3.8.3 The strain concentration tensors

The forth-order strain concentration tensor of subcell $(\beta\gamma)$, $\mathbf{A}^{(\beta\gamma)}$, relates the strain tensor within the subcell (local), $\boldsymbol{\varepsilon}^{(\beta\gamma)}$, to the externally applied strain (global), $\bar{\boldsymbol{\varepsilon}}$, as follows:

$$
\boldsymbol{\varepsilon}^{(\beta\gamma)} = \mathbf{A}^{(\beta\gamma)} \bar{\boldsymbol{\varepsilon}} \quad \text{or} \quad \varepsilon_{ij}^{(\beta\gamma)} = A_{ijkl}^{(\beta\gamma)} \bar{\varepsilon}_{kl}, \quad i, j, k, l = 1, 2, 3
$$

$$(3.92)$$

For the subcell normal strains, these tensors are determined in the following manner. The first relation in (3.86) provides, in conjunction with (3.88), the following four equations for the axial strains in the subcells:

$$\varepsilon_{11}^{(11)} = \varepsilon_{11}^{(12)} = \varepsilon_{11}^{(21)} = \varepsilon_{11}^{(22)} = \bar{\varepsilon}_{11} \tag{3.93}$$

For the normal transverse strains in the subcells, Eqs. (3.80) and (3.82) yield, in conjunction with the second and third relation in Eq. (3.86), that:

$$h_1\varepsilon_{22}^{(11)} + h_2\varepsilon_{22}^{(21)} = (h_1 + h_2)\bar{\varepsilon}_{22} \tag{3.94}$$

$$h_1\varepsilon_{22}^{(12)} + h_2\varepsilon_{22}^{(22)} = (h_1 + h_2)\bar{\varepsilon}_{22} \tag{3.95}$$

$$l_1\varepsilon_{33}^{(11)} + l_2\varepsilon_{33}^{(12)} = (l_1 + l_2)\bar{\varepsilon}_{33} \tag{3.96}$$

$$l_1\varepsilon_{33}^{(21)} + l_2\varepsilon_{33}^{(22)} = (l_1 + l_2)\bar{\varepsilon}_{33} \tag{3.97}$$

As for the subcell transverse shear strains $\varepsilon_{23}^{(\beta\gamma)}$ in the subcells, they are determined from Eq. (3.87) in the following form:

$$h_1l_1\varepsilon_{23}^{(11)} + h_1l_2\varepsilon_{23}^{(12)} + h_2l_1\varepsilon_{23}^{(21)} + h_2l_2\varepsilon_{23}^{(22)} = (h_1 + h_2)(l_1 + l_2)\bar{\varepsilon}_{23} \tag{3.98}$$

Finally, the subcell axial shear strains $\varepsilon_{12}^{(\beta\gamma)}$ are determined from Eq. (3.80) with $i=1$ which provides in conjunction with the sixth relation in (3.86) that

$$h_1\varepsilon_{12}^{(11)} + h_2\varepsilon_{12}^{(21)} = (h_1 + h_2)\bar{\varepsilon}_{12} \tag{3.99}$$

$$h_1\varepsilon_{12}^{(12)} + h_2\varepsilon_{12}^{(22)} = (h_1 + h_2)\bar{\varepsilon}_{12} \tag{3.100}$$

Similarly, Eq. (3.82) with $i=1$ provides in conjunction with the fifth relation in (3.86) that

$$l_1\varepsilon_{13}^{(11)} + l_2\varepsilon_{13}^{(12)} = (l_1 + l_2)\bar{\varepsilon}_{13} \tag{3.101}$$

$$l_1\varepsilon_{13}^{(21)} + l_2\varepsilon_{13}^{(22)} = (l_1 + l_2)\bar{\varepsilon}_{13} \tag{3.102}$$

Equations (3.93)–(3.102) form a system of 13 equations with 24 unknowns, consisting of six components of $\varepsilon_{ij}^{(\beta\gamma)}$ for each of the four subcells $(\beta\gamma)$.

The additional 11 equations that are required to solve the system are given by the traction continuity conditions between the subcells:

$$\sigma_{2i}^{(1\gamma)} = \sigma_{2i}^{(2\gamma)}, \qquad \sigma_{3i}^{(\beta 1)} = \sigma_{3i}^{(\beta 2)} \tag{3.103}$$

Thus, for the normal transverse stresses:

$$\sigma_{22}^{(11)} = \sigma_{22}^{(21)}, \qquad \sigma_{22}^{(12)} = \sigma_{22}^{(22)} \tag{3.104}$$

$$\sigma_{33}^{(11)} = \sigma_{33}^{(12)}, \qquad \sigma_{33}^{(21)} = \sigma_{33}^{(22)} \tag{3.105}$$

whereas, for the transverse and axial shear stresses

$$\sigma_{23}^{(11)} = \sigma_{23}^{(12)} = \sigma_{23}^{(21)} = \sigma_{23}^{(22)} \tag{3.106}$$

$$\sigma_{12}^{(11)} = \sigma_{12}^{(21)}, \qquad \sigma_{12}^{(12)} = \sigma_{12}^{(22)} \tag{3.107}$$

$$\sigma_{13}^{(11)} = \sigma_{13}^{(12)}, \qquad \sigma_{13}^{(21)} = \sigma_{13}^{(22)} \tag{3.108}$$

Using the subcell constitutive Eq. (3.89) these 11 traction continuity equations can be expressed in terms of subcell strains. The full set of 24 MOC equations that relate the subcell strains to the global strains are given in Table 3.3.

It should be noted that, referring to the system equation numbers in Table 3.3 (first column), the four axial strain equations (system Eqs. 1 - 4), the eight transverse normal strain and stress equations (system Eqs. 5 - 12), the four 23 shear strain and stress equations (system Eqs. 13 - 16), the four 12 shear strain and stress equations (system Eqs. 17 - 20), and the four 13 shear strain and stress equations (system Eqs. 21 - 24) are uncoupled. Thus, if desired, every one of these systems can be solved independently and in closed-form (see Aboudi, 1989).

Alternatively, the 24 algebraic equations in the 24 unknown strains $\boldsymbol{\varepsilon}^{(\beta\gamma)}$ in the four subcells expressed in Table 3.3 can be written in matrix form as,

$$\mathbf{M}\,\boldsymbol{\varepsilon}_s = \mathbf{R} \tag{3.109}$$

where $\boldsymbol{\varepsilon}_s$ is a vector containing all 24 subcell strain components (six components per each of the four subcells) ordered by subcell, that is,

$$\boldsymbol{\varepsilon}_s = \left[\, \varepsilon_{11}^{(11)} \;\; \cdots \;\; \varepsilon_{12}^{(11)} \; \varepsilon_{11}^{(12)} \;\; \cdots \;\; \varepsilon_{12}^{(12)} \; \varepsilon_{11}^{(21)} \;\; \cdots \;\; \varepsilon_{12}^{(21)} \; \varepsilon_{11}^{(22)} \;\; \cdots \;\; \varepsilon_{12}^{(21)} \,\right]^{Tr} \tag{3.110}$$

$\mathbf{M}$ is the matrix of coefficients representing the left hand sides of the 24 equations, and $\mathbf{R}$ is a vector of the right hand sides of the 24 equations. These are given explicitly in Fig. 3.6.

The above 24 algebraic equations in the 24 unknown strains $\boldsymbol{\varepsilon}^{(\beta\gamma)}$ in the four subcells can be solved in terms of the global strains $\bar{\boldsymbol{\varepsilon}}$ to yield the strain concentration tensors $\mathbf{A}^{(\beta\gamma)}$ of all subcells. This is performed as follows. Apply the global strain component $\bar{\varepsilon}_{11} = 1$, whereas all other global strain components are set equal to zero. The solution of the above system of equations then yields the components $A_{ij11}^{(\beta\gamma)}$ (which in contracted matrix notation is the first column). Next, apply $\bar{\varepsilon}_{22} = 1$ whereas all other global strain components are set equal to zero. The solution provides the components $A_{ij22}^{(\beta\gamma)}$ (which in contracted matrix notation is the second column). By continuing the application of this procedure, all components of $\mathbf{A}^{(\beta\gamma)}$ can be established. It can be verified that the general form of $\mathbf{A}^{(\beta\gamma)}$, which relates the subcell and global strains, can be represented in contracted matrix form as follows:

Table 3.3 24 MOC equations relating the subcell strains to the global strains.

System Eq. #	Source	Equation	From Chapter Equation number(s)
1	Axial Strains	$\varepsilon_{11}^{(11)} = \bar{\varepsilon}_{11}$	(3.93)
2		$\varepsilon_{11}^{(12)} = \bar{\varepsilon}_{11}$	(3.93)
3		$\varepsilon_{11}^{(21)} = \bar{\varepsilon}_{11}$	(3.93)
4		$\varepsilon_{11}^{(22)} = \bar{\varepsilon}_{11}$	(3.93)
5	Transverse Normal Strains	$h_1\varepsilon_{22}^{(11)} + h_2\varepsilon_{22}^{(21)} = (h_1 + h_2)\bar{\varepsilon}_{22}$	(3.94)
6		$h_1\varepsilon_{22}^{(12)} + h_2\varepsilon_{22}^{(22)} = (h_1 + h_2)\bar{\varepsilon}_{22}$	(3.95)
7		$l_1\varepsilon_{33}^{(11)} + l_2\varepsilon_{33}^{(12)} = (l_1 + l_2)\bar{\varepsilon}_{33}$	(3.96)
8		$l_1\varepsilon_{33}^{(21)} + l_2\varepsilon_{33}^{(22)} = (l_1 + l_2)\bar{\varepsilon}_{33}$	(3.97)
9	Transverse Normal Stresses	$C_{12}^{f}\varepsilon_{11}^{(11)} + C_{22}^{f}\varepsilon_{22}^{(11)} + C_{23}^{f}\varepsilon_{33}^{(11)} - C_{12}^{m}\varepsilon_{11}^{(21)} - C_{22}^{m}\varepsilon_{22}^{(21)} - C_{23}^{m}\varepsilon_{33}^{(21)} = 0$	(3.104), (3.89)
10		$C_{12}^{m}\varepsilon_{11}^{(12)} + C_{22}^{m}\varepsilon_{22}^{(12)} + C_{23}^{m}\varepsilon_{33}^{(12)} - C_{12}^{m}\varepsilon_{11}^{(22)} - C_{22}^{m}\varepsilon_{22}^{(22)} - C_{23}^{m}\varepsilon_{33}^{(22)} = 0$	(3.104), (3.89)
11		$C_{12}^{f}\varepsilon_{11}^{(11)} + C_{23}^{f}\varepsilon_{22}^{(11)} + C_{33}^{f}\varepsilon_{33}^{(11)} - C_{12}^{m}\varepsilon_{11}^{(12)} - C_{23}^{m}\varepsilon_{22}^{(12)} - C_{33}^{m}\varepsilon_{33}^{(12)} = 0$	(3.105), (3.89)
12		$C_{12}^{m}\varepsilon_{11}^{(21)} + C_{23}^{m}\varepsilon_{22}^{(21)} + C_{33}^{m}\varepsilon_{33}^{(21)} - C_{12}^{m}\varepsilon_{11}^{(22)} - C_{23}^{m}\varepsilon_{22}^{(22)} - C_{33}^{m}\varepsilon_{33}^{(22)} = 0$	(3.105), (3.89)
13	23 Shear Strains	$h_1l_1\varepsilon_{23}^{(11)} + h_1l_2\varepsilon_{23}^{(12)} + h_2l_1\varepsilon_{23}^{(21)} + h_2l_2\varepsilon_{23}^{(22)} = (h_1 + h_2)(l_1 + l_2)\bar{\varepsilon}_{23}$	(3.98)
14	23 Shear Stresses	$2C_{44}^{f}\varepsilon_{23}^{(11)} - 2C_{44}^{m}\varepsilon_{23}^{(12)} = 0$	(3.106), (3.89)
15		$2C_{44}^{m}\varepsilon_{23}^{(12)} - 2C_{44}^{m}\varepsilon_{23}^{(21)} = 0$	(3.106), (3.89)
16		$2C_{44}^{m}\varepsilon_{23}^{(21)} - 2C_{44}^{m}\varepsilon_{23}^{(22)} = 0$	(3.106), (3.89)
17	12 Shear Strains	$h_1\varepsilon_{12}^{(11)} + h_2\varepsilon_{12}^{(21)} = (h_1 + h_2)\bar{\varepsilon}_{12}$	(3.99)
18		$h_1\varepsilon_{12}^{(12)} + h_2\varepsilon_{12}^{(22)} = (h_1 + h_2)\bar{\varepsilon}_{12}$	(3.100)
19	12 Shear Stresses	$2C_{66}^{f}\varepsilon_{12}^{(11)} - 2C_{66}^{m}\varepsilon_{12}^{(21)} = 0$	(3.107), (3.89)
20		$2C_{66}^{m}\varepsilon_{12}^{(12)} - 2C_{66}^{m}\varepsilon_{12}^{(22)} = 0$	(3.107), (3.89)
21	13 Shear Strains	$l_1\varepsilon_{13}^{(11)} + l_2\varepsilon_{13}^{(12)} = (l_1 + l_2)\bar{\varepsilon}_{13}$	(3.101)
22		$l_1\varepsilon_{13}^{(21)} + l_2\varepsilon_{13}^{(22)} = (l_1 + l_2)\bar{\varepsilon}_{13}$	(3.102)
23	13 Shear Stresses	$2C_{55}^{f}\varepsilon_{13}^{(11)} - 2C_{55}^{m}\varepsilon_{13}^{(12)} = 0$	(3.108), (3.89)
24		$2C_{55}^{m}\varepsilon_{13}^{(21)} - 2C_{55}^{m}\varepsilon_{13}^{(22)} = 0$	(3.108), (3.89)

$$
\begin{bmatrix}
\varepsilon_{11}^{(\beta\gamma)} \\
\varepsilon_{22}^{(\beta\gamma)} \\
\varepsilon_{33}^{(\beta\gamma)} \\
\varepsilon_{23}^{(\beta\gamma)} \\
\varepsilon_{13}^{(\beta\gamma)} \\
\varepsilon_{12}^{(\beta\gamma)}
\end{bmatrix}
=
\begin{bmatrix}
1 & 0 & 0 & 0 & 0 & 0 \\
A_{21}^{(\beta\gamma)} & A_{22}^{(\beta\gamma)} & A_{23}^{(\beta\gamma)} & 0 & 0 & 0 \\
A_{31}^{(\beta\gamma)} & A_{32}^{(\beta\gamma)} & A_{33}^{(\beta\gamma)} & 0 & 0 & 0 \\
0 & 0 & 0 & A_{44}^{(\beta\gamma)} & 0 & 0 \\
0 & 0 & 0 & 0 & A_{55}^{(\beta\gamma)} & 0 \\
0 & 0 & 0 & 0 & 0 & A_{66}^{(\beta\gamma)}
\end{bmatrix}
\begin{bmatrix}
\bar{\varepsilon}_{11} \\
\bar{\varepsilon}_{22} \\
\bar{\varepsilon}_{33} \\
\bar{\varepsilon}_{23} \\
\bar{\varepsilon}_{13} \\
\bar{\varepsilon}_{12}
\end{bmatrix}
\qquad (3.111)
$$

	M Matrix																								**R** Vector
	Subcell (11)						Subcell (12)						Subcell (21)						Subcell (22)						
Axial Strain Eqs.	1	0	0	0	0	0	0	0	0	0	0	0	0	0	0	0	0	0	0	0	0	0	0	0	$\bar{\varepsilon}_{11}$
	0	0	0	0	0	0	1	0	0	0	0	0	0	0	0	0	0	0	0	0	0	0	0	0	$\bar{\varepsilon}_{11}$
	0	0	0	0	0	0	0	0	0	0	0	0	1	0	0	0	0	0	0	0	0	0	0	0	$\bar{\varepsilon}_{11}$
	0	0	0	0	0	0	0	0	0	0	0	0	0	0	0	0	0	0	1	0	0	0	0	0	$\bar{\varepsilon}_{11}$
Transverse Normal Strain Eqs.	0	h_1	0	0	0	0	0	0	0	0	0	0	0	h_2	0	0	0	0	0	0	0	0	0	0	$(h_1+h_2)\bar{\varepsilon}_{22}$
	0	0	0	0	0	0	0	h_1	0	0	0	0	0	0	0	0	0	0	0	h_2	0	0	0	0	$(h_1+h_2)\bar{\varepsilon}_{22}$
	0	0	l_1	0	0	0	0	0	l_2	0	0	0	0	0	0	0	0	0	0	0	0	0	0	0	$(l_1+l_2)\bar{\varepsilon}_{33}$
	0	0	0	0	0	0	0	0	0	0	0	0	0	0	l_1	0	0	0	0	0	l_2	0	0	0	$(l_1+l_2)\bar{\varepsilon}_{33}$
Transverse Normal Stress Eqs.	C_{12}^f	C_{22}^f	C_{23}^f	0	0	0	0	0	0	0	0	0	$-C_{12}^m$	$-C_{22}^m$	$-C_{23}^m$	0	0	0	0	0	0	0	0	0	0
	0	0	0	0	0	0	C_{12}^m	C_{22}^m	C_{23}^m	0	0	0	0	0	0	0	0	0	$-C_{12}^m$	$-C_{22}^m$	$-C_{23}^m$	0	0	0	0
	C_{13}^f	C_{23}^f	C_{33}^f	0	0	0	$-C_{13}^m$	$-C_{23}^m$	$-C_{33}^m$	0	0	0	0	0	0	0	0	0	0	0	0	0	0	0	0
	0	0	0	0	0	0	0	0	0	0	0	0	C_{13}^m	C_{23}^m	C_{33}^m	0	0	0	$-C_{13}^m$	$-C_{23}^m$	$-C_{33}^m$	0	0	0	0
23 Shear Strain Eq.	0	0	0	h_1l_1	0	0	0	0	0	h_1l_2	0	0	0	0	0	h_2l_1	0	0	0	0	0	h_2l_2	0	0	$(h_1+h_2)(l_1+l_2)\bar{\varepsilon}_{23}$
23 Shear Stress Eqs.	0	0	0	$2C_{44}^f$	0	0	0	0	0	$-2C_{44}^m$	0	0	0	0	0	0	0	0	0	0	0	0	0	0	0
	0	0	0	0	0	0	0	0	0	$2C_{44}^m$	0	0	0	0	0	$-2C_{44}^m$	0	0	0	0	0	0	0	0	0
	0	0	0	0	0	0	0	0	0	0	0	0	0	0	0	$2C_{44}^m$	0	0	0	0	0	$-2C_{44}^m$	0	0	0
12 Shear Strain Eqs.	0	0	0	0	0	h_1	0	0	0	0	0	0	0	0	0	0	0	h_2	0	0	0	0	0	0	$(h_1+h_2)\bar{\varepsilon}_{12}$
	0	0	0	0	0	0	0	0	0	0	0	h_1	0	0	0	0	0	0	0	0	0	0	0	h_2	$(h_1+h_2)\bar{\varepsilon}_{12}$
12 Shear Stress Eqs.	0	0	0	0	0	$2C_{66}^f$	0	0	0	0	0	0	0	0	0	0	0	$-2C_{66}^m$	0	0	0	0	0	0	0
	0	0	0	0	0	0	0	0	0	0	0	$2C_{66}^m$	0	0	0	0	0	0	0	0	0	0	0	$2C_{66}^m$	0
13 Shear Strain Eqs.	0	0	0	0	l_1	0	0	0	0	0	l_2	0	0	0	0	0	0	0	0	0	0	0	0	0	$(l_1+l_2)\bar{\varepsilon}_{13}$
	0	0	0	0	0	0	0	0	0	0	0	0	0	0	0	0	l_1	0	0	0	0	0	l_2	0	$(l_1+l_2)\bar{\varepsilon}_{13}$
13 Shear Stress Eqs.	0	0	0	0	$2C_{55}^f$	0	0	0	0	0	$-2C_{55}^m$	0	0	0	0	0	0	0	0	0	0	0	0	0	0
	0	0	0	0	0	0	0	0	0	0	0	0	0	0	0	0	$2C_{55}^m$	0	0	0	0	0	$-2C_{55}^m$	0	0

Figure 3.6 The MOC **M** matrix and **R** vector with non-zero entries highlighted.

This equation is the contracted form of Eq. (3.92) and is not symmetric. The explicit expressions of the components of $A_{ij}^{(\beta\gamma)}$ in terms of the geometrical dimensions h_1, h_2, l_1, and l_2, and the subcell material stiffness components $C_{ij}^{(\beta\gamma)}$ are given by Brayshaw (1994) and Aboudi et al. (2013).

3.8.4 The effective moduli

Once the strain concentration tensors $\mathbf{A}^{(\beta\gamma)}$ in the four subcells have been established, one can proceed to determine the effective moduli of the composite as follows. Let us write Eq. (3.89) in the form:

$$\boldsymbol{\sigma}^{(\beta\gamma)} = \mathbf{C}^{(\beta\gamma)}\boldsymbol{\varepsilon}^{(\beta\gamma)} \quad \text{or} \quad \sigma_{ij}^{(\beta\gamma)} = C_{ijkl}^{(\beta\gamma)}\varepsilon_{kl}^{(\beta\gamma)} \tag{3.112}$$

By employing Eq. (3.92), the following relation results:

$$\boldsymbol{\sigma}^{(\beta\gamma)} = \mathbf{C}^{(\beta\gamma)}\mathbf{A}^{(\beta\gamma)}\bar{\boldsymbol{\varepsilon}}^{(\beta\gamma)} \quad \text{or} \quad \sigma_{ij}^{(\beta\gamma)} = C_{ijkl}^{(\beta\gamma)}A_{klmn}^{(\beta\gamma)}\bar{\varepsilon}_{mn}^{(\beta\gamma)} \tag{3.113}$$

Hence, Eq. (3.90) yields

$$\bar{\boldsymbol{\sigma}} = \mathbf{C}^{*}\bar{\boldsymbol{\varepsilon}} \quad \text{or} \quad \bar{\sigma}_{ij} = C_{ijkl}^{*}\bar{\varepsilon}_{kl} \tag{3.114}$$

where C_{ijkl}^{*} are the components of the effective stiffness tensor $\mathbf{C}^{*}$ of the composite, which are given by

$$\mathbf{C}^{*} = \frac{1}{(h_1 + h_2)(l_1 + l_2)} \sum_{\beta,\gamma=1}^{2} h_\beta l_\gamma \mathbf{C}^{(\beta\gamma)}\mathbf{A}^{(\beta\gamma)}$$

or   (3.115)

$$C_{ijkl}^{*} = \frac{1}{(h_1 + h_2)(l_1 + l_2)} \sum_{\beta,\gamma=1}^{2} h_\beta l_\gamma C_{ijmn}^{(\beta\gamma)}A_{mnkl}^{(\beta\gamma)}$$

It can be seen that Eq. (3.115) is a generalization of Eq. (3.17) by recognizing that the volume fraction of each subcell is $h_\beta l_\gamma/(h_1 + h_2)(l_1 + l_2)$.

3.8.5 The stress concentration tensors

Given the local subcells strain $\boldsymbol{\varepsilon}^{(\beta\gamma)}$, the subcell stress $\boldsymbol{\sigma}^{(\beta\gamma)}$ can be computed by employing the constitutive relations, Eq. (3.112). Hence, using the conjugate form of Eq. (3.114), $\bar{\boldsymbol{\varepsilon}} = \mathbf{S}^{*}\bar{\boldsymbol{\sigma}}$, along with Eq. (3.92) yields:

$$\boldsymbol{\sigma}^{(\beta\gamma)} = \mathbf{C}^{(\beta\gamma)}\mathbf{A}^{(\beta\gamma)}\mathbf{S}^{*}\bar{\boldsymbol{\sigma}} \quad \text{or} \quad \sigma_{ij}^{(\beta\gamma)} = C_{ijkl}^{(\beta\gamma)}A_{klmn}^{(\beta\gamma)}S_{mnpq}^{*}\bar{\sigma}_{pq} \tag{3.116}$$

where S^*_{mnpq} are the components of the effective compliance tensor $\mathbf{S}^* = \mathbf{C}^{*-1}$. Therefore,

$$\boldsymbol{\sigma}^{(\beta\gamma)} = \mathbf{B}^{(\beta\gamma)}\bar{\boldsymbol{\sigma}} \quad \text{or} \quad \sigma_{ij}^{(\beta\gamma)} = B_{ijkl}^{(\beta\gamma)}\bar{\sigma}_{kl} \tag{3.117}$$

where

$$\mathbf{B}^{(\beta\gamma)} = \mathbf{C}^{(\beta\gamma)}\mathbf{A}^{(\beta\gamma)}\mathbf{S}^* \quad \text{or} \quad B_{ijpq}^{(\beta\gamma)} = C_{ijkl}^{(\beta\gamma)}A_{klmn}^{(\beta\gamma)}S^*_{mnpq} \tag{3.118}$$

and $\mathbf{B}^{(\beta\gamma)}$ is the fourth-order stress concentration tensor, which relates the subcell (local) stress $\boldsymbol{\sigma}^{(\beta\gamma)}$ to the global stress $\bar{\boldsymbol{\sigma}}$. Equation (3.118) is analogous to the general equation for the stress concentration tensor given in Eq. (3.24)

3.8.6   Strain concentration tensors - transverse isotropy

For a square RUC and a "square" fiber (i.e., $h_1 = l_1$ and $h_2 = l_2$), $C^*_{12} = C^*_{13}$, $C^*_{22} = C^*_{33}$, and $C^*_{55} = C^*_{66}$. This leaves six independent elastic constants rather than the five independent elastic constants present in the transversely isotropic case (see Section 2.2.2). An averaging procedure can be applied within the MOC that results in five independent effective moduli required for composite level transverse isotropy. Such an averaging approach has been presented in Brayshaw (1994). It is based on the averaging of the strain concentration tensor $\mathbf{A}^{(\beta\gamma)}$ as follows:

$$\hat{\mathbf{A}}^{(\beta\gamma)} = \frac{2}{\pi} \int\limits_{-\pi/4}^{\pi/4} \mathbf{A}_\xi^{(\beta\gamma)}d\xi \tag{3.119}$$

where $\mathbf{A}_\xi^{(\beta\gamma)}$ is obtained by rotating $\mathbf{A}^{(\beta\gamma)}$ by an angle ξ about the x_1- (fiber) direction. The components of $\mathbf{A}_\xi^{(\beta\gamma)}$ are given by

$$A_{ijkl(\xi)}^{(\beta\gamma)} = T_{ip}T_{jq}T_{kr}T_{ls}A_{pqrs}^{(\beta\gamma)} \tag{3.120}$$

Where all indices run from 1 to 3 and $\mathbf{T}$ is the coordinate transformation matrix, which is given by

$$\mathbf{T} = \begin{bmatrix} 1 & 0 & 0 \\ 0 & \cos\xi & \sin\xi \\ 0 & -\sin\xi & \cos\xi \end{bmatrix} \tag{3.121}$$

The resulting expressions are:

$$\hat{A}_{11}^{(\beta\gamma)} = A_{11}^{(\beta\gamma)}$$
$$\hat{A}_{21}^{(\beta\gamma)} = \left(\tfrac{1}{2} + \tfrac{1}{\pi}\right)A_{21}^{(\beta\gamma)} + \left(\tfrac{1}{2} - \tfrac{1}{\pi}\right)A_{31}^{(\beta\gamma)}$$
$$\hat{A}_{31}^{(\beta\gamma)} = \left(\tfrac{1}{2} - \tfrac{1}{\pi}\right)A_{21}^{(\beta\gamma)} + \left(\tfrac{1}{2} + \tfrac{1}{\pi}\right)A_{31}^{(\beta\gamma)}$$
$$\hat{A}_{22}^{(\beta\gamma)} = \left(\tfrac{3}{8} + \tfrac{1}{\pi}\right)A_{22}^{(\beta\gamma)} + \left(\tfrac{3}{8} - \tfrac{1}{\pi}\right)A_{33}^{(\beta\gamma)} + \tfrac{1}{8}\left(A_{23}^{(\beta\gamma)} + A_{32}^{(\beta\gamma)}\right) + \tfrac{1}{4}A_{44}^{(\beta\gamma)}$$
$$\hat{A}_{32}^{(\beta\gamma)} = \left(\tfrac{3}{8} + \tfrac{1}{\pi}\right)A_{32}^{(\beta\gamma)} + \left(\tfrac{3}{8} - \tfrac{1}{\pi}\right)A_{23}^{(\beta\gamma)} + \tfrac{1}{8}\left(A_{22}^{(\beta\gamma)} + A_{33}^{(\beta\gamma)}\right) - \tfrac{1}{4}A_{44}^{(\beta\gamma)}$$
$$\hat{A}_{23}^{(\beta\gamma)} = \left(\tfrac{3}{8} + \tfrac{1}{\pi}\right)A_{23}^{(\beta\gamma)} + \left(\tfrac{3}{8} - \tfrac{1}{\pi}\right)A_{32}^{(\beta\gamma)} + \tfrac{1}{8}\left(A_{22}^{(\beta\gamma)} + A_{33}^{(\beta\gamma)}\right) - \tfrac{1}{4}A_{44}^{(\beta\gamma)}$$
$$\hat{A}_{33}^{(\beta\gamma)} = \left(\tfrac{3}{8} + \tfrac{1}{\pi}\right)A_{33}^{(\beta\gamma)} + \left(\tfrac{3}{8} - \tfrac{1}{\pi}\right)A_{22}^{(\beta\gamma)} + \tfrac{1}{8}\left(A_{23}^{(\beta\gamma)} + A_{32}^{(\beta\gamma)}\right) + \tfrac{1}{4}A_{44}^{(\beta\gamma)}$$
$$\hat{A}_{44}^{(\beta\gamma)} = \tfrac{1}{4}\left(A_{22}^{(\beta\gamma)} + A_{33}^{(\beta\gamma)}\right) - \tfrac{1}{4}\left(A_{23}^{(\beta\gamma)} + A_{32}^{(\beta\gamma)}\right) + \tfrac{1}{2}A_{44}^{(\beta\gamma)}$$
$$\hat{A}_{55}^{(\beta\gamma)} = \left(\tfrac{1}{2} + \tfrac{1}{\pi}\right)A_{55}^{(\beta\gamma)} + \left(\tfrac{1}{2} - \tfrac{1}{\pi}\right)A_{66}^{(\beta\gamma)}$$
$$\hat{A}_{66}^{(\beta\gamma)} = \left(\tfrac{1}{2} - \tfrac{1}{\pi}\right)A_{55}^{(\beta\gamma)} + \left(\tfrac{1}{2} + \tfrac{1}{\pi}\right)A_{66}^{(\beta\gamma)}$$

$$(3.122)$$

The effective moduli of a transversely isotropic unidirectional composite are calculated by the MOC through Eq. (3.115) in which the strain concentration tensor $\mathbf{A}^{(\beta\gamma)}$ is replaced by $\hat{\mathbf{A}}^{(\beta\gamma)}$, i.e.,

$$\mathbf{C}^* = \frac{1}{(h_1 + h_2)(l_1 + l_2)} \sum_{\beta,\gamma=1}^{2} h_\beta l_\gamma \mathbf{C}^{(\beta\gamma)} \hat{\mathbf{A}}^{(\beta\gamma)} \tag{3.123}$$

This equation can be expanded as,

$$\mathbf{C}^* = \frac{h_1 l_1 \mathbf{C}^{(11)}\hat{\mathbf{A}}^{(11)} + h_1 l_2 \mathbf{C}^{(12)}\hat{\mathbf{A}}^{(12)} + h_2 l_1 \mathbf{C}^{(21)}\hat{\mathbf{A}}^{(21)} + h_2 l_2 \mathbf{C}^{(22)}\hat{\mathbf{A}}^{(22)}}{(h_1 + h_2)(l_1 + l_2)}$$

$$(3.124)$$

As stated, when $h_1 = l_1$ and $h_2 = l_2$ (square RUC), Eq. (3.124) will provide effective stiffness tensor components such that $C_{12}^* = C_{13}^*$, $C_{22}^* = C_{33}^*$, $C_{55}^* = C_{66}^*$, and $C_{44}^* = \tfrac{1}{2}\left(C_{22}^* - C_{23}^*\right)$, resulting in five independent elastic constants and transversely isotropic behavior. Similarly, the corresponding transversely isotropic stress concentration tensor is given by

$$\hat{\mathbf{B}}^{(\beta\gamma)} = \mathbf{C}^{(\beta\gamma)}\hat{\mathbf{A}}^{(\beta\gamma)}\mathbf{S}^* \tag{3.125}$$

The MATLAB implementation of the MOC theory is given in MATLAB Function 3.4.

MATLAB Function 3.4 – `MOC.m`

This function performs micromechanics calculations for the Method of Cells (MOC). Results that include averaging to produce transversely isotropic behavior and unaveraged results (MOCu) are produced.

```
function [Cstar,CTEs,As,Cstar_unavg,CTEs_unavg,As_unavg] = MOC(Fconstit, Mconstit, Vf)
% +++++++++++++++++++++++++++++++++++++++++++++++++++++++++++++++++++++++++++++++++++++++
% Copyright 2020 United States Government as represented by the Administrator of the
% National Aeronautics and Space Administration. No copyright is claimed in the
% United States under Title 17, U.S. Code. All Other Rights Reserved. By downloading
% or using this software you agree to the terms of NOSA v1.3. This code was prepared
% by Drs. B.A. Bednarcyk and S.M. Arnold to complement the book "Practical
% Micromechanics of Composite Materials" during the course of their government work.
% +++++++++++++++++++++++++++++++++++++++++++++++++++++++++++++++++++++++++++++++++++++++
% Purpose: Calculate the effective stiffness, effective CTES, and the mechanical and
%          thermal strain concentration tensors for the Method of Cells (MOC)
%          micromechanics theory (both with and without averaging for transverse
%          isotropy)
% Input:
% - Fconstit: Struct containing fiber material properties
% - Mconstit: Struct containing matrix material properties
% - Vf: Fiber volume fraction
% Output:
% - Cstar: Effective stiffness matrix (averaged)
% - CTEs: Effective CTEs (averaged)
% - As: Struct containing subcell concentration tensors (averaged)
% - Cstar_unavg: Effective stiffness matrix (unaveraged)
% - CTEs_unavg: Effective CTEs (unaveraged)
% - As_unavg: Struct containing subcell concentration tensors (unaveraged)
% +++++++++++++++++++++++++++++++++++++++++++++++++++++++++++++++++++++++++++++++++++++++

% -- Get fiber stiffness matrix
Cf = GetCFromProps(Fconstit);

ALPHAf=[Fconstit.aL; Fconstit.aT; Fconstit.aT; 0; 0; 0];   % -- Fiber CTE vector

% -- Get matrix stiffness matrix
Cm = GetCFromProps(Mconstit);

ALPHAm=[Mconstit.aL; Mconstit.aL; Mconstit.aL; 0; 0; 0];   % -- Matrix CTE vector

% -- Get subcell dimensions from vf - square RUC
h1 = sqrt(Vf);
h2 = 1 - h1;
l1 = h1;
l2 = h2;

% -- Define matrix of coefficients for local subcell strains M(24,24) in Eq. 3.109
M = MOCM(Cm, Cf, h1, h2, l1, l2);

% -- Initialize subcell strain concentration tensors
A11 = zeros(6);
A12 = zeros(6);
A21 = zeros(6);
A22 = zeros(6);

% -- Apply consecutively each far-field average strain component
for I = 1: 6

    % -- Vector of 6 global strain components (eps_bar)
    EPSB = zeros(6,1);
    EPSB(I) = 1;

    % -- Determine the right hand side vector R(24) in Eq. 3.109
    R = MOCR(EPSB, h1, h2, l1, l2);
```

```
% -- Solve Eq. 3.109 for the vector of subcell strains (stored as X)
X = M\R; % -- Equivalent to X = inv(M)*R

% -- Unaveraged strain concentration tensors in the four subcells
A11(1, I) = X(1 );
A11(2, I) = X(2 );
A11(3, I) = X(3 );
A11(4, I) = X(4 );
A11(5, I) = X(5 );
A11(6, I) = X(6 );

A12(1, I) = X(7 );
A12(2, I) = X(8 );
A12(3, I) = X(9 );
A12(4, I) = X(10);
A12(5, I) = X(11);
A12(6, I) = X(12);

A21(1, I) = X(13);
A21(2, I) = X(14);
A21(3, I) = X(15);
A21(4, I) = X(16);
A21(5, I) = X(17);
A21(6, I) = X(18);

A22(1, I) = X(19);
A22(2, I) = X(20);
A22(3, I) = X(21);
A22(4, I) = X(22);
A22(5, I) = X(23);
A22(6, I) = X(24);

end  % -- End of loop over 6 applied global strain components

% -- Copy into struct
As_unavg(1,1).A = A11;
As_unavg(1,2).A = A12;
As_unavg(2,1).A = A21;
As_unavg(2,2).A = A22;

% -- Unaveraged effective stiffness matrix (Eq. 3.115)
Cstar_unavg = (h1*l1*Cf*A11 + h1*l2*Cm*A12 + h2*l1*Cm*A21 + h2*l2*Cm*A22)...
              /((h1+h2)*(l1+l2));

% -- Unaveraged effective compliance matrix
SS = inv(Cstar_unavg);

% -- Unaveraged stress concentration matrices (Eq. 3.118)
B11 = Cf*A11*SS;
B12 = Cm*A12*SS;
B21 = Cm*A21*SS;
B22 = Cm*A22*SS;

% -- Transpose of unaveraged stress concentration matrices
BT11 = transpose(B11);
BT12 = transpose(B12);
BT21 = transpose(B21);
BT22 = transpose(B22);

% -- Unaveraged CTEs (Eq. 3.165)
CTEs_unavg =(h1*l1*BT11*ALPHAf + h1*l2*BT12*ALPHAm + h2*l1*BT21*ALPHAm...
            + h2*l2*BT22*ALPHAm)/((h1 + h2)*(l1 + l2));

%+++++++++++++++++++++++++++++++++++++++++++++++++++++++++++++++++++++
% -- Perform averaging to obtain transversely isotropy
%+++++++++++++++++++++++++++++++++++++++++++++++++++++++++++++++++++++

%  -- Averaged subcell strain concentration matrices (Eqs. 3.122)
COF1 = 1./2 + 1/pi;
COF2 = 1./2 - 1/pi;
COF3 = 3./8 + 1/pi;
COF4 = 3./8 - 1/pi;
COF5 = 1./2 + 1/pi;
COF6 = 1./2 - 1/pi;
```

```
AH11(1,1) = A11(1,1);
AH11(2,1) = COF1*A11(2,1) + COF2*A11(3,1);
AH11(3,1) = COF2*A11(2,1) + COF1*A11(3,1);
AH11(2,2) = COF3*A11(2,2) + COF4*A11(3,3) + (A11(2,3) + A11(3,2))/8 + A11(4,4)/4;
AH11(3,2) = COF3*A11(3,2) + COF4*A11(2,3) + (A11(2,2) + A11(3,3))/8 - A11(4,4)/4;
AH11(2,3) = COF3*A11(2,3) + COF4*A11(3,2) + (A11(2,2) + A11(3,3))/8 - A11(4,4)/4;
AH11(3,3) = COF3*A11(3,3) + COF4*A11(2,2) + (A11(2,3) + A11(3,2))/8 + A11(4,4)/4;
AH11(4,4) = (A11(2,2) + A11(3,3))/4 - (A11(2,3) + A11(3,2))/4 + A11(4,4)/2;
AH11(5,5) = COF5*A11(5,5) + COF6*A11(6,6);
AH11(6,6) = COF6*A11(5,5) + COF5*A11(6,6);

AH12(1,1) = A12(1,1);
AH12(2,1) = COF1*A12(2,1) + COF2*A12(3,1);
AH12(3,1) = COF2*A12(2,1) + COF1*A12(3,1);
AH12(2,2) = COF3*A12(2,2) + COF4*A12(3,3) + (A12(2,3) + A12(3,2))/8 + A12(4,4)/4;
AH12(3,2) = COF3*A12(3,2) + COF4*A12(2,3) + (A12(2,2) + A12(3,3))/8 - A12(4,4)/4;
AH12(2,3) = COF3*A12(2,3) + COF4*A12(3,2) + (A12(2,2) + A12(3,3))/8 - A12(4,4)/4;
AH12(3,3) = COF3*A12(3,3) + COF4*A12(2,2) + (A12(2,3) + A12(3,2))/8 + A12(4,4)/4;
AH12(4,4) = (A12(2,2) + A12(3,3))/4 - (A12(2,3) + A12(3,2))/4 + A12(4,4)/2;
AH12(5,5) = COF5*A12(5,5) + COF6*A12(6,6);
AH12(6,6) = COF6*A12(5,5) + COF5*A12(6,6);

AH21(1,1) = A21(1,1);
AH21(2,1) = COF1*A21(2,1) + COF2*A21(3,1);
AH21(3,1) = COF2*A21(2,1) + COF1*A21(3,1);
AH21(2,2) = COF3*A21(2,2) + COF4*A21(3,3) + (A21(2,3) + A21(3,2))/8 + A21(4,4)/4;
AH21(3,2) = COF3*A21(3,2) + COF4*A21(2,3) + (A21(2,2) + A21(3,3))/8 - A21(4,4)/4;
AH21(2,3) = COF3*A21(2,3) + COF4*A21(3,2) + (A21(2,2) + A21(3,3))/8 - A21(4,4)/4;
AH21(3,3) = COF3*A21(3,3) + COF4*A21(2,2) + (A21(2,3) + A21(3,2))/8 + A21(4,4)/4;
AH21(4,4) = (A21(2,2) + A21(3,3))/4 - (A21(2,3) + A21(3,2))/4 + A21(4,4)/2;
AH21(5,5) = COF5*A21(5,5) + COF6*A21(6,6);
AH21(6,6) = COF6*A21(5,5) + COF5*A21(6,6);

AH22(1,1) = A22(1,1);
AH22(2,1) = COF1*A22(2,1) + COF2*A22(3,1);
AH22(3,1) = COF2*A22(2,1) + COF1*A22(3,1);
AH22(2,2) = COF3*A22(2,2) + COF4*A22(3,3) + (A22(2,3) + A22(3,2))/8 + A22(4,4)/4;
AH22(3,2) = COF3*A22(3,2) + COF4*A22(2,3) + (A22(2,2) + A22(3,3))/8 - A22(4,4)/4;
AH22(2,3) = COF3*A22(2,3) + COF4*A22(3,2) + (A22(2,2) + A22(3,3))/8 - A22(4,4)/4;
AH22(3,3) = COF3*A22(3,3) + COF4*A22(2,2) + (A22(2,3) + A22(3,2))/8 + A22(4,4)/4;
AH22(4,4) = (A22(2,2) + A22(3,3))/4 - (A22(2,3) + A22(3,2))/4 + A22(4,4)/2;
AH22(5,5) = COF5*A22(5,5) + COF6*A22(6,6);
AH22(6,6) = COF6*A22(5,5) + COF5*A22(6,6);

% -- Copy into struct
As(1,1).A = AH11;
As(1,2).A = AH12;
As(2,1).A = AH21;
As(2,2).A = AH22;

% -- Averaged effective stiffness matrix (Eq. 3.124)
Cstar = (h1*l11*Cf*AH11 + h1*l12*Cm*AH12 + h2*l11*Cm*AH21 + h2*l12*Cm*AH22) ...
        /((h1 + h2)*(l11 + l12));

% -- Averaged effective compliance matrix
SS = inv(Cstar);

% -- Averaged stress concentration matrices (Eq. 3.125)
BH11 = Cf*AH11*SS;
BH12 = Cm*AH12*SS;
BH21 = Cm*AH21*SS;
BH22 = Cm*AH22*SS;

% -- Transpose of averaged stress concentration matrices
```

```
BHT11=transpose(BH11);
BHT12=transpose(BH12);
BHT21=transpose(BH21);
BHT22=transpose(BH22);

% -- Averaged CTEs (Eq. 3.165)
CTEs = (h1*l1*BHT11*ALPHAf + h1*l2*BHT12*ALPHAm + h2*l1*BHT21*ALPHAm + h2*l2...
       *BHT22*ALPHAm)/((h1 + h2)*(l1 + l2));

% -- Thermal strain concentration tensors

% -- Fiber and matrix thermal stress tensors (Eq. 3.128)
GAMf = Cf*ALPHAf;
GAMm = Cm*ALPHAm;

% -- R thermal (Eq. 3.172)
RT = zeros(24,1);
RT(9)  = GAMf(2) - GAMm(2);
RT(11) = GAMf(2) - GAMm(2);

% -- Subcell strains solution for EPSB = 0, DT = 1
X = M\RT;

% -- Extract subcell thermal strain concentration tensors from X
AT11 = zeros(6,1);
AT11(1) = X(1 );
AT11(2) = X(2 );
AT11(3) = X(3 );
AT11(4) = X(4 );
AT11(5) = X(5 );
AT11(6) = X(6 );

AT12 = zeros(6,1);
AT12(1) = X(7 );
AT12(2) = X(8 );
AT12(3) = X(9 );
AT12(4) = X(10);
AT12(5) = X(11);
AT12(6) = X(12);

AT21 = zeros(6,1);
AT21(1) = X(13);
AT21(2) = X(14);
AT21(3) = X(15);
AT21(4) = X(16);
AT21(5) = X(17);
AT21(6) = X(18);

AT22 = zeros(6,1);
AT22(1) = X(19);
AT22(2) = X(20);
AT22(3) = X(21);
AT22(4) = X(22);
AT22(5) = X(23);
AT22(6) = X(24);

% -- Copy into struct
As(1,1).AT = AT11;
As(1,2).AT = AT12;
As(2,1).AT = AT21;
As(2,2).AT = AT22;

% -- Copy into struct (unaverage AT = averaged AT)
As_unavg(1,1).AT = AT11;
As_unavg(1,2).AT = AT12;
As_unavg(2,1).AT = AT21;
As_unavg(2,2).AT = AT22;

end

%-------------------------------------------------------------------------
%-------------------------------------------------------------------------
```

```
function [M] = MOCM(Cm, Cf, h1, h2, l1, l2)

% -- Define matrix of coefficients for local subcell strains M(24,24) in Eq. 3.109
   M = zeros(24,24);

% -- See Table 3.3 & Fig. 3.6 - Axial Strains (System Eqs 1-4)
   M(1 ,1 ) = 1;
   M(2 ,7 ) = 1;
   M(3 ,13) = 1;
   M(4 ,19) = 1;

% -- See Table 3.3 & Fig. 3.6 - Transverse Normal Strains (System Eqs 5-8)
   M(5 ,2 ) = h1;
   M(5 ,14) = h2;

   M(6 ,8 ) = h1;
   M(6 ,20) = h2;

   M(7 ,3 ) = l1;
   M(7 ,9 ) = l2;

   M(8 ,15) = l1;
   M(8 ,21) = l2;

% -- See Table 3.3 & Fig. 3.6 - Transverse Normal Stresses (System Eqs 9-12)
   M(9 ,1 ) = Cf(1,2);
   M(9 ,2 ) = Cf(2,2);
   M(9 ,3 ) = Cf(2,3);
   M(9 ,13) =-Cm(1,2);
   M(9 ,14) =-Cm(2,2);
   M(9 ,15) =-Cm(2,3);

   M(10,7 ) = Cm(1,2);
   M(10,8 ) = Cm(2,2);
   M(10,9 ) = Cm(2,3);
   M(10,19) =-Cm(1,2);
   M(10,20) =-Cm(2,2);
   M(10,21) =-Cm(2,3);

   M(11,1 ) = Cf(1,3);
   M(11,2 ) = Cf(2,3);
   M(11,3 ) = Cf(3,3);
   M(11,7 ) =-Cm(1,3);
   M(11,8 ) =-Cm(2,3);
   M(11,9 ) =-Cm(3,3);

   M(12,13) = Cm(1,3);
   M(12,14) = Cm(2,3);
   M(12,15) = Cm(3,3);
   M(12,19) =-Cm(1,3);
   M(12,20) =-Cm(2,3);
   M(12,21) =-Cm(3,3);

% -- See Table 3.3 & Fig. 3.6 - 23 Shear Strains (System Eq 13)
   M(13,4 ) = h1*l1;
   M(13,10) = h1*l2;
   M(13,16) = h2*l1;
   M(13,22) = h2*l2;

% -- See Table 3.3 & Fig. 3.6 - 23 Shear Stresses (System Eq 14-16)
   M(14,4 ) = 2*Cf(4,4);
   M(14,10) =-2*Cm(4,4);

   M(15,10) = 2*Cm(4,4);
   M(15,16) =-2*Cm(4,4);

   M(16,16) = 2*Cm(4,4);
   M(16,22) =-2*Cm(4,4);

% -- See Table 3.3 & Fig. 3.6 - 12 Shear Strains (System Eq 17-18)
   M(17,6 ) = h1;
   M(17,18) = h2;
   M(18,12) = h1;
   M(18,24) = h2;
```

```
% -- See Table 3.3 & Fig. 3.6 - 12 Shear Stresses (System Eq 19-20)
    M(19,6 ) = 2*Cf(6,6);
    M(19,18) =-2*Cm(6,6);

    M(20,12) = 2*Cm(6,6);
    M(20,24) =-2*Cm(6,6);

% -- See Table 3.3 & Fig. 3.6 - 13 Shear Strains (System Eq 21-22)
    M(21,5 ) = 11;
    M(21,11) = 12;

    M(22,17) = 11;
    M(22,23) = 12;

% -- See Table 3.3 & Fig. 3.6 - 13 Shear Stresses (System Eq 23-24)
    M(23,5 ) = 2*Cf(5,5);
    M(23,11) =-2*Cm(5,5);

    M(24,17) = 2*Cm(5,5);
    M(24,23) =-2*Cm(5,5);

end

%-------------------------------------------------------------------------
%-------------------------------------------------------------------------

function [R] = MOCR(EPSB, h1, h2, 11, 12)

% -- Determine the right hand side vector R(24) in Eq. 3.109

% -- See Table 3.3 & Fig. 3.6
    R(1 ,1) = EPSB(1); % -- Axial Strains (System Eqs 1-4)
    R(2 ,1) = EPSB(1);
    R(3 ,1) = EPSB(1);
    R(4 ,1) = EPSB(1);

    R(5 ,1) = (h1 + h2)*EPSB(2); % -- Transverse Normal Strains (System Eqs 5-8)
    R(6 ,1) = (h1 + h2)*EPSB(2);
    R(7 ,1) = (11 + 12)*EPSB(3);
    R(8 ,1) = (11 + 12)*EPSB(3);

    R(9 ,1) = 0; % -- Transverse Normal Stresses (System Eqs 9-12)
    R(10,1) = 0;
    R(11,1) = 0;
    R(12,1) = 0;

    R(13,1) = (h1 + h2)*(11 + 12)*EPSB(4); % -- 23 Shear Strains (System Eq 13)

    R(14,1) = 0; % -- 23 Shear Stresses (System Eq 14-16)
    R(15,1) = 0;
    R(16,1) = 0;

    R(17,1) = (h1 + h2)*EPSB(6); % -- 12 Shear Strains (System Eq 17-18)
    R(18,1) = (h1 + h2)*EPSB(6);

    R(19,1) = 0; % -- 12 Shear Stresses (System Eq 19-20)
    R(20,1) = 0;

    R(21,1) = (11 + 12)*EPSB(5); % -- 13 Shear Strains (System Eq 21-22)
    R(22,1) = (11 + 12)*EPSB(5);

    R(23,1) = 0; % -- 13 Shear Stresses (System Eq 23-24)
    R(24,1) = 0;

end
```

3.9 Thermomechanical effects

The previous sections in this chapter dealt with purely linear elastic mechanical micromechanics analysis. Thermomechanical effects were initially not included in order
to keep the presentation of the micromechanics theories as clear and straightforward
as possible. Now, in this section, the closed-form micromechanics methods discussed
earlier are extended to include thermomechanical effects resulting from a uniform
temperature change. It is important to keep in mind that the constituents within a
composite material are not free to expand individually when subjected to a temperature
change. Rather, the constituents restrain each other, resulting in mechanical strains and
stresses that are induced due to mismatches in the thermomechanical properties and the
temperature change.

Extension of the micromechanics theories to include thermomechanical effects
involves two parts: (1) the calculation of the composite effective coefficients of thermal
expansion (CTEs) and (2) the determination of the local fields in the composite constituents resulting from an applied uniform temperature change. Levin (1967) showed
that the effective CTE can be calculated directly from the mechanical strain concentration tensor. Interestingly, one can thus obtain the effective CTEs without determining
the local stress and strain fields in the composite. However, part (2) of the thermomechanical problem requires calculation of an additional thermal strain concentration
tensor (which is specific to each micromechanics model) in order to determine the local
fields in the composite constituents. Consequently, in this section, the two parts of the
thermomechanical problem are first discussed in general and then specialized to each
micromechanics theory.

3.9.1 The effective coefficients of thermal expansion via the Levin theorem

When considering thermal effects in composites, the elastic constitutive relations in a
material can be generalized to represent thermoelastic material behavior as follows

$$\sigma_{ij} = C_{ijkl}\varepsilon_{kl} - \Gamma_{ij}\Delta T \quad \text{or} \quad \boldsymbol{\sigma} = \mathbf{C}\boldsymbol{\varepsilon} - \boldsymbol{\Gamma}\Delta T \tag{3.126}$$

or alternatively,

$$\varepsilon_{ij} = S_{ijkl}\sigma_{kl} + \alpha_{ij}\Delta T \quad \text{or} \quad \boldsymbol{\varepsilon} = \mathbf{S}\boldsymbol{\sigma} + \boldsymbol{\alpha}\Delta T \tag{3.127}$$

where the thermal stress tensor $\boldsymbol{\Gamma}$ is given in terms of the coefficients of thermal
expansion $\boldsymbol{\alpha}$ by

$$\Gamma_{ij} = C_{ijkl}\alpha_{kl} \quad \text{or} \quad \boldsymbol{\Gamma} = \mathbf{C}\boldsymbol{\alpha} \tag{3.128}$$

and ΔT is the temperature deviation from a reference temperature at which the material is undeformed. Note that, as in Chapter 2, ΔT will be treated as constant throughout the composite.

Levin's (1967) theorem immediately provides the effective CTEs of a composite once the mechanical stress concentration tensors, $\mathbf{B}_k$, have been established. For this, consider a composite subjected to two types of boundary conditions:

1. In the first case, under the isothermal situation, homogeneous traction boundary conditions are imposed in the form given by Eq. (3.4). The resulting local (constituent level) elastic stress and strain field variations in the composite are denoted by $\boldsymbol{\sigma}(\mathbf{x})$ and $\boldsymbol{\varepsilon}(\mathbf{x})$, respectively. The global stress is given by $\bar{\boldsymbol{\sigma}} = \boldsymbol{\sigma}^0$.
2. In the second case, traction-free boundary conditions are imposed, but with a temperature change of ΔT. This gives rise to local stress and strain field variations, which are denoted by $\boldsymbol{\sigma}'(\mathbf{x})$ and $\boldsymbol{\varepsilon}'(\mathbf{x}) = \boldsymbol{\alpha}(\mathbf{x})\Delta T$, respectively, where $\boldsymbol{\alpha}(\mathbf{x})$ is the local CTE tensor. Here however, the resulting global fields are: $\bar{\boldsymbol{\sigma}}' = 0$ and $\bar{\boldsymbol{\varepsilon}}' = \boldsymbol{\alpha}^* \Delta T$, where $\boldsymbol{\alpha}^*$ is the effective CTE tensor of the composite (such that the global constitutive equation is $\bar{\boldsymbol{\sigma}} = \mathbf{C}^* (\bar{\boldsymbol{\varepsilon}} - \boldsymbol{\alpha}^* \Delta T)$).

From the Hill-Mandel relation (see Aboudi et al., 2013) one obtains that

$$\frac{1}{V} \int_V \boldsymbol{\sigma}(\mathbf{x}) \boldsymbol{\varepsilon}'(\mathbf{x}) \, dV = \boldsymbol{\sigma}^0 \bar{\boldsymbol{\varepsilon}}' \tag{3.129}$$

Alternatively,

$$\frac{1}{V} \int_V \boldsymbol{\sigma}'(\mathbf{x}) \boldsymbol{\varepsilon}(\mathbf{x}) \, dV = 0 \tag{3.130}$$

Substituting $\boldsymbol{\varepsilon}'(\mathbf{x})$ and $\bar{\boldsymbol{\varepsilon}}'$ into Eq. (3.129) yields

$$\frac{1}{V} \int_V \boldsymbol{\sigma}(\mathbf{x}) \boldsymbol{\alpha}(\mathbf{x}) \Delta T \, dV = \boldsymbol{\sigma}^0 \boldsymbol{\alpha}^* \Delta T \tag{3.131}$$

Since $\boldsymbol{\alpha}$ and $\boldsymbol{\sigma}$ are both symmetric second-order tensors, $(\boldsymbol{\sigma}\,\boldsymbol{\alpha})^{\text{Tr}} = \boldsymbol{\alpha}^{\text{Tr}}\boldsymbol{\sigma}^{\text{Tr}} = \boldsymbol{\alpha}\boldsymbol{\sigma}$, thus taking the transpose (Tr) of both sides of Eq. (3.131), one obtains,

$$\int_V \boldsymbol{\alpha}(\mathbf{x}) \boldsymbol{\sigma}(\mathbf{x}) \, dV = V \boldsymbol{\alpha}^* \boldsymbol{\sigma}^0 \tag{3.132}$$

Given that $\boldsymbol{\alpha}(\mathbf{x})$ and $\boldsymbol{\sigma}(\mathbf{x})$ are both constant per constituent material, the left hand side of Eq. (3.132) becomes,

$$\int_V \boldsymbol{\alpha}(\mathbf{x}) \boldsymbol{\sigma}(\mathbf{x}) \, dV = \sum_{k=1}^{N} \boldsymbol{\alpha}^{(k)} \int_{V_k} \boldsymbol{\sigma}(\mathbf{x}) \, dV = \sum_{k=1}^{N} \boldsymbol{\alpha}^{(k)} \bar{\boldsymbol{\sigma}}^{(k)} V_k \tag{3.133}$$

where $\bar{\sigma}^{(k)}$ is the average of σ over the kth phase, which has a volume V_k. Therefore,

$$\sum_{k=1}^{N} \alpha^{(k)}\bar{\sigma}^{(k)}V_k = V\alpha^*\sigma^0 \tag{3.134}$$

For a composite consisting of N homogeneous thermoelastic phases (constituent materials), the constitutive equation of each phase k, $k=1,\ldots,N$, is given by

$$\sigma^{(k)} = \mathbf{C}^{(k)}\varepsilon^{(k)} - \mathbf{\Gamma}^{(k)}\Delta T \quad \text{or} \quad \varepsilon^{(k)} = \mathbf{S}^{(k)}\sigma^{(k)} + \alpha^{(k)}\Delta T \tag{3.135}$$

where $\sigma^{(k)}$, $\varepsilon^{(k)}$, $\mathbf{C}^{(k)}$, $\mathbf{S}^{(k)}$, $\alpha^{(k)}$, $\mathbf{\Gamma}^{(k)}$ are the stress, strain, stiffness, compliance, CTE, and thermal stress tensors in the kth phase, respectively.

In the purely mechanical problem, the strain $\mathbf{A}^{(k)}$ and stress $\mathbf{B}^{(k)}$ concentration tensors relate, according to Eqs. (3.14) and (3.18), the local strain and stress in the kth phase to the average external strain and stress:

$$\bar{\varepsilon}^{(k)} = \mathbf{A}_k\,\varepsilon^0, \qquad \bar{\sigma}^{(k)} = \mathbf{B}_k\,\sigma^0 \tag{3.136}$$

By substituting the second relation in Eq. (3.136) into the left-hand side of Eq. (3.134), one obtains

$$\left[\alpha^* - \sum_{k=1}^{N} v_k\alpha^{(k)}\mathbf{B}_k\right]\sigma^0 = 0 \tag{3.137}$$

where $v_k = V_k/V$ is the volume fraction of the kth phase. Since σ^0 is arbitrary, the following relation is obtained:

$$\alpha^* = \sum_{k=1}^{N} v_k\alpha^{(k)}\mathbf{B}_k = \sum_{k=1}^{N} v_k[\mathbf{B}_k]^{\text{Tr}}\alpha^{(k)} \tag{3.138}$$

where $[\mathbf{B}_k]^{\text{Tr}}$ is the transpose of $\mathbf{B}_k$. This relation shows that the effective CTE of the composite can be readily determined from the knowledge of its mechanical stress concentration tensors of the phases. It is also applicable to any micromechanics theory that can provide the strain and stress concentration tensors.

3.9.2   Thermomechanical local fields

In the presence of thermal strains due to a temperature change, ΔT, Eq. (3.14) (generalized for k phases), is extended as,

$$\bar{\varepsilon}^{(k)} = \mathbf{A}_k\bar{\varepsilon} + \mathbf{A}_k^T\Delta T \tag{3.139}$$

where $\mathbf{A}_k^T$ is the thermal-strain concentration tensor for phase k. Just like $\mathbf{A}_k$ is specific to a given micromechanics model, so too is $\mathbf{A}_k^T$. Similarly, a thermal-stress concentration tensor can be introduced such that,

$$\bar{\sigma}^{(k)} = \mathbf{B}_k \bar{\sigma} + \mathbf{B}_k^T \Delta T \tag{3.140}$$

where the thermal stress concentration tensor, $\mathbf{B}_k^T$, is also micromechanics theory specific. It should be noted that $\mathbf{A}_k^T \Delta T$ and $\mathbf{B}_k^T \Delta T$ are not simply the thermal strains and thermal stresses for material k. Rather, they account for thermomechanical interaction effects of the constituents. Consider the case where a composite material is subject to free thermal expansion, such that $\bar{\sigma} = \mathbf{0}$ with $\Delta T \neq 0$. Then, according to Eq. (3.140), the average local stress induced in each constituent k is equal to $\mathbf{B}_k^T \Delta T$. Similarly, for the case where a composite material is fully constrained such that $\bar{\varepsilon} = \mathbf{0}$ with $\Delta T \neq 0$, according to Eq. (3.139), the average local strain in constituent k is equal to $\mathbf{A}_k^T \Delta T$. Therefore, $\mathbf{A}_k^T$ can be determined by solving for the constituent strains under the condition $\bar{\varepsilon} = \mathbf{0}$ with $\Delta T \neq 0$.

Benveniste and Dvorak (1990) established a relationship between $\mathbf{A}_k^T$ and $\mathbf{A}_k$ for a two-phase thermoelastic composite through superposition of two problems resulting in the desired condition of, $\bar{\varepsilon} = \mathbf{0}$ with $\Delta T \neq 0$. Problem I is thermomechanical, with $\Delta T \neq 0$, in which a uniform strain, $\hat{\varepsilon} = \hat{\varepsilon}^{(f)} = \hat{\varepsilon}^{(m)}$, is determined that results in a uniform stress throughout the composite, that is,

$$\bar{\sigma}^{(f)} = \bar{\sigma}^{(m)} = \bar{\sigma} \tag{3.141}$$

This uniform strain, $\hat{\varepsilon}$, is obtained by substituting the first of Eq. (3.135) into Eq. (3.141) yielding,

$$\mathbf{C}^{(f)}\hat{\varepsilon} - \mathbf{\Gamma}^{(f)}\Delta T = \mathbf{C}^{(m)}\hat{\varepsilon} - \mathbf{\Gamma}^{(m)}\Delta T \tag{3.142}$$

Solving for $\hat{\varepsilon}$,

$$\hat{\varepsilon}_I = \left(\mathbf{C}^{(f)} - \mathbf{C}^{(m)}\right)^{-1}\left(\mathbf{\Gamma}^{(f)} - \mathbf{\Gamma}^{(m)}\right)\Delta T \tag{3.143}$$

where the subscript I refers to problem I.

The second problem (problem II), is purely mechanical, with $\Delta T = 0$, wherein $\bar{\varepsilon} = -\hat{\varepsilon}_I$ is imposed. Consequently, from Eq. (3.139), the second problem results in the following constituent strains,

$$\bar{\varepsilon}_{II}^{(f)} = -\mathbf{A}_f \hat{\varepsilon}_I, \qquad \bar{\varepsilon}_{II}^{(m)} = -\mathbf{A}_m \hat{\varepsilon}_I, \tag{3.144}$$

Then, superimposing the global strains associated with the two problems results in, $\bar{\varepsilon} = \hat{\varepsilon}_I - \hat{\varepsilon}_I = \mathbf{0}$, and superimposing the constituent strains results in,

$$\bar{\varepsilon}^{(f)} = \hat{\varepsilon}_I + \hat{\varepsilon}_{II}^f = \left(\mathbf{I} - \mathbf{A}_f\right)\left(\mathbf{C}^{(f)} - \mathbf{C}^{(m)}\right)^{-1}\left(\mathbf{\Gamma}^{(f)} - \mathbf{\Gamma}^{(m)}\right)\Delta T \tag{3.145}$$

$$\bar{\varepsilon}^{(m)} = \hat{\varepsilon}_I + \hat{\varepsilon}_{II}^m = (\mathbf{I} - \mathbf{A}_m)(\mathbf{C}^{(f)} - \mathbf{C}^{(m)})^{-1}(\mathbf{\Gamma}^{(f)} - \mathbf{\Gamma}^{(m)})\Delta T \qquad (3.146)$$

Consequently, from Eq. (3.139), since $\bar{\varepsilon} = \mathbf{0}$, $\bar{\varepsilon}^{(k)} = \mathbf{A}_k^T \Delta T$. Therefore, from Eqs. (3.145) and (3.146), the thermal strain concentration tensors can be identified as,

$$\mathbf{A}_f^T = (\mathbf{I} - \mathbf{A}_f)(\mathbf{C}^{(f)} - \mathbf{C}^{(m)})^{-1}(\mathbf{\Gamma}^{(f)} - \mathbf{\Gamma}^{(m)}) \qquad (3.147)$$

$$\mathbf{A}_m^T = (\mathbf{I} - \mathbf{A}_m)(\mathbf{C}^{(f)} - \mathbf{C}^{(m)})^{-1}(\mathbf{\Gamma}^{(f)} - \mathbf{\Gamma}^{(m)}) \qquad (3.148)$$

A similar procedure can provide the $\mathbf{B}_k^T$ in terms of $\mathbf{B}_k$,

$$\mathbf{B}_f^T = (\mathbf{I} - \mathbf{B}_f)(\mathbf{S}^{(f)} - \mathbf{S}^{(m)})^{-1}(\boldsymbol{\alpha}^{(m)} - \boldsymbol{\alpha}^{(f)}) \qquad (3.149)$$

$$\mathbf{B}_m^T = (\mathbf{I} - \mathbf{B}_m)(\mathbf{S}^{(f)} - \mathbf{S}^{(m)})^{-1}(\boldsymbol{\alpha}^{(m)} - \boldsymbol{\alpha}^{(f)}) \qquad (3.150)$$

3.9.3   Thermomechanical Voigt approximation

For the Voigt approximation, where $\mathbf{A}_f^{Voigt} = \mathbf{A}_m^{Voigt} = \mathbf{I}$, and Eq. (3.24) provides the stress concentration tensors as,

$$\mathbf{B}_f^{Voigt} = \mathbf{C}^{(f)}\mathbf{I}\mathbf{C}^{*-1}, \qquad \mathbf{B}_m^{Voigt} = \mathbf{C}^{(m)}\mathbf{I}\mathbf{C}^{*-1} \qquad (3.151)$$

Eq. (3.138) then becomes,

$$\boldsymbol{\alpha}_{Voigt}^* = v_m\left[\mathbf{C}^{(m)}\mathbf{C}^{*-1}\right]^{Tr}\boldsymbol{\alpha}^{(m)} + v_f\left[\mathbf{C}^{(f)}\mathbf{C}^{*-1}\right]^{Tr}\boldsymbol{\alpha}^{(f)} \qquad (3.152)$$

From Eqs. (3.147) and (3.148), it is clear that thermal strain concentration tensor for the Voigt approximation is $\mathbf{0}$,

$$\left(\mathbf{A}_k^T\right)^{Voigt} = \mathbf{0} \qquad (3.153)$$

Thus, the full thermomechanical Voigt approximation is complete as the effective stiffness is given by Eq. (3.26), the effective CTE by Eq. (3.152), all local strains are known since they are equal to the global strains, and the local stresses can be obtained from the local constitutive Eq. (3.126).

3.9.4   Thermomechanical Reuss approximation

For the Reuss approximation, where $\mathbf{B}_{Reuss}^{(f)} = \mathbf{B}_{Reuss}^{(m)} = \mathbf{I}$, Eq. (3.138) becomes the rule of mixtures,

$$\boldsymbol{\alpha}_{Reuss}^* = v_m\boldsymbol{\alpha}^{(m)} + v_f\boldsymbol{\alpha}^{(f)} \qquad (3.154)$$

It is clear from Eqs. (3.149) and (3.150) that the thermal stress concentration tensor for the Reuss approximation is $\mathbf{0}$,

$$\left(\mathbf{B}_k^T\right)^{Reuss} = \mathbf{0} \tag{3.155}$$

To obtain the thermal strain concentration tensors, Eq. (3.24) provides,

$$\mathbf{A}_f^{Reuss} = \mathbf{C}^*\mathbf{I}\mathbf{C}^{(f)-1}, \qquad \mathbf{A}_m^{Reuss} = \mathbf{C}^*\mathbf{I}\mathbf{C}^{(m)-1} \tag{3.156}$$

Then, from Eqs. (3.147) and (3.148),

$$(\mathbf{A}_f^T)^{Reuss} = (\mathbf{I} - \mathbf{C}^*\mathbf{C}^{(f)-1})(\mathbf{C}^{(f)} - \mathbf{C}^{(m)})^{-1}(\boldsymbol{\Gamma}^{(f)} - \boldsymbol{\Gamma}^{(m)}) \tag{3.157}$$

$$(\mathbf{A}_m^T)^{Reuss} = (\mathbf{I} - \mathbf{C}^*\mathbf{C}^{(m)-1})(\mathbf{C}^{(f)} - \mathbf{C}^{(m)})^{-1}(\boldsymbol{\Gamma}^{(f)} - \boldsymbol{\Gamma}^{(m)}) \tag{3.158}$$

Thus, the full thermomechanical Reuss approximation is complete as the effective stiffness is given by Eq. (3.28), the effective CTE by Eq. (3.154), all local stresses are known since they are equal to the global stresses, and the local strains are given by Eq. (3.139).

3.9.5   Thermomechanical Mori-Tanaka method

For the Mori-Tanaka (MT) method, Eq. (3.138) provides,

$$\boldsymbol{\alpha}_{MT}^* = v_m\left[\mathbf{B}_m^{MT}\right]^{\mathrm{Tr}}\boldsymbol{\alpha}^{(m)} + v_f\left[\mathbf{B}_f^{MT}\right]^{\mathrm{Tr}}\boldsymbol{\alpha}^{(f)} \tag{3.159}$$

Given $\mathbf{A}_f^{MT}$ from Eq. (3.57), $\mathbf{B}_f^{MT}$ is obtained from Eq. (3.24),

$$\mathbf{B}_f^{MT} = \mathbf{C}^{(f)}\mathbf{A}_f^{MT}\mathbf{C}^{*-1} \tag{3.160}$$

Using Eq. (3.24) provides $\mathbf{B}_m^{MT}$ as,

$$\mathbf{B}_m^{MT} = \mathbf{C}^{(m)}\mathbf{A}_m^{MT}\mathbf{C}^{*-1} \tag{3.161}$$

where $\mathbf{A}_m^{MT}$ is obtained from Eq. (3.44),

$$\mathbf{A}_m^{MT} = \frac{1}{v_m}\left(\mathbf{I} - v_f\mathbf{A}_f^{MT}\right) \tag{3.162}$$

Using Eqs. (3.147) and (3.148), the thermal strain concentration tensors for the MT method are obtained as,

$$\left(\mathbf{A}_f^T\right)^{MT} = \left(\mathbf{I} - \mathbf{A}_f^{MT}\right)\left(\mathbf{C}^{(f)} - \mathbf{C}^{(m)}\right)^{-1}\left(\boldsymbol{\Gamma}^{(f)} - \boldsymbol{\Gamma}^{(m)}\right) \tag{3.163}$$

$$\left(\mathbf{A}_m^T\right)^{MT} = \left(\mathbf{I} - \mathbf{A}_m^{MT}\right)\left(\mathbf{C}^{(f)} - \mathbf{C}^{(m)}\right)^{-1}\left(\boldsymbol{\Gamma}^{(f)} - \boldsymbol{\Gamma}^{(m)}\right) \tag{3.164}$$

Thus, the full thermomechanical MT method is complete as the effective stiffness is given by Eq. (3.60), the effective CTE by Eq. (3.159), and the local strains by Eqs. (3.139), (3.57), (3.163), and (3.164).

3.9.6   Thermomechanical method of cells

For the method of cells (MOC), Eq. (3.138) takes the form,

$$\alpha_{MOC}^*$$
$$= \frac{h_1 l_1 \left[\mathbf{B}^{(11)}\right]^{\mathrm{Tr}} \alpha^{(11)} + h_1 l_2 \left[\mathbf{B}^{(12)}\right]^{\mathrm{Tr}} \alpha^{(12)} + h_2 l_1 \left[\mathbf{B}^{(21)}\right]^{\mathrm{Tr}} \alpha^{(21)} + h_2 l_2 \left[\mathbf{B}^{(22)}\right]^{\mathrm{Tr}} \alpha^{(22)}}{(h_1 + h_2)(l_1 + l_2)}$$

$$(3.165)$$

where the subcell stress concentration tensors are given by Eq. (3.118).

In the framework of the MOC, which considers four subcells rather than just two subvolumes, the Benveniste and Dvorak (1990) superposition problem for binary composites (used for the simpler methods above) cannot be used. Instead the following approach is used to establish the thermal concentration tensor. To obtain the MOC thermal strain concentration tensor, Eq. (3.109) must be extended to include thermal effects,

$$\mathbf{M}\varepsilon_s = \mathbf{R} + \mathbf{R}^T \Delta T \tag{3.166}$$

where $\mathbf{R}^T$ contains the required additional thermal terms.

Towards this end, the subcell constitutive Eq. (3.89) is written as,

$$
\begin{bmatrix}
\sigma_{11}^{(\beta\gamma)} \\
\sigma_{22}^{(\beta\gamma)} \\
\sigma_{33}^{(\beta\gamma)} \\
\sigma_{23}^{(\beta\gamma)} \\
\sigma_{13}^{(\beta\gamma)} \\
\sigma_{12}^{(\beta\gamma)}
\end{bmatrix}
=
\begin{bmatrix}
C_{11}^{(\beta\gamma)} & C_{12}^{(\beta\gamma)} & C_{13}^{(\beta\gamma)} & 0 & 0 & 0 \\
C_{12}^{(\beta\gamma)} & C_{22}^{(\beta\gamma)} & C_{23}^{(\beta\gamma)} & 0 & 0 & 0 \\
C_{13}^{(\beta\gamma)} & C_{23}^{(\beta\gamma)} & C_{33}^{(\beta\gamma)} & 0 & 0 & 0 \\
0 & 0 & 0 & C_{44}^{(\beta\gamma)} & 0 & 0 \\
0 & 0 & 0 & 0 & C_{55}^{(\beta\gamma)} & 0 \\
0 & 0 & 0 & 0 & 0 & C_{66}^{(\beta\gamma)}
\end{bmatrix}
\begin{bmatrix}
\varepsilon_{11}^{(\beta\gamma)} \\
\varepsilon_{22}^{(\beta\gamma)} \\
\varepsilon_{33}^{(\beta\gamma)} \\
2\varepsilon_{23}^{(\beta\gamma)} \\
2\varepsilon_{13}^{(\beta\gamma)} \\
2\varepsilon_{12}^{(\beta\gamma)}
\end{bmatrix}
$$

$$
-
\begin{bmatrix}
\Gamma_1^{(\beta\gamma)} \\
\Gamma_2^{(\beta\gamma)} \\
\Gamma_3^{(\beta\gamma)} \\
0 \\
0 \\
0
\end{bmatrix}
\Delta T
\tag{3.167}
$$

Referring to Table 3.3, the only equations that are impacted by this change in the constitutive equation are the transverse normal stress equations (system equations 9 – 12 in Table 3.3), which now take the form,

$$C_{12}^f \varepsilon_{11}^{(11)} + C_{22}^f \varepsilon_{22}^{(11)} + C_{23}^f \varepsilon_{33}^{(11)} - C_{12}^m \varepsilon_{11}^{(21)} - C_{22}^m \varepsilon_{22}^{(21)} - C_{23}^m \varepsilon_{33}^{(21)}$$
$$= \left(\Gamma_2^{(11)} - \Gamma_2^{(21)} \right) \Delta T = \left(\Gamma_2^{(f)} - \Gamma_2^{(m)} \right) \Delta T \tag{3.168}$$

$$C_{12}^m \varepsilon_{11}^{(12)} + C_{22}^m \varepsilon_{22}^{(12)} + C_{23}^m \varepsilon_{33}^{(12)} - C_{12}^m \varepsilon_{11}^{(22)} - C_{22}^m \varepsilon_{22}^{(22)} - C_{23}^m \varepsilon_{33}^{(22)}$$
$$= \left(\Gamma_2^{(12)} - \Gamma_2^{(22)} \right) \Delta T = \left(\Gamma_2^{(m)} - \Gamma_2^{(m)} \right) \Delta T = 0 \tag{3.169}$$

$$C_{12}^f \varepsilon_{11}^{(11)} + C_{23}^f \varepsilon_{22}^{(11)} + C_{33}^f \varepsilon_{33}^{(11)} - C_{12}^m \varepsilon_{11}^{(12)} - C_{23}^m \varepsilon_{22}^{(12)} - C_{33}^m \varepsilon_{33}^{(12)}$$
$$= \left(\Gamma_3^{(11)} - \Gamma_3^{(12)} \right) \Delta T = \left(\Gamma_3^{(f)} - \Gamma_3^{(m)} \right) \Delta T \tag{3.170}$$

$$C_{12}^m \varepsilon_{11}^{(21)} + C_{23}^m \varepsilon_{22}^{(21)} + C_{33}^m \varepsilon_{33}^{(21)} - C_{12}^m \varepsilon_{11}^{(22)} - C_{23}^m \varepsilon_{22}^{(22)} - C_{33}^m \varepsilon_{33}^{(22)}$$
$$= \left(\Gamma_3^{(21)} - \Gamma_3^{(22)} \right) \Delta T = \left(\Gamma_3^{(m)} - \Gamma_3^{(m)} \right) \Delta T = 0 \tag{3.171}$$

Eqs. (3.169) and (3.171) are equal to zero because Γ is equal in all three matrix subcells (12, 21, 22), as can be seen from Eq. (3.128). Therefore,

$$\mathbf{R}^T = \begin{bmatrix} 0 & 0 & 0 & 0 & 0 & 0 & 0 & 0 & \Gamma_2^{(f)} - \Gamma_2^{(m)} & 0 & \Gamma_3^{(f)} - \Gamma_3^{(m)} & 0 & \cdots \\ 0 & 0 & 0 & 0 & 0 & 0 & 0 & 0 & 0 & 0 & 0 & 0 \end{bmatrix}^{\text{Tr}} \tag{3.172}$$

where, from Eq. (3.128),

$$\Gamma_2^{(k)} = C_{12}^{(k)} \alpha_1^{(k)} + C_{22}^{(k)} \alpha_2^{(k)} + C_{23}^{(k)} \alpha_3^{(k)}, \qquad k = m, f \tag{3.173}$$

$$\Gamma_3^{(k)} = C_{13}^{(k)} \alpha_1^{(k)} + C_{23}^{(k)} \alpha_2^{(k)} + C_{33}^{(k)} \alpha_3^{(k)}, \qquad k = m, f \tag{3.174}$$

As was done with the other micromechanics theories presented, the thermal strain concentration tensor is obtained from Eq. (3.139), which can be written for the method of cells as,

$$\varepsilon^{(\beta\gamma)} = \mathbf{A}^{(\beta\gamma)} \bar{\varepsilon} + \mathbf{A}_T^{(\beta\gamma)} \Delta T \tag{3.175}$$

where $\mathbf{A}_T^{(\beta\gamma)}$ is the thermal strain concentration tensor for subcell $(\beta\gamma)$. Under the conditions $\bar{\varepsilon} = 0$ with $\Delta T = 1$, $\mathbf{A}_T^{(\beta\gamma)} = \varepsilon^{(\beta\gamma)}$. Consequently, solving Eq. (3.166) for ε_s with applied $\bar{\varepsilon} = 0$ and $\Delta T = 1$ provides $\mathbf{A}_T^{(\beta\gamma)}$. It should be noted that, as was the case with $\mathbf{A}^{(\beta\gamma)}$, $\mathbf{A}_T^{(\beta\gamma)}$ could also be determined analytically (see Aboudi et al., 2013).

Further, the effective CTEs and thermal strain concentration tensors are not affected by averaging to obtain transversely isotropic behavior as they are already transversely isotropic (i.e., $\alpha_{22}^* = \alpha_{33}^*$, $\alpha_{23}^* = \alpha_{13}^* = \alpha_{12}^* = 0$).

Thus, the full MOC thermomechanical problem is complete, with the effective stiffness given by Eq. (3.124), the effective CTE by Eq. (3.165), the local strains from Eq. (3.175), where the strain concentration tensors are determined as described, and the local stresses from the constitutive Eq. (3.167).

3.10 Numerical methods

Multiple fully numerical methods (finite element analysis (FEA), boundary element, and finite difference) have been applied in the literature to predict not only effective elastic properties, but also local and macro stress and strain fields for all types of composites, see for example, Banerjee and Butterfield (1981), Sun and Vaidya (1996), Kwon et al. (2008), Fish (2009), Kanoute et al. (2009), Galvanetto and Aliabadi (2010), Vaz et al. (2011), and Bednarcyk et al. (2015). The finite element method appears to be the most popular of these numerical methods (see Fig. 3.7) and has been used extensively to predict the effective elastic properties and inelastic response of fibrous composites under thermomechanical loading [e.g., see Feyel, and Chaboche (2000)]. The earliest application of FEA for micromechanics purposes appears to be that of Foye (1966a,b), who studied the effective elastic properties, inelastic response, and stress distributions of a unidirectional boron/epoxy system. Since then, numerous detailed (high fidelity) finite element studies have been performed to examine a variety of pertinent micromechanics issues in composites, for example the influence of fiber packing architecture on the inelastic response of metal matrix composites (see references in Arnold et., 1996). A key aspect impacting the accuracy of FEA micromechanics models is the correct application of periodic boundary conditions on the RUC or RVE considered. Stier et al. (2015) provide the correct periodic boundary conditions for 3D finite element models.

The finite-difference technique was employed by Adams and Doner (1967a,b) to investigate the effects of material properties, fiber shape, spacing, volume fraction, and periodic arrangement on the elastic response under longitudinal shear and transverse loading. Lastly, the boundary element method has been utilized to determine effective properties and composite response since the mid-1990s, see Banerjee et al. (1997), Rizzo et al (1998), Liu et al (2005). Although, numerical methods typically provide the highest level of fidelity in both local and global fields, one must always balance this increase in accuracy with the additional computation and manpower resources required to achieve this increment in accuracy over the ultra-efficient but lower fidelity methods. Chapters 5 and 6 provide generalizations to the MOC to increase accuracy while still maintaining improved efficiency compared to numerical methods. These generalizations, along with the original MOC presented in this chapter are collectively referred to as the "MOC Family" of micromechanics theories (see Fig. 3.7).

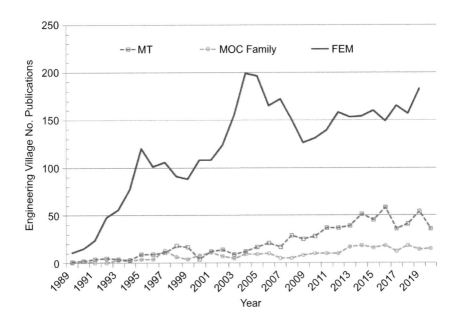

Figure 3.7 Comparison of the number of publications involving micromechanics of composites and numerical (FEM) and analytical/semianalytical (MT and MOC Family) micromechanics methods as obtained from a search of publications titles and abstracts on Engineering Village (2020).

3.11 MATLAB implementation of stand-alone micromechanics and micromechanics-based CLT

The MATLAB code provided with this text is available from:

https://github.com/nasa/Practical-Micromechanics

In this section, the provided MATLAB (2018) functions and scripts for solving (1) stand-alone micromechanics problems (material point) and (2) micromechanics-based laminate problems are discussed. A flow chart of the main stand-alone micromechanics problem script (`MicroAnalysis.m`) is shown in Fig. 3.8A, whereas the script itself is shown in MATLAB Script 3.5. As shown in Fig. 3.8A, a new function to define stand-alone micromechanics problems has been introduced, `MicroProblemDef.m` (see MATLAB Function 3.6), which is similar to `LamProblemDef.m` described in Chapter 2. Now, however, only a material (`MicroMat{NP}`) and the applied loading (in terms of stresses and strains) need be defined per problem (`NP`). All micromechanics-based materials are assigned a unique material number that is greater than or equal to 100 in `GetEffProps.m` (see MATLAB Function 3.7). In this function, for each

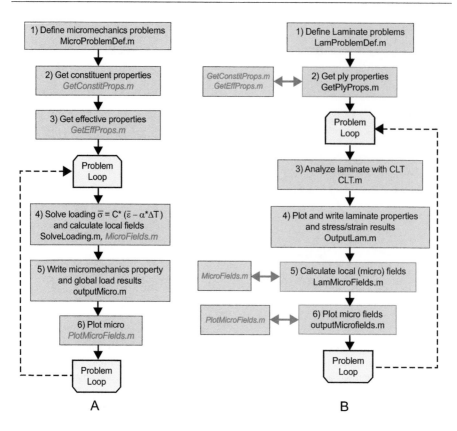

Figure 3.8 Flow charts describing steps involved in (A) solving a stand-alone composite micromechanics problem in the MATLAB Script 3.5 – MicroAnalysis.m and (B) solving a multiscale laminate problem, wherein the ply materials are based on micromechanics, in the MATLAB Script 2.6 – LamAnalysis.m. In (B), steps 2, 5, and 6, outlined in red, indicate coupling with micromechanics functions to calculate effective ply properties, determine the micro scale fields, and report results.

micromechanics-based material, the fiber volume fraction and the fiber and matrix constituent materials are specified, and then the selected micromechanics theory is called (RunMicro.m, see MATLAB Function 3.8) to calculate the effective properties of the material. All constituent materials (fiber and matrix) are defined in GetConstit-Props.m (see MATLAB Function 3.9). The constituent properties considered in this chapter are given in Table 3.4, where IM7 and glass are fibers and 8552 and epoxy are matrix resins.

GetConstitProps.m and GetEffProps.m are called in steps 2 and 3 in MicroAnal-ysis.m, as shown in Fig. 3.8A. Then, each micromechanics problem is solved, as indicated by Problem Loop. This entails solving for the unknown global stress and strain components, calculating the local fields in each constituent (MicroFields.m,

Table 3.4 Constituent material properties.

Mat	Material Name	E_1 (GPa)	E_2 (GPa)	G_{12} (GPa)	ν_{12}	ν_{23}	α_1 $(10^{-6}/°C)$	α_2 $(10^{-6}/°C)$
1	IM7	262.2	11.8	18.9	0.17	0.21	−0.9	9.0
2	Glass	73.0	73.0	29.918	0.22	0.22	5.0	5.0
3	8552	4.67	4.67	1.610	0.45	0.45	42.0	42.0
4	Epoxy	3.45	3.45	1.278	0.35	0.35	54.0	54.0

MATLAB Function 3.10) based on the global strain, $\bar{\varepsilon}$, and the strain concentration tensors, and reporting output. Note that, for the MOC, pseudocolor plots are generated wherein the subcell stress and strain fields are plotted over the RUC (see Fig. 3.5B). For the MT, Reuss, and Voigt methods, in order to visualize the local fields similarly, an RUC that captures the correct fiber volume fraction is generated and used to make pseudocolor plots.

The loading for stand-alone micromechanics problems is in terms of global stress and strain components (see MATLAB Function 3.6) applied to a continuum point of the composite material (e.g., the RUC in the case of MOC). For example, to apply a global strain in the x_2-direction of 2 percent, $\bar{\varepsilon}_{22} = 0.02$, while specifying all other global stress components to be zero, $\bar{\sigma}_{11} = \bar{\sigma}_{33} = \bar{\sigma}_{23} = \bar{\sigma}_{13} = \bar{\sigma}_{12} = 0$, one would specify;

```
Loads{NP}.Type = [S, E, S, S, S, S];
Loads{NP}.Value = [0, 0.02, 0, 0, 0, 0];
```

where E indicates an applied global strain component and S indicates an applied global stress component, and the location in the `Loads{NP}.Type` vector corresponds to the six components (ordered as 11, 22, 33, 23, 13, and 12), and specified shear strains are engineering quantities.

The micromechanics-based laminate solution flow chart (still based on the MATLAB Script 2.6 – `LamAnalysis.m`), shown in Fig. 3.8B, is the same as the ply-based lamination theory flow chart shown in Fig. 2.7A, except for the new connections/steps required to access the lower length scale (constituent material) information (indicated in red). First, `GetPlyProps.m` has been modified to obtain the effective ply properties as calculated by the micromechanics theories. This entails calling `GetConstitProps.m` and `GetEffProps.m` and involves only minor additions to `GetPlyProps.m` as shown in MATLAB Function 2.1. Then, after the lamination theory calculations have been performed, the local (constituent material) fields must be calculated per ply based on the strains in each ply and the strain concentration tensors. As shown in `LamMicroFields.m` (see MATLAB Function 3.11), this entails first calculating the ply level through-thickness (x_3-direction) strain, ε_{33}^k, that results in a state of plane stress for the ply [see Eq. (2.76)]. This component, along with the in-plane strain components, are then passed to the function `MicroFields.m` as the global composite strains, $\bar{\varepsilon}$. Note that $\bar{\gamma}_{13} = \bar{\gamma}_{23} = 0$. Finally, in step 6, the micro scale stress and strain fields are reported.

MATLAB Script 3.5 – `MicroAnalysis.m`

This script performs stand-alone micromechanics analysis of a unidirectional material. The composite material and micromechanics model are defined in `GetEffProps.m` (MATLAB Function 3.7) and the details of the problem to be analyzed are specified in `MicroProblemDef.m` (MATLAB Function 3.6).

```matlab
% ++++++++++++++++++++++++++++++++++++++++++++++++++++++++++++++++++++++++++++
% Copyright 2020 United States Government as represented by the Administrator of the
% National Aeronautics and Space Administration. No copyright is claimed in the
% United States under Title 17, U.S. Code. All Other Rights Reserved. By downloading
% or using this software you agree to the terms of NOSA v1.3. This code was prepared
% by Drs. B.A. Bednarcyk and S.M. Arnold to complement the book "Practical
% Micromechanics of Composite Materials" during the course of their government work.
% ++++++++++++++++++++++++++++++++++++++++++++++++++++++++++++++++++++++++++++
%
% Purpose: This script runs stand-alone micromechanics problems treating the
%          composite as a material point.  The composite material is defined in
%          GetEffProps.m and the loading and problem setup is specified in
%          MicroProblemDef.m
%
% ++++++++++++++++++++++++++++++++++++++++++++++++++++++++++++++++++++++++++++

% -- Clear memory and close files
clear;
close all;
fclose('all');
clc;

% -- Add needed function locations to the path
addpath('Functions/Utilities');
addpath('Functions/WriteResults');
addpath('Functions/Micromechanics');
addpath('Functions/Margins');

%-----------------------------------------------------------------
% 1) Define Micromechanics Problems
%-----------------------------------------------------------------
[NProblems, OutInfo, MicroMat, Loads] = MicroProblemDef();

%-----------------------------------------------------------------
% 2) Get constituent properties
%-----------------------------------------------------------------
[constitprops] = GetConstitProps;

%-----------------------------------------------------------------
% 3) Get effective properties from micromechanics
%-----------------------------------------------------------------

% -- Preallocate
effprops = cell(1, 200);
Results = cell(1, NProblems);

% -- Determine which mats are used in any problem
for NP = 1: NProblems
    mat = MicroMat{NP};
    effprops{mat}.used = true;
end

[effprops] = GetEffProps(constitprops, effprops);

% -- Loop through problems
for NP = 1: NProblems

    mat = MicroMat{NP};

    % -- Check for missing problem name
    if (ismissing(OutInfo.Name(NP)))
        OutInfo.Name(NP) = string(['Problem ', char(num2str(NP))]);

    end

    % -- Echo problem info to command window
    disp(['Micro Problem #',num2str(NP),' - ', char(OutInfo.Name(NP))]);
    disp(['   Material Number ',num2str(mat)]);

    % -- Option to quit if just getting eff props in GetEffProps
    if isfield(effprops{mat}, 'Quit')
        if effprops{mat}.Quit
            disp('   Completed calculation of effective properties -- quitting');
            disp(['   *** Problem ',char(num2str(NP)),' Completed ***'])
            disp(' ');
            continue;
        end
    end
```

```
% -- Check that loads are specified for this problem
if ~isfield(Loads{NP}, 'Type') || ~isfield(Loads{NP}, 'Value')
    error(strcat('Problem #', num2str(NP), ' Loads not properly defined'));
end
if ~isfield(Loads{NP}, 'DT')
    Loads{NP}.DT = 0;
end

% -- Check that the problem's ply material has been defined
if ~isfield(effprops{mat}, 'name')
    error(strcat('effective material #', num2str(mat), ...
                 ' undefined ... check GetEffProps'));
end

%-----------------------------------------------------------------
% 4) Solve loading and calculate local fields for micromechanics
%-----------------------------------------------------------------

% -- Calculate global thermal strains
epsth = Loads{NP}.DT * [effprops{mat}.a1; effprops{mat}.a2; effprops{mat}.a3; ...
                        0; 0; 0];
% -- Solve for unknown strains and stresses in SG = C*FullGlobalStrain + B
B = -effprops{mat}.Cstar*epsth;
[SG, FullGlobalStrain] = SolveLoading(6, effprops{mat}.Cstar, B, Loads{NP});

% -- Calculate micro scale (constituent level) fields
[Results{NP}] = MicroFields(FullGlobalStrain, Loads{NP}.DT, effprops{mat});

%-----------------------------------------------------------------
% 5) Write micromechanics property and global load results
%-----------------------------------------------------------------

effprops{mat}.Mat = mat;
[OutInfo] = OutputMicro(OutInfo, NP, effprops{mat}, Loads{NP}, SG, ...
                        FullGlobalStrain, epsth);

%-----------------------------------------------------------------
% 6) Plot micro fields
%-----------------------------------------------------------------
if (~isfield(OutInfo,'Format'))
    OutInfo.Format = "txt";
end

if OutInfo.MakePlots
    PlotMicroFields(OutInfo, effprops{mat}, Results{NP});
end

disp(['  *** Problem ',char(num2str(NP)),' Completed ***'])
disp(' ');
close all;

end
```

MATLAB Function 3.6 – `MicroProblemDef.m`

This function defines any number of stand-alone micromechanics problems.

```
function [NP, OutInfo, MicroMat, Loads] = MicroProblemDef()
% +++++++++++++++++++++++++++++++++++++++++++++++++++++++++++++++++++++++++++++++
% Copyright 2020 United States Government as represented by the Administrator of the
% National Aeronautics and Space Administration. No copyright is claimed in the
% United States under Title 17, U.S. Code. All Other Rights Reserved. By downloading
% or using this software you agree to the terms of NOSA v1.3. This code was prepared
% by Drs. B.A. Bednarcyk and S.M. Arnold to complement the book "Practical
% Micromechanics of Composite Materials" during the course of their government work.
% +++++++++++++++++++++++++++++++++++++++++++++++++++++++++++++++++++++++++++++++
%
% Purpose: Sets up and defines stand-alone micromechanics problems.  Specifies
%          problem name, material to be analyzed, loading, and failure criteria.
% Output:
% - NP: Number of problems
% - OutInfo: Struct containing output information
% - MicroMat: Effective material number (from GetEff props) to be analyzed
%             per problem
% - Loads: Cell array containing problem loading information per problem
% +++++++++++++++++++++++++++++++++++++++++++++++++++++++++++++++++++++++++++++++
```

```
NProblemsMax = 200; % -- To preallocate cells
NP = 0; % -- Problem number counter
E = 1; % -- Applied loading type identifier (strains)
S = 2; % -- Applied loading type identifier (stresses)

% -- Preallocate cell
Loads = cell(1,NProblemsMax);

% -- Type of output requested
OutInfo.Format = "txt"; % --Text format
%OutInfo.Format = "doc"; % -- MS Word format (Windows only, MUCH slower)

% -- Include or exclude plots from output
OutInfo.MakePlots = true;
%OutInfo.MakePlots = false;

%==================================================================
% -- Define all problems below
%==================================================================

%% -- Chapter 3 Examples

NP = NP + 1;
OutInfo.Name(NP) = "Sect 3.12.1&2 - Voigt IM7-8552 Vf=0.55";
MicroMat{NP} = 100;
Loads{NP}.DT = 0.0;
Loads{NP}.Type  = [S,  S,  S,  S,  S,  S];
Loads{NP}.Value = [0,  1,  0,  0,  0,  0];

NP = NP + 1;
OutInfo.Name(NP) = "Sect 3.12.1&2 - Reuss IM7-8552 Vf=0.55";
MicroMat{NP} = 101;
Loads{NP}.DT = 0.0;
Loads{NP}.Type  = [S,  S,  S,  S,  S,  S];
Loads{NP}.Value = [0,  1,  0,  0,  0,  0];

NP = NP + 1;
OutInfo.Name(NP) = "Sect 3.12.1&2 - MT IM7-8552 Vf=0.55";
MicroMat{NP} = 102;
Loads{NP}.DT = 0.0;
Loads{NP}.Type  = [S,  S,  S,  S,  S,  S];
Loads{NP}.Value = [0,  1,  0,  0,  0,  0];

NP = NP + 1;
OutInfo.Name(NP) = "Sect 3.12.1&2 - MOC IM7-8552 Vf=0.55";
MicroMat{NP} = 103;
Loads{NP}.DT = 0.0;
Loads{NP}.Type  = [S,  S,  S,  S,  S,  S];
Loads{NP}.Value = [0,  1,  0,  0,  0,  0];

NP = NP + 1;
OutInfo.Name(NP) = "Sect 3.12.1&2 - MOCu IM7-8552 Vf=0.55";
MicroMat{NP} = 104;
Loads{NP}.DT = 0.0;
Loads{NP}.Type  = [S,  S,  S,  S,  S,  S];
Loads{NP}.Value = [0,  1,  0,  0,  0,  0];

NP = NP + 1;
OutInfo.Name(NP) = "Sect 3.12.3 - All theories IM7-8552 all Vf";
MicroMat{NP} = 105;

NP = NP + 1;
OutInfo.Name(NP) = "Sect 3.12.3 - All theories glass-epoxy all Vf";
MicroMat{NP} = 106;

%%

%==================================================================
% -- End of problem definitions
%==================================================================

% -- Store applied loading type identifiers
for I = 1:NP
    Loads{I}.E = E;
    Loads{I}.S = S;
end

end
```

MATLAB Function 3.7 – *GetEffProps.m*

This function defines micromechanics-based composite materials for use in `Micro-ProblemDef.m` or `LamProblemDef.m`.

```matlab
function [props] = GetEffProps(constitprops, props)
% ++++++++++++++++++++++++++++++++++++++++++++++++++++++++++++++++++++++++++++++++
% Copyright 2020 United States Government as represented by the Administrator of the
% National Aeronautics and Space Administration. No copyright is claimed in the
% United States under Title 17, U.S. Code. All Other Rights Reserved. By downloading
% or using this software you agree to the terms of NOSA v1.3. This code was prepared
% by Drs. B.A. Bednarcyk and S.M. Arnold to complement the book "Practical
% Micromechanics of Composite Materials" during the course of their government work.
% ++++++++++++++++++++++++++++++++++++++++++++++++++++++++++++++++++++++++++++++++
%
% Purpose: Specifies material number, name, micromechanics theory, constituent volume
%          fractions, and constituent materials for micromechanics-based composite
%          materials, and runs the micromechanics theory to obtain effective props
%          and concentration tensors.
%          ** Any Mat not used in a Problem will not be evaluated **
% Input:
% - constitprops: Cell/struct containing constituent properties
% - props: Cell/struct containing effective composite properties
% Output:
% - props: Updated cell/struct containing effective composite properties
% ++++++++++++++++++++++++++++++++++++++++++++++++++++++++++++++++++++++++++++++++

disp('*** Evaluating Micromechanics-Based Materials ***');

%================================================================
% -- Define all composite materials below
% -- *** Begin numbering at 100 ***
%================================================================

Mat = 100;
name = "Voigt IM7-8552 Vf = 0.55";
Theory = 'Voigt';
Vf = 0.55;
Constits.Fiber = constitprops{1};
Constits.Matrix = constitprops{3};
[props{Mat}] = RunMicro(Theory, name, Vf, Constits, props{Mat});

Mat = 101;
name = "Reuss IM7-8552 Vf = 0.55";
Theory = 'Reuss';
Vf = 0.55;
Constits.Fiber = constitprops{1};
Constits.Matrix = constitprops{3};
[props{Mat}] = RunMicro(Theory, name, Vf, Constits, props{Mat});

Mat = 102;
name = "MT IM7-8552 Vf = 0.55";
Theory = 'MT';
Vf = 0.55;
Constits.Fiber = constitprops{1};
Constits.Matrix = constitprops{3};
[props{Mat}] = RunMicro(Theory, name, Vf, Constits, props{Mat});

Mat = 103;
name = "MOC IM7-8552 Vf = 0.55";
Theory = 'MOC';
Vf = 0.55;
Constits.Fiber = constitprops{1};
Constits.Matrix = constitprops{3};
[props{Mat}] = RunMicro(Theory, name, Vf, Constits, props{Mat});

Mat = 104;
name = "MOCu IM7-8552 Vf = 0.55";
Theory = 'MOCu';
Vf = 0.55;
Constits.Fiber = constitprops{1};
Constits.Matrix = constitprops{3};
[props{Mat}] = RunMicro(Theory, name, Vf, Constits, props{Mat});

Mat = 105;
name = "All theories IM7-8552 Vf = x";
Constits.Fiber = constitprops{1};
Constits.Matrix = constitprops{3};
% -- Call function (contained in this file) to get eff props across all Vf
RunOverVf(name, Constits, props, Mat);
props{Mat}.Quit = true; % -- Tells MicroAnalysis to quit after getting eff props
```

```
Mat = 106;
name = "All theories glass-epoxy Vf = x";
Constits.Fiber = constitprops{2};
Constits.Matrix = constitprops{4};
% -- Call function (contained in this file) to get eff props across all Vf
RunOverVf(name, Constits, props, Mat);
props{Mat}.Quit = true; % -- Tells MicroAnalysis to quit after getting eff props

end

%==================================================================
% -- End composite material definitions
%==================================================================

%------------------------------------------------------------------
%------------------------------------------------------------------

function RunOverVf(name, Constits, props, Mat)

% -- Run Voigt, Reuss, MT, MOC, and MOCu micromechanics theories over all
%    Vf for the specifed Constits, write out eff prop results to file

Fact1 = 1000; % -- Factor - MPa to GPa
Fact2 = 1E6; % -- Factor - /C to 1.E-6/C

VfResults = cell(5);
% -- Loop over theory
for k = 1:5
    switch k
        case 1
            Theory = 'Voigt';
            Name(k) = "Voigt       ";
        case 2
            Theory = 'Reuss';
            Name(k) = "Reuss       ";
        case 3
            Theory = 'MT';
            Name(k) = "MT          ";
        case 4
            Theory = 'MOC';
            Name(k) = "MOC         ";
        case 5
            Theory = 'MOCu';
            Name(k) = "MOCu        ";
    end

    N_Vf = 100;
    for i = 1: N_Vf + 1
        Vf = (i - 1)/N_Vf;
        if exist('RUCid','var')
            [props{Mat}] = RunMicro(Theory, name, Vf, Constits, props{Mat}, RUCid);
        else
            [props{Mat}] = RunMicro(Theory, name, Vf, Constits, props{Mat});
        end
        if ~props{Mat}.used
            return;
        end
        VfResults{k}.Vf(i) = Vf;
        VfResults{k}.EffProps(1,i) = props{Mat}.E1/Fact1;
        VfResults{k}.EffProps(2,i) = props{Mat}.E2/Fact1;
        VfResults{k}.EffProps(3,i) = props{Mat}.E3/Fact1;
        VfResults{k}.EffProps(4,i) = props{Mat}.G23/Fact1;
        VfResults{k}.EffProps(5,i) = props{Mat}.G13/Fact1;
        VfResults{k}.EffProps(6,i) = props{Mat}.G12/Fact1;
        VfResults{k}.EffProps(7,i) = props{Mat}.v12;
        VfResults{k}.EffProps(8,i) = props{Mat}.v13;
        VfResults{k}.EffProps(9,i) = props{Mat}.v23;
        VfResults{k}.EffProps(10,i) = props{Mat}.a1*Fact2;
        VfResults{k}.EffProps(11,i) = props{Mat}.a2*Fact2;
        VfResults{k}.EffProps(12,i) = props{Mat}.a3*Fact2;
    end
end

% -- Print to file for Excel plotting
[~,~,~] = mkdir('Output');
tttt = datetime(datetime,'Format','yyyy-MMM-dd HH.mm.ss');
OutFile = ['Output/Props vs Vf - ',char(name),char(tttt),'.txt'];
fid = fopen(OutFile,'wt');
Fmet = '%E \t %E \t %E \t %E \t %E \t %E \t %E \t %E \t %E \t %E \n';
Fmst = '%s \t %s \t %s \t %s \t %s \t %s \t %s \t %s \t %s \t %s \n';
Fmnt = '\n %s \n';
Props = ["E1", "E2", "E3", "G23", "G13", "G12", "V12", "V13", "V23", ...
         "alpha1", "alpha2", "alpha3"];

for j = 1:12
    TSProp = ['---- ', char(Props(j)),' ----'];
    fprintf(fid, Fmnt, TSProp);
    fprintf(fid, Fmst, ["Vf          ", Name]);
    for i = 1: N_Vf + 1
        fprintf(fid, Fmet, VfResults{1}.Vf(i), VfResults{1}.EffProps(j,i), ...
                VfResults{2}.EffProps(j,i), VfResults{3}.EffProps(j,i), ...
                VfResults{4}.EffProps(j,i), VfResults{5}.EffProps(j,i));
    end
end

fclose(fid);

end
```

MATLAB Function 3.8 – `RunMicro.m`

This function runs the chosen micromechanics theory to obtain the effective material properties and the strain concentration tensors.

```
function [plyprops] = RunMicro(Theory, name, Vol, Constits, plyprops, RUCid)
% +++++++++++++++++++++++++++++++++++++++++++++++++++++++++++++++++++++++++++++++
% Copyright 2020 United States Government as represented by the Administrator of the
% National Aeronautics and Space Administration. No copyright is claimed in the
% United States under Title 17, U.S. Code. All Other Rights Reserved. By downloading
% or using this software you agree to the terms of NOSA v1.3. This code was prepared
% by Drs. B.A. Bednarcyk and S.M. Arnold to complement the book "Practical
% Micromechanics of Composite Materials" during the course of their government work.
% +++++++++++++++++++++++++++++++++++++++++++++++++++++++++++++++++++++++++++++++
% Purpose: Runs the chosen micromechanics theory to obtain the effective properties
%                   and strain concentration tensors
% Input:
% - Theory: Chosen micromechanics theory
% - name: Micromechanics-based material name
% - Vol: Contains either Vf or is struct containing Vf and Vi
% - Constits: Struct containing constituent material properties
% - plyprops: Struct containing effective properties
% - RUCid: RUC i.d. number or full RUC struct containing RUC info
%               Optional argument, only used for GMC and HFGMC (Ch. 5 & 6)
% Output:
% - plyprops: Updated struct containing effective properties
% +++++++++++++++++++++++++++++++++++++++++++++++++++++++++++++++++++++++++++++++

% -- Do not run micromechanics if this material not used in any problems
if ~isfield(plyprops, 'used')
    plyprops.used = false;
    return
elseif ~plyprops.used
    return
end

% -- Extract fiber and interface vol fractions (Vol can be Vf or struct)
if isstruct(Vol)
    Vf = Vol.Vf;
    if isfield(Vol, 'Vi')
        Vi = Vol.Vi;
    else
        Vi = 0;
    end
else
    Vf = Vol;
    Vi = 0;
end

% -- Run the specified micromechanics theory
switch(Theory)
    case 'Voigt'
        [Cstar, CTEs, Af, Am, ATf, ATm] = Voigt(Constits.Fiber, Constits.Matrix, Vf);
        plyprops.micro = "Voigt";

    case 'Reuss'
        [Cstar, CTEs, Af, Am, ATf, ATm] = Reuss(Constits.Fiber, Constits.Matrix, Vf);
        plyprops.micro = "Reuss";

    case 'MT'
        [Cstar, CTEs, Af, Am, ATf, ATm] = MT(Constits.Fiber, Constits.Matrix, Vf);
        plyprops.micro = "MT";

    case 'MOC'
        [Cstar,CTEs,As,Cstaru,CTEsu,Asu] = MOC(Constits.Fiber, Constits.Matrix, Vf);
        plyprops.micro = "MOC";
        [RUC] = GetRUC(Constits, Vf, Vi, 2);

    case 'MOCu'
        [Cstar,CTEs,As,Cstaru,CTEsu,Asu] = MOC(Constits.Fiber, Constits.Matrix, Vf);
        Cstar = Cstaru;
        CTEs = CTEsu;
        As = Asu;
        plyprops.micro = "MOCu";

    otherwise
        error(['Invalid Theory Specified in RunMicro - Theory = ', Theory])

end
```

```
% -- Store info in plyprops struct
plyprops.name = name;
plyprops.Vf = Vf;
plyprops.Vi = Vi;
plyprops.Fiber = Constits.Fiber;
plyprops.Matrix = Constits.Matrix;
if isfield(Constits,'Interface')
    plyprops.Interface = Constits.Interface;
end

[Effprops] = GetEffPropsFromC(Cstar);
plyprops.E1 = Effprops.E1;
plyprops.E2 = Effprops.E2;
plyprops.E3 = Effprops.E3;
plyprops.v12 = Effprops.v12;
plyprops.v13 = Effprops.v13;
plyprops.v23 = Effprops.v23;
plyprops.G12 = Effprops.G12;
plyprops.G13 = Effprops.G13;
plyprops.G23 = Effprops.G23;

plyprops.a1 = CTEs(1);
plyprops.a2 = CTEs(2);
plyprops.a3 = CTEs(3);

plyprops.Cstar = Cstar;

switch(Theory)
    case {'Voigt', 'Reuss', 'MT'}
        plyprops.Af = Af;
        plyprops.Am = Am;
        plyprops.ATf = ATf;
        plyprops.ATm = ATm;
    case {'MOC', 'MOCu'}
        plyprops.As = As;
    otherwise
        plyprops.As = As;
        plyprops.RUC = RUC;
end
```

MATLAB Function 3.9 – *GetConstitProps.m*

This function is used to specify the properties of constituent materials for use in micromechanics calculations.

```
function [constitprops] = GetConstitProps
% ++++++++++++++++++++++++++++++++++++++++++++++++++++++++++++++++++++++++++++++
% Copyright 2020 United States Government as represented by the Administrator of the
% National Aeronautics and Space Administration. No copyright is claimed in the
% United States under Title 17, U.S. Code. All Other Rights Reserved. By downloading
% or using this software you agree to the terms of NOSA v1.3. This code was prepared
% by Drs. B.A. Bednarcyk and S.M. Arnold to complement the book "Practical
% Micromechanics of Composite Materials" during the course of their government work.
% ++++++++++++++++++++++++++++++++++++++++++++++++++++++++++++++++++++++++++++++
% Purpose: Specifies properties of composite constituent materials
% Input: None
% Output:
% - constitprops: Cell/struct containing constituent properties
% ++++++++++++++++++++++++++++++++++++++++++++++++++++++++++++++++++++++++++++++

% -- Preallocate
NconstitMax = 20;
constitprops = cell(1,NconstitMax);

%=============================================================================
% -- Define all constituent materials below
%=============================================================================

% -- Constituent Mat#1 - IM7 (Transversely Isotropic)
Mat = 1;
constitprops{Mat}.name = "IM7";
constitprops{Mat}.constID = Mat;
constitprops{Mat}.EL = 262.2E3;
constitprops{Mat}.ET = 11.8E3;
constitprops{Mat}.vL = 0.17;
constitprops{Mat}.vT = 0.21;
constitprops{Mat}.GL = 18.9E3;
constitprops{Mat}.aL = -0.9E-06;
constitprops{Mat}.aT = 9.E-06;

% -- Constituent Mat#2 - Glass (Isotropic)
Mat = 2;
constitprops{Mat}.name = "Glass";
constitprops{Mat}.constID = Mat;
constitprops{Mat}.EL = 73.E3;
constitprops{Mat}.ET = 73.E3;
constitprops{Mat}.vL = 0.22;
constitprops{Mat}.vT = 0.22;
constitprops{Mat}.GL = constitprops{Mat}.EL/(2*(1+constitprops{Mat}.vL)); % -- isotropic
constitprops{Mat}.aL = 5.E-06;
constitprops{Mat}.aT = 5.E-06;

% -- Constituent Mat#3 - 8552 (Isotropic)
Mat = 3;
constitprops{Mat}.name = "8552";
constitprops{Mat}.constID = Mat;
constitprops{Mat}.EL = 4.67E3;
constitprops{Mat}.ET = 4.67E3;
constitprops{Mat}.vL = 0.45;
constitprops{Mat}.vT = 0.45;
constitprops{Mat}.GL = constitprops{Mat}.EL/(2*(1+constitprops{Mat}.vL)); % -- isotropic
constitprops{Mat}.aL = 42.E-06;
constitprops{Mat}.aT = 42.E-06;

% -- Constituent Mat#4 - Epoxy (Isotropic) - Allowables based on MT
Mat = 4;
constitprops{Mat}.name = "Epoxy";
constitprops{Mat}.constID = Mat;
constitprops{Mat}.EL = 3.45E3;
constitprops{Mat}.ET = 3.45E3;
constitprops{Mat}.vL = 0.35;
constitprops{Mat}.vT = 0.35;
constitprops{Mat}.GL = constitprops{Mat}.EL/(2*(1+constitprops{Mat}.vL)); % -- isotropic
constitprops{Mat}.aL = 54.e-06;
constitprops{Mat}.aT = 54.e-06;

%=============================================================================
% -- End of constituent material definitions
%=============================================================================

end
```

MATLAB Function 3.10 – `MicroFields.m`

This function calculates the micromechanics-based local (constituent scale) stress and strain fields based on the strain concentration tensors and the global (composite scale) strains.

```
function [Results] = MicroFields(FullGlobalStrain, DT, props, Results)
% ++++++++++++++++++++++++++++++++++++++++++++++++++++++++++++++++++++++++++++++++
% Copyright 2020 United States Government as represented by the Administrator of the
% National Aeronautics and Space Administration. No copyright is claimed in the
% United States under Title 17, U.S. Code. All Other Rights Reserved. By downloading
% or using this software you agree to the terms of NOSA v1.3. This code was prepared
% by Drs. B.A. Bednarcyk and S.M. Arnold to complement the book "Practical
% Micromechanics of Composite Materials" during the course of their government work.
% ++++++++++++++++++++++++++++++++++++++++++++++++++++++++++++++++++++++++++++++++
% Purpose: Calculate the local (constituent level) stress and strain fields in the
%          composite given the concentration tensors and global fields and
%          temperature change
% Input:
% - FullGlobalStrain: Vector of global total strain components
% - DT: Applied temperature change
% - props: Struct containing effective composite properties
% - Results: Struct containing micromechanics analysis results
% Output:
% - Results: Updated struct containing micromechanics analysis results
% ++++++++++++++++++++++++++++++++++++++++++++++++++++++++++++++++++++++++++++++++

% -- Obtain fiber and matrix stiffness tensors
Cf = GetCFromProps(props.Fiber);
Cm = GetCFromProps(props.Matrix);

% -- Micromechanics theories with only average fiber and matrix fields
if (props.micro == "MT" || props.micro == "Voigt" || props.micro == "Reuss")
    ThStrainF = [props.Fiber.aL; props.Fiber.aT; props.Fiber.aT; 0; 0; 0]*DT;
    MicroStrainF = props.Af*FullGlobalStrain + props.ATf*DT;
    MicroStressF = Cf*(MicroStrainF - ThStrainF);
    ThStrainM = [props.Matrix.aL; props.Matrix.aT; props.Matrix.aT; 0; 0; 0]*DT;
    MicroStrainM = props.Am*FullGlobalStrain + props.ATm*DT;
    MicroStressM = Cm*(MicroStrainM - ThStrainM);
    Results.Type = "FM";
    Results.MicroFieldsF.thstrain = ThStrainF;
    Results.MicroFieldsF.strain = MicroStrainF;
    Results.MicroFieldsF.stress = MicroStressF;
    Results.MicroFieldsM.thstrain = ThStrainM;
    Results.MicroFieldsM.strain = MicroStrainM;
    Results.MicroFieldsM.stress = MicroStressM;

% -- MOC with 2x2 subcell RUC
elseif props.micro == "MOC" || props.micro == "MOCu"
    for g = 1:2
        for b = 1:2
            if b == 1 && g == 1
                C = Cf;
                ThStrain = [props.Fiber.aL; props.Fiber.aT; props.Fiber.aT; ...
                            0; 0; 0]*DT;
            else
                C = Cm;
                ThStrain = [props.Matrix.aL; props.Matrix.aT; props.Matrix.aT; ...
                            0; 0; 0]*DT;
            end
            MicroFields(b,g).strain = props.As(b,g).A*FullGlobalStrain + ...
                                      props.As(b,g).AT*DT;
            MicroFields(b,g).stress = C*(MicroFields(b,g).strain - ThStrain);
            MicroFields(b,g).thstrain = ThStrain;
            Results.Type = "bg";
            Results.MicroFields(b,g) = MicroFields(b,g);
        end
    end

else
    txt = ['plyprops{',char(num2str(mat)),'}.micro = ', char(props.micro)];
    error(['** Incorrect micromechanics theory name specified: ', txt])

end

end
```

MATLAB Function 3.11 – *LamMicroFields.m*

This function calls `MicroFields.m` to calculate the local (constituent scale) fields at the top and bottom of each ply in a micromechanics-based laminate analysis.

```
function [LamResults] = LamMicroFields(Geometry, Loads, plyprops, LamResults)
% ++++++++++++++++++++++++++++++++++++++++++++++++++++++++++++++++++++++++++++++++++
% Copyright 2020 United States Government as represented by the Administrator of the
% National Aeronautics and Space Administration. No copyright is claimed in the
% United States under Title 17, U.S. Code. All Other Rights Reserved. By downloading
% or using this software you agree to the terms of NOSA v1.3. This code was prepared
% by Drs. B.A. Bednarcyk and S.M. Arnold to complement the book "Practical
% Micromechanics of Composite Materials" during the course of their government work.
% ++++++++++++++++++++++++++++++++++++++++++++++++++++++++++++++++++++++++++++++++++
% Purpose: Calculate the local (constituent level) stress and strain fields at the
%          top and bottom of each ply in a composite laminate by first determining
%          the full 3D strain at each point and then calling MicroFields.m
% Input:
% - Geometry: Struct containing laminate definition variables
% - Loads: Struct containing problem loading information
% - plyprops: Cell/struct containing effective composite material properties
% - LamResults: Struct containing laminate analysis results
% Output:
% - LamResults: Struct containing laminate analysis results updated to include local
%               micromechanics result ("Micro(kk)") per ply location
% ++++++++++++++++++++++++++++++++++++++++++++++++++++++++++++++++++++++++++++++++++

DT = Loads.DT;

% -- Calculate and plot microfields per ply top and bottom location (kk)
for kk = 1: 2*Geometry.N

    k = round(kk/2);  % -- Ply number
    mat = Geometry.plymat(k);

    if isfield(plyprops{mat},'micro')

        % -- Calculate eps_zz = eps_33 (Eq. 2.76)
        Cstar = plyprops{mat}.Cstar;
        alpha = [plyprops{mat}.a1; plyprops{mat}.a2; plyprops{mat}.a3];
        Ptstrain(:) = LamResults.MCstrain(:, kk);
        epsz = -(Ptstrain(1) - alpha(1)*DT)*Cstar(1,3)/Cstar(3,3) ...
               -(Ptstrain(2) - alpha(2)*DT)*Cstar(2,3)/Cstar(3,3) ...
               + alpha(3)*DT;

        FullGlobalStrain = [Ptstrain(1); Ptstrain(2); epsz; 0; 0; Ptstrain(3)];

        % -- Call MicroFields.m for this ply location
        if ~isfield(LamResults, 'Micro')
            [LamResults.Micro(kk)] = MicroFields(FullGlobalStrain, DT, plyprops{mat});
        elseif length(LamResults.Micro) < kk
            [LamResults.Micro(kk)] = MicroFields(FullGlobalStrain, DT, plyprops{mat});
        else
            [LamResults.Micro(kk)] = MicroFields(FullGlobalStrain, DT, ...
                                      plyprops{mat}, LamResults.Micro(kk));
        end

    end

end

end
```

3.12   Example micromechanics problems

In this section, the two most important features of micromechanics will be demonstrated; 1) the prediction of effective composite properties through homogenization of the constituent properties and 2) the prediction of constituent level stress and strain fields through localization of applied composite level loading (stresses or strains). These features will be demonstrated through stand-alone micromechanics problems and a micromechanics-based laminate problem.

3.12.1   Stand-alone micromechanics effective properties

The four micromechanics methods presented in this chapter, the Voigt approximation, the Reuss approximation, MT, and MOC, all predict effective properties for composite materials based on the constituent properties and the fiber volume fraction (v_f). In this example, the computed effective properties are compared for a unidirectional IM7/8552 composite with a fiber volume fraction of 0.55. To execute these examples, the following MATLAB functions are used to provide the required input

1. `GetConstitProps.m` (MATLAB Function 3.9) – Specifies the constituent material properties (see Table 3.4).

   For the IM7 fiber:
   ```
   % -- Constituent Mat#1 - IM7 (Transversely Isotropic)
   Mat = 1;
   constitprops{Mat}.name = "IM7";
   constitprops{Mat}.constID = Mat;
   constitprops{Mat}.EL = 262.2E3;
   constitprops{Mat}.ET = 11.8E3;
   constitprops{Mat}.vL = 0.17;
   constitprops{Mat}.vT = 0.21;
   constitprops{Mat}.GL = 18.9E3;
   constitprops{Mat}.aL = -0.9E-06;
   constitprops{Mat}.aT = 9.E-06;
   ```

   For the 8552 resin matrix:
   ```
   % -- Constituent Mat#3 - 8552 (Isotropic)
   Mat = 3;
   constitprops{Mat}.name = "8552";
   constitprops{Mat}.constID = Mat;
   constitprops{Mat}.EL = 4.67E3;
   constitprops{Mat}.ET = 4.67E3;
   constitprops{Mat}.vL = 0.45;
   constitprops{Mat}.vT = 0.45;
   constitprops{Mat}.GL = constitprops{Mat}.EL/(2*(1+constitprops{Mat}.vL)); % -- isotropic
   constitprops{Mat}.aL = 42.E-06;
   constitprops{Mat}.aT = 42.E-06;
   ```

2. `GetEffProps.m` (MATLAB Function 3.7) – Defines the micromechanics-based composite material (micromechanics theory, fiber volume fraction, and constituent materials). Note that the numbering of these micromechanics-based materials starts at 100. For example:
   ```
   Mat = 100;
   name = "Voigt IM7-8552 Vf = 0.55";
   Theory = 'Voigt';
   Vf = 0.55;
   Constits.Fiber = constitprops{1};
   Constits.Matrix = constitprops{3};
   [props{Mat}] = RunMicro(Theory, name, Vf, Constits, props{Mat});
   ```

3. `MicroProblemDef.m` (see MATLAB Function 3.6) – Specifies which micromechanics-based composite material is being analyzed per problem (e.g., `MicroMat{NP} = 100`) and the applied loading (in terms of global, composite level, stresses and strains plus an optional temperature change). Note that mechanical loading must be specified for each problem, even if only effective property results are needed. For example:
   ```
   NP = NP + 1;
   OutInfo.Name(NP) = "Sect 3.12.1&2 - Voigt IM7-8552 Vf=0.55";
   MicroMat{NP} = 100;
   Loads{NP}.DT = 0.0;
   Loads{NP}.Type  = [S, S, S, S, S, S];
   Loads{NP}.Value = [0, 1, 0, 0, 0, 0];
   ```

Table 3.5 Predicted effective mechanical properties of a 0.55 fiber volume fraction unidirectional IM7/8552 composite by different micromechanics theories. Note that this corresponds to the output of the MATLAB code with respect to significant figures.

Theory	E1 (MPa)	E2 (MPa)	E3 (MPa)	G12 (MPa)	G13 (MPa)	G23 (MPa)	Nu12	Nu13	Nu23
Voigt	147135.36	10394.23	10394.23	11119.66	11119.66	3406.47	0.3485	0.3485	0.5257
Reuss	10156.68	6994.48	6994.48	3241.03	3241.03	2549.46	0.444	0.444	0.3718
MT	146420.02	8733.25	8733.25	4394.31	4394.31	2809.55	0.3029	0.3029	0.5542
MOC	146404.54	8764.9	8764.9	4129.95	4129.95	2824.54	0.3019	0.3019	0.5516
MOCu	146404.54	9412.97	9412.97	4129.95	4129.95	2549.46	0.3019	0.3019	0.5184

Table 3.6 Predicted effective CTEs of a 0.55 fiber volume fraction unidirectional IM7/8552 composite by different micromechanics theories.

Theory	alpha1 (1/°C)	alpha2 (1/°C)	alpha3 (1/°C)
Voigt	7.5871e-07	4.1765e-05	4.1765e-05
Reuss	1.8405e-05	2.385e-05	2.385e-05
MT	−1.4582e-07	3.36e-05	3.36e-05
MOC	−1.6549e-07	3.3423e-05	3.3423e-05
MOCu	−1.6549e-07	3.3423e-05	3.3423e-05

Once the input is specified, the script `MicroAnalysis.m` (MATLAB Script 3.5) is executed. Output (text-based or MS Word) is written to the `Output` folder.

The effective property results of this example, for all of the micromechanics theories, are shown in Table 3.5 and Table 3.6. MOCu refers to the unaveraged version of MOC, whereas the MOC results involve averaging to obtain transversely isotropic effective properties (see Section 3.8.6). It is clear from Table 3.5 that both of these versions of MOC predict E2 and E3 that are identical, but calculating the transversely isotropic shear modulus from the transverse Young's modulus and Poisson's ratio [see Eq. (2.18)],

$$G_{23} = \frac{E_2}{2(1 + \nu_{23})} \tag{3.176}$$

yields for MOC 2824.5 MPa and for MOCu 3099.6 MPa. For MOC, this matches the predicted G_{23} in Table 3.5, whereas it does not match MOCu. As discussed in Section 3.8.6, this is because MOCu yields six rather than five independent elastic constants. The averaging in the MOC results only affects E_2, E_3, G_{23}, and Nu_{23}. Like the MOC results, Voigt, Reuss, and MT all provide transversely isotropic effective properties. In addition, the MOC and MT results are in reasonably good agreement for all properties, whereas Voigt and Reuss are quite different. The predicted properties of the theories will be compared across all fiber volume fractions in Section 3.12.3 below.

3.12.2 Stand-alone micromechanics local fields

The four micromechanics methods presented in this chapter are also capable of predicting the local stresses and strains within the composite constituent materials because all four calculate strain concentration tensors, which relate the local fields to the applied global loading. The Voigt, Reuss, and MT methods provide average fiber and matrix strain concentration tensors, which enable the calculation of the average strains (and stresses) in each constituent material. In contrast, the MOC predicts distinct strain (and stress) values for each of the four subcells. For consistency in visualization of the local fields, the Voigt, Reuss, and MT results are plotted on an RUC identical to the MOC RUC. The code will produce MATLAB pseudocolor plots of the local stress and strain fields (over the composite geometry) provided the flag `OutInfo.MakePlots` is set equal to `true` in `MicroProblemDef.m` (MATLAB Function 3.6).

In this example, the local fields predicted by the methods are compared for the same unidirectional composite considered above, a 0.55 fiber volume fraction unidirectional IM7/8552 composite, whose constituent properties are given in Table 3.4, when subjected to unit transverse stress loading, $\bar{\sigma}_{22} = 1$ MPa, with all other global stress components set equal to zero (loading specified in `MicroProblemDef.m`). The normal stress fields are plotted in Fig. 3.9, while the normal strain fields are plotted in Fig. 3.10. All shear stress and strain components are identically zero for the given load case. Note that each component is plotted on the same scale across all theories. This is accomplished by specifying `caxis` for each plot in `PlotMicroFields.m`. In this function, these `caxis` lines are commented out by default in the provided code.

Examining the Voigt and Reuss stress and strain results, the iso-strain assumption of Voigt is clear in Fig. 3.10 as each strain component is constant in and across both the fiber and matrix. Similarly, the Reuss iso-stress assumption is apparent in Fig. 3.9, with each stress component constant in the entire composite. Obviously, given that the fiber and matrix stiffnesses differ, the Reuss strain predictions must be different between these phases. Similarly, Voigt predicts distinct stresses for each constituent. Note that the Reuss σ_{22} is exactly 1 MPa, which is exactly the applied global stress magnitude as required by the average stress theorem, Eq. (3.1). This theorem also requires that the volume average of the σ_{22} fields for all other theories equals 1 MPa. Furthermore, the ε_{11} fields for all models, with the exception of Reuss, are constant in both phases. This is because, for continous reinforcements, MT and MOC assume iso-strain in the fiber direction, just like the Voigt approximation (see Eq. (3.93) for MOC).

Examining the stress and strain fields predicted by MT, it is clear that distinct average stress and strain components are predicted between the fiber and matrix (with the exception of ε_{11}). In contrast, MOC and MOCu both provide additional stress and strain field variations, enabling both MOC versions to approximate gradients in the matrix. While MT is capable of predicting the average matrix stress and strain accurately, in reality, stress and strain gradients are present in the matrix, thus motivating the use of the MOC approach. The ability to capture gradients in the matrix is particularly important when modeling nonlinear matrix behavior (e.g., inelasticity and damage), see Aboudi et al. (2013) along with Chapter 7.

Comparing the MOC and MOCu stress and strain fields in Fig. 3.9 and Fig. 3.10, it should be noted that MOCu predicts constant σ_{22} along columns of subcells (in the x_2-direction) and constant σ_{33} along rows of subcells (in the x_3-direction). This is because of the traction continuity applied in an average sense in the formulation combined with the linear displacement field per subcell. However, the averaging performed on the strain concentration tensors to obtain transversely isotropic effective properties relieves the strict traction continuity to some degree, as evident in the MOC fields. As such, the

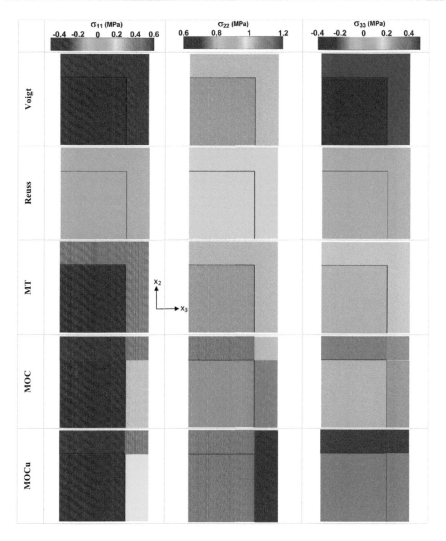

Figure 3.9 Comparison of predicted $v_f = 0.55$ IM7/8552 local stresses in response to an applied global stress of $\bar{\sigma}_{22} = 1$ MPa for the Voigt, Reuss, MT, MOC, and unaveraged MOC (MOCu) micromechanics models.

averaging process for MOC enables the stress gradients in the matrix to be retained while obtaining an average stress state that is associated with transverse isotropy.

3.12.3 Micromechanics theory predictions across fiber volume fraction

In this example, the effective properties predicted by Voigt, Reuss, MT, MOC, and MOCu are compared across all fiber volume fractions (Vf), from zero to one, for unidirectional IM7/8552 and glass/epoxy composites. While there are a variety of

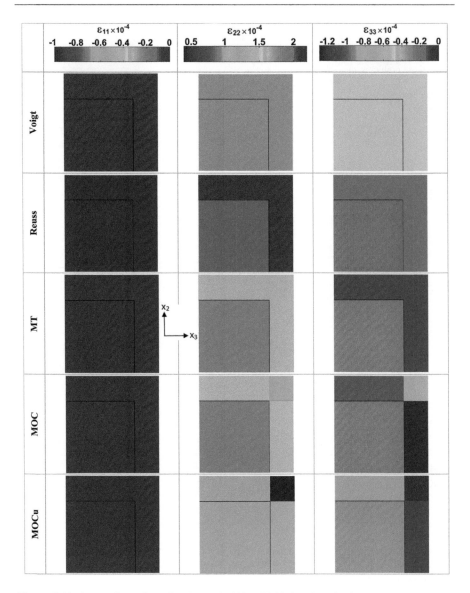

Figure 3.10 Comparison of predicted $v_f = 0.55$ IM7/8552 local strains in response to an applied global stress of $\bar{\sigma}_{22} = 1$MPa for the Voigt, Reuss, MT, MOC, and unaveraged MOC (MOCu) micromechanics models.

ways to obtain quickly the theories' effective property predictions over the range of fiber volume fractions, herein, a function has been added to GetEffProps.m, called RunOverVf (see MATLAB Function 3.7). Mat = 105 represents IM7/8552 and Mat = 106 represents glass/epoxy. Then RunOverVf is called from within GetEffProps.m to run all micromechanics theories over all fiber volume fractions and write the results to a text file ("`Props vs. Vf ... .txt`"). Plots comparing each property among the theories can then be made using any desired software.

To run these problems, the following code defines the problems in `MicroProblemDef.m`:

```
NP = NP + 1;
OutInfo.Name(NP) = "Sect 3.12.3 - All theories IM7-8552 all Vf";
MicroMat{NP} = 105;

NP = NP + 1;
OutInfo.Name(NP) = "Sect 3.12.3 - All theories glass-epoxy all Vf";
MicroMat{NP} = 106;
```

The composite materials are defined, and the calls to `RunOverVf` are made, in `GetEffProps.m`:

```
Mat = 105;
name = "All theories IM7-8552 Vf = x";
Constits.Fiber = constitprops{1};
Constits.Matrix = constitprops{3};
% -- Call function (contained in this file) to get eff props across all Vf
RunOverVf(name, Constits, props, Mat);
props{Mat}.Quit = true; % -- Tells MicroAnalysis to quit after getting eff props

Mat = 106;
name = "All theories glass-epoxy Vf = x";
Constits.Fiber = constitprops{2};
Constits.Matrix = constitprops{4};
% -- Call function (contained in this file) to get eff props across all Vf
RunOverVf(name, Constits, props, Mat);
props{Mat}.Quit = true; % -- Tells MicroAnalysis to quit after getting eff props
```

As shown, a flag, `props{Mat}.Quit`, has been added to tell the MATLAB code to stop after the effective properties are obtained, so no further analysis is performed and no further output is written.

Fig. 3.11 shows the axial and transverse Young's moduli predicted by the four micromechanics methods as a function of Vf. As before, for MOC, results that have employed averaging to obtain transversely isotropic effective properties (see Section 3.8.6) are denoted by "MOC", whereas, the unaveraged predictions are denoted by "MOCu". The dotted horizontal lines represent the ply properties for IM7/8552 and glass/epoxy used in Chapter 2. Obviously, in all cases, for $Vf = 0$, the effective properties correspond to the properties of the matrix constituent material, whereas, for $Vf = 1$, the effective properties correspond to the properties of the fiber constituent material. Also, the Voigt and Reuss approximations represent upper and lower bounds for the composite stiffness tensor components. However, when dealing with engineering constants, as done here, these will not always provide such bounds.

Fig. 3.11A and B show that all methods aside from Reuss predict nearly linear Vf dependence of the axial Young's modulus (E_1). The Reuss E_1 prediction is highly nonlinear with Vf and significantly lower in magnitude compared to the other methods. In contrast to Fig. 3.11A and B, Fig. 3.11C and D show that there is a significant qualitative difference in the E_2 plots between material systems. This qualitative difference is driven by the mismatch ratio of the fiber and matrix E_2, which is approximately 2.5 in the case of IM7/8552 and 21 in the case of glass/epoxy. It is also clear that across all Vf values, the MT and MOC predictions are in close agreement, whereas

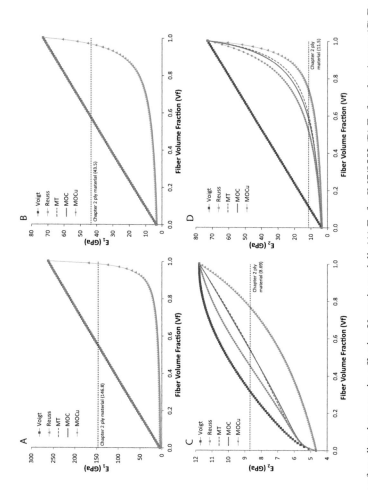

Figure 3.11 Comparison of predicted composite effective Young's moduli. (A) E_1 for IM7/8552 (B) E_1 for glass/epoxy (C) E_2 for IM7/8552 (D) E_2 for glass epoxy. Note the differences in the axis scales.

the MOCu always overpredicts the E_2 compared to these models, particularly at non-dilute fiber volume fractions (> 0.2). The Voigt and Reuss predictions for E_2 provide wide bounds compared to the results of the other methods. Note that the Voigt E_1 and E_2 predictions for glass/epoxy are identical (compare Fig. 3.11B and D) as this theory predicts isotropic composite properties when the constituent materials are isotropic. The same is true for the Reuss theory.

For the shear moduli predictions shown in Fig. 3.12, the Voigt and Reuss predictions are truly upper and lower bounds because the shear moduli correspond to stiffness tensor components. The G_{12} predictions (Fig. 3.12A and B) show that the MT and MOC predictions are in excellent agreement for both material systems, with MOCu corresponding exactly with MOC (because the averaging does not affect G_{12}). Similarly, for G_{23} (Fig. 3.12C and D), the MT and MOC predictions are in excellent agreement for IM7/8852, but with some differences for glass/epoxy for Vf values between 0.6 and 0.9. As expected, the MOCu predictions for G_{23} agree identically with the Reuss predictions, because of the MOCu 23 shear iso-stress condition represented by Eq. (3.106).

Fig. 3.13, which shows the Poisson ratio predictions, clearly indicates that the Voigt and Reuss predictions are not upper and lower bounds. Because the averaging in MOC has no effect on the ν_{12} predictions (see Fig. 3.13A and B), the MOC and MOCu predictions are identical, and both of these are in excellent agreement with MT. For ν_{23} (Fig. 3.13C and D), as one would expect, the averaging within MOC has a significant effect. For the IM7/8552 composite, the MT, MOC, and MOCu predictions are still in reasonably good agreement. However, in the case of glass/epoxy, the predictions of all three models are vastly different in both magnitude and character. It is also interesting to note that these differences are very pronounced in the range of realistic Vf (0.45 – 0.65). Fortunately, this out of plane property does not influence classical lamination theory predictions (although it can impact delamination and damage).

For the predicted coefficients of thermal expansion (CTEs) (see Fig. 3.14), the MOC and MOCu predictions are identical because the MOCu CTEs are already transversely isotropic, therefore the averaging to obtain transverse isotropy has no effect. These models' CTE predictions are in excellent agreement with those of MT for both material systems. As in the case of the effective Poisson ratios, the Voigt and Reuss approximations do not provide bounds for the CTEs.

As mentioned, the dotted lines in Fig. 3.11 through Fig. 3.14 correspond to the ply material properties given in Chapter 2 for the unidirectional IM7/8552 and glass/epoxy composites. It is thus possible to assess the consistency of the micromechanics model predictions with each other, and with respect to the Chapter 2 ply properties, by determining the fiber volume fraction at which each micromechanics model matches a given Chapter 2 ply property (within 0.01). This is the fiber volume fraction at which the horizontal dotted line intersects the curve associated with a model in Fig. 3.11 through Fig. 3.14. The matching fiber volume fractions are given in Table 3.7 and Table 3.8.

Table 3.7 and Table 3.8 show that, in all cases, the MT and MOC volume fractions are in close agreement, indicating that the predictions of these methods are consistent with each other. The MOCu agrees with MT and MOC for all properties except E_2.

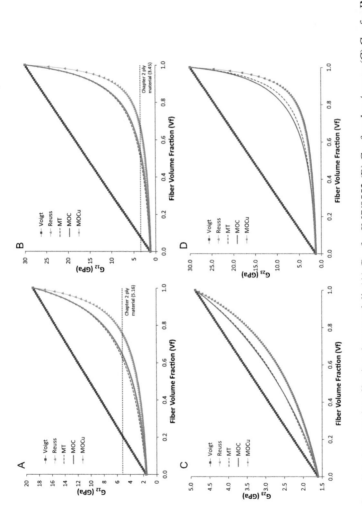

Figure 3.12 Comparison of predicted composite effective shear moduli. (A) G_{12} for IM7/8552 (B) G_{12} for glass/epoxy (C) G_{23} for IM7/8552 (D) G_{23} for glass/epoxy.

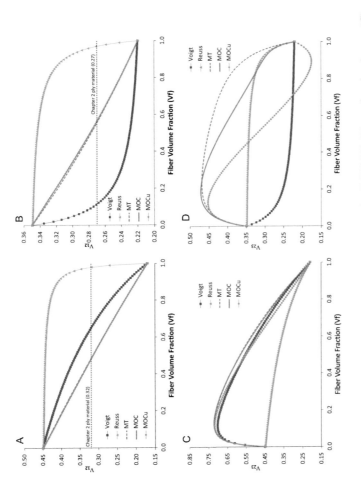

Figure 3.13 Comparison of predicted composite effective Poisson's ratios. (A) ν_{12} for IM7/8552 (B) ν_{12} for glass/epoxy (C) ν_{23} for IM7/8552 (D) ν_{23} for glass/epoxy. Note the differences in the axis scales.

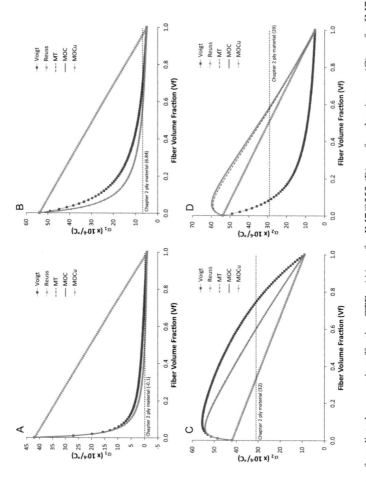

Figure 3.14 Comparison of predicted composite effective CTEs. (A) α_1 for IM7/8552 (B) α_1 for glass/epoxy (C) α_2 for IM7/8552 (D) α_2 for glass epoxy. Note the differences in the axis scales.

Table 3.7 Fiber volume fractions required to match IM7/8552 ply properties from Chapter 2.

	E_1 (GPa)	E_2 (GPa)	G_{12} (GPa)	ν_{12}	α_1 ($\times 10^{-6}/°C$)	α_2 ($\times 10^{-6}/°C$)
Voigt	0.55	0.32	0.21	0.65	0.79	0.73
Reuss	0.98	0.77	0.75	0.98	0.98	0.30
MT	0.55	0.54	0.62	0.49	0.53	0.58
MOC	0.55	0.54	0.64	0.48	0.53	0.58
MOCu	0.55	0.45	0.64	0.48	0.53	0.58

Table 3.8 Fiber volume fractions required to match glass/epoxy ply properties from Chapter 2.

	E_1 (GPa)	E_2 (GPa)	G_{12} (GPa)	ν_{12}	α_1 ($\times 10^{-6}/°C$)	α_2 ($\times 10^{-6}/°C$)
Voigt	0.57	0.11	0.08	0.11	0.69	0.08
Reuss	0.97	0.73	0.66	0.97	0.96	0.51
MT	0.58	0.61	0.50	0.56	0.58	0.57
MOC	0.58	0.58	0.53	0.57	0.57	0.58
MOCu	0.58	0.52	0.53	0.57	0.57	0.58

The Voigt and Reuss predictions vary widely and are thus clearly inconsistent with each other and with the other models.

Experimentally measured ply properties are always associated with a single average fiber volume fraction. Clearly, from Table 3.7 and Table 3.8, none of the considered methods can exactly match all properties for either material system using one fiber volume fraction value. There are three potential sources for this inconsistency: (1) errors in the measured ply properties; (2) errors in the constituent properties; and (3) limitations of the theories behind micromechanics methods. Only if the ply properties, constituent properties, and micromechanics theory were perfect would one expect a single fiber volume fraction to match every property for a given ply material. Although the Chapter 2 ply properties are not taken from a set of experiments with known fiber volume fraction, they are representative of structural PMCs with fiber volume fractions in the range of 0.55 to 0.60. Examining the MT and MOC fiber volume fractions in Table 3.7 and Table 3.8, it is clear that they are reasonably consistent with this range of fiber volume fractions. In practice, both measurements and predictions of E_1 almost always are associated with the lowest level of uncertainty (both for ply materials and for constituents). Consequently, the choice of a single value of fiber volume fraction should be based on matching E_1, while the errors this would introduce in the other properties must be accepted. Quantifying the significance of these errors is an additional problem for which micromechanics is well suited.

3.12.4 Micromechanics-based cross-ply laminate

In this example, the same IM7/8552 $[90/0]_s$ cross-ply laminate considered in Section 2.5.2 is analyzed, but now using MOC to predict the ply properties (rather than using effective ply properties). This follows the procedure shown in Fig. 3.8B. The only difference between a micromechanics-based laminate analysis and a ply-based

Table 3.9 Comparison of IM7/8552 ply properties predicted by MOC with the ply properties from Chapter 2.

	E_1 (GPa)	E_2 (GPa)	G_{12} (GPa)	ν_{12}	α_1 ($\times 10^{-6}/°C$)	α_2 ($\times 10^{-6}/°C$)
Chapter 2 IM7/8552 Ply Properties	146.8	8.69	5.16	0.32	−0.1	31
MOC Predicted Properties (Vf = 0.55)	146.4	8.765	4.130	0.302	−0.165	33.42

laminate analysis involves assigning a micromechanics-based material to the plies in `LamProblemDef.m` (MATLAB Function 2.7). In this example, the IM7/8552 MOC material defined previously (`Mat = 103` in `GetEffProps.m`, MATLAB Function 3.7):

```
Mat = 103;
name = "MOC IM7-8552 Vf = 0.55";
Theory = 'MOC';
Vf = 0.55;
Constits.Fiber = constitprops{1};
Constits.Matrix = constitprops{3};
[props{Mat}] = RunMicro(Theory, name, Vf, Constits, props{Mat});
```

is assigned to all four plies in `LamProblemDef.m`:

```
NP = NP + 1;
OutInfo.Name(NP) = "Sect 3.12.4 - cross-ply IM7-8552 MOC";
Geometry{NP}.Orient = [90,0,0,90];
nplies = length(Geometry{NP}.Orient);
Geometry{NP}.plymat(1:nplies) = 103;
Geometry{NP}.tply(1:nplies) = 0.15;
Loads{NP}.Type  = [NM,   EK, NM, NM, NM, NM];
Loads{NP}.Value = [ 0, 0.02,  0,  0,  0,  0];
```

As shown, loading in the form of a laminate midplane strain of $\varepsilon_{yy}^0 = 0.02$, with all other force and moment resultants equal to zero, has been applied.

Table 3.9 compares the ply properties predicted by MOC to those specified in Chapter 2. The only major difference is in shear modulus, G_{12}. As shown in Table 3.7, a fiber volume fraction of 0.64 would have been required to obtain a perfect match with the ply properties (which would result in greater differences in the other properties). Note that a better match for G_{12} could be obtained by altering the IM7 and 8552 constituent material properties as well. In fact, all constituent properties could be optimized to obtain the best fit with known ply properties (e.g., Murthy et al., 2018).

Table 3.10 compares the laminate-level effective properties for the cross-ply laminate for the ply-based and micromechanics-based analyses. The laminate effective properties are in good agreement, aside from (again) the shear modulus, G_{xy}. Because the laminate is cross-ply, the ply G_{12} is equal to the laminate G_{xy}.

Fig. 3.15 shows the local stress fields in the 90° and 0° plies in response to an applied laminate midplane strain of $\varepsilon_{yy}^0 = 0.02$, with all other force and moment resultants equal to zero. Note that the local stresses are plotted in local ply coordinates, thus the σ_{11} in the 90° ply fiber (Fig. 3.15A), which is aligned with the applied loading direction,

Table 3.10 Comparison of IM7/8552 cross-ply laminate effective properties for ply-based (see Section 2.5.2) and micromechanics-based CLT.

	E_x (GPa)	E_y (GPa)	G_{xy} (GPa)	ν_{xy}	α_x ($\times 10^{-6}/^\circ C$)	α_y ($\times 10^{-6}/^\circ C$)
Chapter 2 (using ply-based properties)	78.12	78.12	5.160	0.0358	2.115	2.115
Micromechanics-based properties	77.92	77.92	4.130	0.0341	2.223	2.223

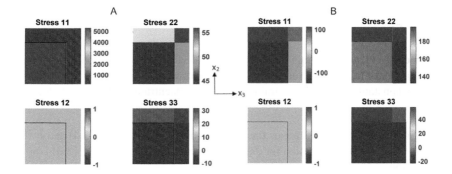

Figure 3.15 Predicted local (constituent) stress fields (MPa) in the (A) 90° plies and (B) 0° plies for a micromechanics-based (MOC) cross-ply IM7/8552 laminate with Vf = 0.55.

is very high. Examining the 0° ply stress fields, (Fig. 3.15B), the σ_{22} is aligned with the loading direction, but the stresses are much lower because the fiber in this ply is transverse to the loading direction. σ_{12} is zero throughout the laminate because there are no off-axis plies. Finally, note that the x_3-direction is aligned with the laminate z-direction (which is through-thickness). Thus the average of the σ_{33} in each ply must be zero to satisfy plane stress conditions.

3.13 Summary

This chapter has presented the fundamental concepts related to micromechanics, most importantly the strain and stress concentration tensors relating the local fields to the applied composite-level loading. Then four classical micromechanics theories [the Voigt approximation, Reuss approximation, MT method, and MOC] were presented in terms of their unique concentration tensors, from which all effective mechanical properties can be determined. While the Voigt, Reuss, and MT theories provide average fields for each constituent, MOC enables prediction of the variations of the stress and strain fields within the matrix constituent. Furthermore, in contrast to MT and MOC, the simple Voigt and Reuss approaches, while illustrative from a theoretical standpoint, do not provide good approximations of the effective composite properties

nor the local fields. All four theories have been extended to accommodate thermomechanical loading, wherein the effective coefficients of thermal expansion (CTEs) were
predicted, along with thermal strain concentration tensors, which enable the calculation
of local fields in the presence of thermal loading. The MATLAB implementation of
these classical micromechanics theories (for solving stand-alone micromechanics and
multiscale (micromechanics-based) laminate analysis problems), along with associated
flowcharts, were provided and discussed. Finally, several example problems comparing
the micromechanics theory results were presented for both stand-alone micromechanics and micromechanics-based laminate simulations.

3.14 Exercises

Exercise 3.1. Generalize Eq. (3.8) through Eq. (3.20) wherein an RVE is discretized
into k number of regions each containing its own material. Assume the stiffness in each
region is constant and a unique strain and stress concentration tensor exists for each
region.

Exercise 3.2. Given homogeneous traction boundary conditions (Eq. 3.4) and $\bar{\sigma}^{(f)} =
\mathbf{B}_f^{\text{dilute}}\, \bar{\sigma}^{(m)}$, derive the expression for the effective compliance tensor, Eq. (3.61) for the
MT method.

Exercise 3.3. Modify the MT MATLAB Function 3.3 to enable prediction of the
effective properties of composite materials containing *spherical* inclusions using the
Mori-Tanaka micromechanics theory. For the glass/epoxy composite (constituent properties given in Table 3.4), with $v_f = 0.5$, calculate the effective mechanical properties
and show that the effective properties are isotropic.

Exercise 3.4. Show the difference between the general 3-D Voigt approximation and
the classical 1-D iso-strain approximation for effective axial thermal expansion, i.e.,

$$\alpha_{11}^* = \frac{v_f E_{11}^f \alpha_{11}^f + (1 - v_f) E_{11}^m \alpha_{11}^m}{v_f E_{11}^f + (1 - v_f) E_{11}^m}$$

by plotting the α_{11}^* predicted by each for an IM7/8552 composite material for fiber
volume fractions from 0 to 1. What drives the difference between the Voigt results and
the 1-D iso-strain approximation?

Exercise 3.5. Voids (or porosity) are a common problem in composites. High quality
composites have a void content under 1%. Use the MT method for spherical inclusions,
incorporated into the MATLAB in Exercise 3.3, to model 8552 resin with a spherical
void. The void can be modeled as the spherical inclusion with material properties
identical to those of 8552, but with Young's moduli and shear moduli reduced by a
factor of 1000. Plot the Young's modulus of the 8552 with voids as a function of the
void volume fraction (for 0 to 1).

Exercise 3.6. Consider a unidirectional 0.55 fiber volume fraction IM7/8552 composite that contains 5% voids by volume (e.g., 55% fiber, 40% matrix, and 5% voids). Using the approach described in Exercise 3.5, first determine the effective properties of the matrix with voids (using MT), and then calculate the effective properties of the composite (using MT with a continuous cylindrical fiber, assuming all voids are in the matrix). Compare the composite effective properties with and without (see Table 3.5) voids in the matrix and calculate the percentage difference.

Exercise 3.7. The currently implemented MOC MATLAB code considers a square fiber packing arrangement ($h_2 = l_2$), see Fig. 3.5B. However, the equations are sufficiently general to allow consideration of rectangular fiber packing. Such packing is shown here:

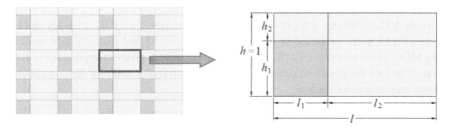

Given this Figure, use the RUC aspect ratio, $AR = l/h$, which express the ratio between the RUC width ($l = l_1 + l_2$) and height ($h = h_1 + h_2 = 1$), to derive expressions for l_1, l_2, h_1, and h_2 based on a known fiber volume fraction, v_f, and aspect ratio, AR. Assume that the RUC total height, $h = 1$. Given that l_2 and h_2 must be positive, determine the max and min limits of AR for a given v_f, in the form,

$$AR_{min} < AR < AR_{max}$$

Exercise 3.8. Implement the rectangular fiber packing arrangement equations derived in Exercise 3.7 by modifying the appropriate section of the MATLAB Function 3.4 – MOC.m. For a unidirectional glass/epoxy composite with fiber volume fractions of 0.25 and 0.5, determine and tabulate the effective mechanical properties (E_1, E_2, E_3, G_{23}, G_{13}, G_{12}, v_{12}, v_{13}, v_{23}) for the two RUC aspect ratios $AR = 0.52632$ and 1.9 using the *unaveraged* MOCu model. Are these composites transversely isotropic? Which effective properties are unaffected by changes in AR? For each v_f, calculate the ratio, $\frac{E_2^{AR=1.9}}{E_2^{AR=0.52632}}$. Note, the MATLAB plots will not reflect the AR (unless this capability is added to PlotMicroFields.m).

Exercise 3.9. Consider a unidirectional 0.55 fiber volume fraction glass/epoxy composite, subjected to an applied $\bar{\sigma}_{22} = 100$ MPa, with all other stress component equal to zero. Calculate the global strain component in the loading direction ($\bar{\varepsilon}_{22}$) and plot the constituent stresses using the Voigt, Reuss, MT, MOC, and MOCu micromechanics theories.

Exercise 3.10. Consider a unidirectional 0.55 fiber volume fraction glass/epoxy composite, subjected to an applied multiaxial stress state $\bar{\sigma}_{11}= \bar{\sigma}_{22}=50$ MPa, $\bar{\sigma}_{12}=-50$ MPa with all other stress components equal to zero. Calculate the global strain components and plot the constituent stresses using the Voigt, Reuss, MT, MOC, and MOCu micromechanics theories.

Exercise 3.11. Consider a unidirectional lamina, [45], 1 mm thick, composed of glass/epoxy with $v_f=0.55$, determine mid-plane strain components given an applied $Nx = 100$ N/mm (with all other force and moment resultants zero). Calculate and plot the constituent stresses for the Voigt, Reuss, MT, MOC, and MOCu micromechanics theories. Why does $\gamma_{xy}^0 = 0$ for the Voigt and Reuss theories? Are the local stresses and global/laminate midplane strains the same or different compared to the results of Exercise 3.10 and why or why not?

Exercise 3.12. Consider a 6 ply quasi-isotropic laminate, $[\pm30/0]_s$, with ply thicknesses of 0.15 mm, composed of IM7/8552 with $v_f=0.55$ and each ply represented using MOC. Determine the mid-plane strains and curvatures given an applied $Nx = 100$ N/mm. Tabulate the ply-level stresses in material coordinates and plot the constituent stresses and strains for each ply.

Exercise 3.13. Consider a 4 ply cross-ply laminate, $[90/0]_s$ with ply thickness of 0.15 mm, composed of IM7/8552 with $v_f=0.55$ and each ply represented using MOC. Determine the mid-plane strains and curvatures given an applied $Mx = 100$ N-mm/mm. Also, tabulate the ply-level stress in material coordinates and plot the constituent stresses and strains for each ply.

Exercise 3.14. Given Eqs. (3.139) and (3.140) and the local and global constitutive equations, establish the following relationship between the thermal strain and stress concentration tensors, $\mathbf{A}_k^T$ and $\mathbf{B}_k^T$:

$$\mathbf{A}_k^T = -\mathbf{A}_k\boldsymbol{\alpha}^* + \mathbf{S}^{(k)}\mathbf{B}_k^T + \boldsymbol{\alpha}^{(k)}$$

Exercise 3.15. Referring to the modification of the MATLAB Function 3.3 – MT.m for spherical inclusions made in Exercise 3.3, for a glass/epoxy composite with $v_f=0.5$, determine the thermal strain concentration tensors, $\mathbf{A}_f^T$ and $\mathbf{A}_m^T$ for both a discontinuous (spherical) and continuous (cylindrical) inclusion. Why are the last three components of the thermal strain concentration tensor zero in all cases? Why is the first (x_1-direction) component zero for the cylindrical inclusion case? Hint: place a break in GetEffProps.m to examine the calculated thermal strain concentration tensors as they are not written out.

Exercise 3.16. Consider a unidirectional composite composed of IM7/8552 subjected to an applied $\Delta T = 100$ °C and all stress components equal to zero. Using the MOC micromechanics theory to represent the composite material, plot the constituent stresses and strains for $v_f=0$, $v_f=0.25$, and $v_f=0.5$. What must the average of each stress component equal for all three cases? Hint: $v_f=0$ denotes monolithic 8552; this can be specified by constructing a composite material where both the fiber and the matrix are the 8552 constituent material.

Exercise 3.17. Plot the local (constituent) residual stresses induced by an applied $\Delta T = -150°C$ (with all stress components equal to zero) for both MT (cylindrical inclusion) and MOC for two material systems: IM7/8552 and glass/epoxy. Consider a unidirectional composite with $v_f = 0.5$. Make a table comparing the maximum and minimum matrix stress component σ_{22} for the two models and the two material systems. Hint: Place a break in `MicroAnalysis.m` to examine the local stress values.

Exercise 3.18. Consider regular quasi-isotropic laminates, $[-45/0/45/90]_S$ with a ply thickness of 0.15 mm composed of IM7/8552 and glass/epoxy where the plies are modeled with both MOC and MT. Apply a $\Delta T = -150° C$ with all force and moment resultants equal to zero. Compare the ply-level σ_{22} to the maximum and minimum matrix σ_{22} values tabulated in Exercise 3.17.

3.15 Appendix: Establishment of Eq. (3.88)

Eq. (3.88) relates the gradients of the subcell centroid displacements to the average RUC strains in MOC,

$$\bar{\varepsilon}_{ij} = \frac{1}{2}\left(\frac{\partial w_i}{\partial x_j} + \frac{\partial w_j}{\partial x_i}\right) \tag{3.88}$$

This can be established in a component by component manner as follows.
For $i=1$ and $j=1$, substituting the first relation in Eq. (3.86) into Eq. (3.87) yields,

$$\bar{\varepsilon}_{11} = \frac{\partial w_1}{\partial x_1} \tag{3A.1}$$

For $i=2$ and $j=2$, multiplication of Eq. (3.80) (with $i=2$) by l_γ and summation over γ, and using Eq. (3.84), yields

$$h_1 l_1 \phi_2^{(11)} + h_1 l_2 \phi_2^{(12)} + h_2 l_1 \phi_2^{(21)} + h_2 l_2 \phi_2^{(22)} = (h_1 + h_2)(l_1 + l_2)\frac{\partial w_2}{\partial x_2} \tag{3A.2}$$

which, by utilizing the second relation in Eq. (3.86), indicates that

$$\sum_{\beta,\gamma=1}^{2} h_\beta l_\gamma \varepsilon_{22}^{(\beta\gamma)} = (h_1 + h_2)(l_1 + l_2)\frac{\partial w_2}{\partial x_2} \tag{3A.3}$$

Then, from Eq. (3.87),

$$\bar{\varepsilon}_{22} = \frac{\partial w_2}{\partial x_2} \tag{3A.4}$$

Following a similar procedure, starting with Eq. (3.82) with $i=3$, multiplying by h_β and summing over β, it can be shown that

$$\bar{\varepsilon}_{33} = \frac{\partial w_3}{\partial x_3} \tag{3A.5}$$

For $i=2$ and $j=3$, Eq. (3.80) (with $i=3$) is multiplied by l_γ and summed over γ . Then, Eq. (3.82) (with $i=3$) is multiplied by h_β and summed over β. Addition of the resulting two equations provides

$$h_1 l_1 \left(\phi_3^{(11)} + \psi_2^{(11)}\right) + h_1 l_2 \left(\phi_3^{(12)} + \psi_2^{(12)}\right) + h_2 l_1 \left(\phi_3^{(21)} + \psi_2^{(21)}\right)$$
$$+ h_2 l_2 \left(\phi_3^{(22)} + \psi_2^{(22)}\right) = (h_1 + h_2)(l_1 + l_2)\left(\frac{\partial w_3}{\partial x_2} + \frac{\partial w_2}{\partial x_3}\right) \tag{3A.6}$$

This yields, according to the fourth relation in Eq. (3.86), that

$$2h_1 l_1 \varepsilon_{23}^{(11)} + 2h_1 l_2 \varepsilon_{23}^{(12)} + 2h_2 l_1 \varepsilon_{23}^{(21)} + 2h_2 l_2 \varepsilon_{23}^{(22)} = (h_1 + h_2)(l_1 + l_2)\left(\frac{\partial w_3}{\partial x_2} + \frac{\partial w_2}{\partial x_3}\right) \tag{3A.7}$$

which, from Eq. (3.87), implies that,

$$\bar{\varepsilon}_{23} = \frac{1}{2}\left(\frac{\partial w_3}{\partial x_2} + \frac{\partial w_2}{\partial x_3}\right) \tag{3A.8}$$

For $i=1$ and $j=2$, Eq. (3.80) (with $i=1$) is multiplied by l_γ and summed over γ, and using Eq. (3.84) yielding

$$h_1 l_1 \phi_1^{(11)} + h_1 l_2 \phi_1^{(12)} + h_2 l_1 \phi_1^{(21)} + h_2 l_2 \phi_1^{(22)} = (h_1 + h_2)(l_1 + l_2)\frac{\partial w_1}{\partial x_2} \tag{3A.9}$$

Next, Eq. (3.84) with $i=1$ and $j=2$ is multiplied by $h_\beta\, l_\gamma$ and summed over β and γ, yielding

$$h_1 l_1 \frac{\partial w_2^{(11)}}{\partial x_1} + h_1 l_2 \frac{\partial w_2^{(12)}}{\partial x_1} + h_2 l_1 \frac{\partial w_2^{(21)}}{\partial x_1} + h_2 l_2 \frac{\partial w_2^{(22)}}{\partial x_1} = (h_1 + h_2)(l_1 + l_2)\frac{\partial w_2}{\partial x_1} \tag{3A.10}$$

Addition of Eqs. (3A.9) and (3A.10) yields,

$$h_1 l_1 \left(\phi_1^{(11)} + \frac{\partial w_2^{(11)}}{\partial x_1}\right) + h_1 l_2 \left(\phi_1^{(12)} + \frac{\partial w_2^{(12)}}{\partial x_1}\right) + h_2 l_1 \left(\phi_1^{(21)} + \frac{\partial w_2^{(21)}}{\partial x_1}\right)$$

$$+ h_2 l_2 \left(\phi_1^{(22)} + \frac{\partial w_2^{(22)}}{\partial x_1} \right) = (h_1 + h_2)(l_1 + l_2) \left(\frac{\partial w_1}{\partial x_2} + \frac{\partial w_2}{\partial x_1} \right) \quad (3A.11)$$

It follows from the sixth relation in Eq. (3.86) that

$$\bar{\varepsilon}_{12} = \frac{1}{2} \left(\frac{\partial w_1}{\partial x_2} + \frac{\partial w_2}{\partial x_1} \right) \quad (3A.12)$$

A similar procedure can be used to establish that

$$\bar{\varepsilon}_{13} = \frac{1}{2} \left(\frac{\partial w_1}{\partial x_3} + \frac{\partial w_3}{\partial x_1} \right) \quad (3A.13)$$

Thus all components of Eq. (3.88) have been established as Eqs. (3A.1, 3A.4, 3A.5, 3A.8, 3A.12, and 3A.13).

References

Aboudi, J., 1989. Micromechanical Analysis of composites by the method of cells. Appl. Mech. Rev. 42 (7), 193–221.

Aboudi, J., 1993. Constitutive Behavior of Multiphase Metal Matrix Composites with Interfacial Damage by the Generalized Cell Model. In: Voyiadjis, G.Z. (Ed.), Damage in Composite Materials. Elsevier, Amsterdam, The Netherlands, p. 3.

Aboudi, J., Arnold, S.M., Bednarcyk, B.A., 2013. Micromechanics of Composite Materials A Generalized Multiscale Analysis Approach. Elsevier, New York, United States.

Adams, D.F., Doner, D.R., 1967a. Longitudinal shear loading of a unidirectional composite. J. Comp. Mater. 1, 4–17.

Adams, D.F., Doner, D.R., 1967b. Transverse normal loading of a unidirectional composite. J. Comp. Mater. 1, 152–164.

Arnold, S.M., Pindera, M.J., Wilt, T.E., 1996. Influence of fiber architecture on the inelastic response of metal matrix composites. Int. J. Plasticity 12 (4), 507–545.

Banerjee, P.K., Henry, D.P., Hopkins, D.A., Goldberg, R.K, 1997. Further developments of BEM for micro and macromechanical analyses of composites : BEST-CMS (Boundary Element Software Technology-Composite Modeling System), User's Manual. NASA-CR-204254.

Banerjee, P.K., Butterfield, R., 1981. Boundary Element Methods in Engineering Science. McGraw-Hill, London, United Kingdom.

Barral, M., Chatzigeorgiou, G., Meraghni, F., Leon, R., 2020. Homogenization using modified Mori-Tanaka and TFA framework for elastoplastic-viscoelastic-viscoplastic composites: theory and numerical validation. Int. J. Plast. 127, 102632.

Bazant, Z.P., 2007. Can multiscale-multiphysics methods predict softening damage and structural failure? Mechanics 36, 5–12.

Bednarcyk, B.A., Stier, B., Simon, J.-W., Reese, S., Pineda, E.J., 2015. Meso and micro-scale modeling of damage in plain weave composites. Compos. Struct. 121, 258–270.

Benveniste, Y., 1987. A new approach to the application of Mori-Tanaka's theory in composite materials. Mech. Mater 6, 147–157.

Benveniste, Y., Dvorak, G.J., 1990. On a Correspondence Between Mechanical and Thermal Effects in Two-Phase Composites. In: Weng, G.J., Taya, M., Abe, H. (Eds.), Micromechanics and Inhomogeneity. Springer-Verlag, New York, United States, pp. 65–81.

Brayshaw, J.B., 1994. Consistent Formulation of the Method of Cells Micromechanics Model for Transversely Isotropic Metal Matrix Composites. Ph.D. Dissertation, University of Virginia, Charlottesville, VA, United States.

Christensen, R.M., 1979. Mechanics of Composite Materials. Wiley, New York, United States.

Drago, A., Pindera, M.-J., 2007. Micro-macromechanical analysis of heterogeneous materials: macroscopically homogeneous vs. periodic microstructures. Comp. Sci. Tech. 67, 1243–1263.

Engineering Village (2020) https://www.engineeringvillage.com/search/quick.url, accessed May 8, 2020.

Eshelby, J.D., 1957. The determination of the elastic field of an Ellipsoidal inclusion, and related problems. Proc. Roy. Soc. London A241, 376–396.

Eshelby, J.D., 1959. The elastic field outside an ellipsoidal inclusion. Proc. Roy. Soc. London A252, 561–569.

Ferrari, M., 1991. Asymmetry and high concentration limit of the Mori-Tanaka medium theory. Mech. Mater. 11, 251–256.

Feyel, F., Chaboche, J.-L., 2000. FE2 multiscale approach for modeling the elastoviscoplastic behavior of long fiber SiC/Ti composite materials. Comput. Methods Appl. Mech. Eng. 183, 309–330.

Fish, J. (Ed.), 2009. Multiscale Methods Bridging the Scales in Science and Engineering. Oxford University Press.

Foye, R.L., 1966a. An evaluation of various engineering estimates of the transverse properties of unidirectional composites. SAMPE 10, 31–42.

Foye, R. L., 1966b, Structural Composites, Quarterly Progress Reports No. I and 2, AFML Contract No. AF33(615)-5150.

Galvanetto, U., Aliabadi, M.H.F. (Eds.), 2010. Multiscale Modeling in Solid Mechanics Computational Approaches. Imperial College Press, London.

Hill, R., 1963. Elastic properties of elastic solids: some theoretical principles. J Mech. Phys. Sol. 11, 357–372.

Kanoute, P., Boso, D.P., Chaboche, J.L., Schrefler, B.A., 2009. Multiscale methods for composites: a review. Arch. Comput. Methods Eng. 16, 31–75.

Kwon, Y.W., Allen, D.H., Talreja, R. (Eds.), 2008. Multiscale Modeling and Simulation of Composite Materials and Structures. *Springer*, New York, NY, United States.

Levin, V.M., 1967. On the coefficients of thermal expansion of. heterogeneous materials, Mech. Solids 2, 58–61.

Liu, C.H., 1988. The Elastic Plastic Behavior of Composite Materials, PhD Dissertation, Yale University.

Liu, Y.J., Nishimura, N., Otani, Y., Takahashi, T., Chen, X.L., 2005. A fast boundary element method for the analysis of fiber-reinforced composites based on a rigid-inclusion model. J. Appl. Mech. 72 (1), 115.

MATLAB, 2018. version R2018. The MathWorks Inc, Natick, Massachusetts, United States.

Mori, T., Tanaka, K., 1973. Average stresses in matrix and average energy of materials with misfitting inclusions. Acta Metall 21, 571–574.

Mura, T., 1982. Micromechanics of Defects in Solids. Martinus Nijhoff, The Hague, Switzerland.

Murthy, P.L.N, Naghipour Ghezeljeh, P., Bednarcyk, B.A, 2018. Development and Application of a Tool for Optimizing Composite Matrix Viscoplastic Material Parameters. NASA-TM-2018-219745.

Onat, E.T., Wright, S.I., 1991. On Representation of Mechanical Behavior and Stereological Measures of Microstructure. NASA CR 187130.

Pineda, E.J., Waas, A.M., Bednarcyk, B.A., Collier, C.S., 2011. A multiscale progressive damage analysis and design tool for advanced composite structures. In: 16th International Conference on Composite Structures June.

Reuss, A., 1929. Berechnung der fließgrenze von mischkristallen auf grund der plastizitätsbedingung für einkristalle. J Appl. Math. Mech. 9, 49–58.

Rizzo, F.J., Pan, L., Adams, D.O., 1998. Boundary element analysis for composite materials and a library of green's functions. Comput. Struct. 66 (5), 685–693.

Stier, B., Bednarcyk, B.A., Simon, J.-W., Reese, S, 2015. Investigation of Micro-Scale Effects on Damage of Composites. NASA Technical Memorandum 218740.

Sun, C.T., Vaidya, R.S., 1996. Prediction of composite properties from a representative volume element. Comp. Sci. Tech. 56, 171–179.

Tanaka, Mori, 1972. Note on volume integrals of the elastic field around an ellipsoidal inclusion. J. Elast. 2, 199–200.

Taya, M., Arsenault, R.J., 1989. Metal Matrix Composites. Pergamon, Oxford, United States.

Voigt, W., 1887. Theoretische studien über die elasticitätsverhältnisse der krystalle. Abhandlungen der Königlichen Gesellschaft der Wissenschaften in Göttingen 34, 3–51.

Vaz, M. Jr., de Souza Neto, E.A., Munoz-Rojas, P.A. (Eds.), 2011. Advanced Computational Materials Modeling From Classical to Multi-Scale Techniques, 2011. WILEY-VCH Verlag & Co. KGaA, Weinheim, Germany.

Failure criteria and margins of safety

4

Chapter outline

Chapter 2 presented classical lamination theory (CLT) to enable calculation of the effective properties of a composite laminate, along with the ply-level stresses and strains. In Chapter 3, the ability of micromechanics theories to provide not only effective composite properties, but also the local stress and strain fields in the composite constituents, was demonstrated. Now, in this chapter, the local stress and strains, be they ply level quantities from CLT or constituent scale quantities from micromechanics, are used for a practical purpose: assessing the ability of the composite or laminate to safely support applied loading. The point at which the applied loading can no longer be safely supported may be thought of as "failure". It may correspond to an actual failure, fracture, or other loss of structural integrity of a specimen in an experiment, but in design, failure more generally refers to exceeding some limit or criterion that has been established to assess the ability to support loads safely.

Assessing the ability of the composite or laminate to safely support applied loading is extremely valuable as it enables structural design. For example, given the force and moment resultants at a point in a structure, one can determine the number of composite plies required to safely carry the loads with no expected failure. Of course, this then requires some quantity that characterizes a material's ability to carry loads. This quantity is called an allowable. Allowables are usually in terms of stresses or strains, and they are, quite simply, intended to provide an upper bound on the magnitude

Practical Micromechanics of Composite Materials. DOI: 10.1016/C2019-0-04193-3

Table 4.1 Nomenclature and symbols used for allowables. Compressive allowables are considered to be negative, and shear allowables are positive.

	Stress Allowables		Strain Allowables	
	Tension	Compression	Tension	Compression
Axial Normal (x_1-direction)	X_T	X_C	$X_{\varepsilon C}$	$X_{\varepsilon T}$
Transverse Normal (x_2-direction)	Y_T	Y_C	$Y_{\varepsilon T}$	$Y_{\varepsilon C}$
Transverse Normal (x_3-direction)	Z_T	Z_C	$Z_{\varepsilon C}$	$Z_{\varepsilon T}$
Shear (2-3 component)	Q		Q_ε	
Shear (1-3 component)	R		R_ε	
Shear (1-2 component)	S		S_ε	

of the stress or strain that the material can safely experience, with no expected loss of structural integrity. Thus, for example, if the ply level stress or strain allowables are known, one can use CLT to determine ply level stresses and strains and compare them to the allowables. If the allowables are exceeded (by some measure) the design is not good and the design (e.g., layup or number of plies) must be altered. Again, it must be emphasized that allowables do not necessarily correspond to a strength, failure point, or fracture in a test. Rather, as the name implies, they are simply a limit on stresses or strains above which there is too high a risk of not being able to support loads safely.

Actual failure of composites is, in general, extremely complex and is the subject of a great deal of active research. Composite failure modes operate at different length scales (e.g., constituent, ply or laminate, and structural level). For instance, at the constituent level there is matrix cracking, fiber breakage, fiber/matrix debonding, and fiber microbuckling; whereas, at the laminate level, there is ply cracking, splitting, fiber bridging, delamination, and laminate fracture. Typically, unless a composite behaves as purely brittle, there is both damage initiation and propagation prior to final composite failure or fracture. In the present chapter, failure is treated only through determination of the point at which allowables are exceeded, which does not directly model any true failure or damage mechanisms. Damage initiation, progression, and ultimate failure, through micromechanics-based simulations, are the subject of Chapter 7.

In composite laminates, the plies are typically highly anisotropic, and therefore the allowables are typically different in the different directions and also in tension vs. compression. Herein, the nomenclature shown in Table 4.1 has been adopted for the composite stress and strain allowables. It is important to note that the *compression allowables are negative*. For transversely isotropic materials (with x_2-x_3 plane of isotropy) it is often assumed that $Z_T = Y_T$, $Z_C = Y_C$, $R = S$, $Z_{\varepsilon T} = Y_{\varepsilon T}$, $Z_{\varepsilon T} = Y_{\varepsilon T}$, and $R_\varepsilon = S_\varepsilon$.

Allowables used in design are determined via a great deal of careful mechanical testing to determine statistically-based quantities. For example, A-Basis allowables are intended to provide stress or strain magnitudes that would not cause failure (defined as loss of structural integrity) 99% of the time (with a 95% confidence interval). B-Basis allowables are intended to provide stress or strain magnitudes that would not cause failure (defined as loss of structural integrity) 90% of the time (with a 95% confidence interval). The reader is referred to (MIL-HDBK-17-1F (2002)) for details on material allowables and how they are determined from statistical test data. Determining safe, but not overly conservative, allowables is extremely important for enabling efficient design with composite materials. As such the required testing can be quite expensive ($> \$1$ million), and companies often treat their allowables as proprietary.

As mentioned, to determine if a given point in a structure is safe, one can compare the stresses or strains at that point to the allowables. Simply doing this on a component by component basis constitutes the maximum stress or maximum strain criterion. That is, for example, if any stress component exceeds the associated allowable stress, the point is not safe (or not good). There are however, more complex failure criteria that involve interaction of the stress (or strain) components. These interactive failure criteria attempt to capture the effects of multiaxial stress or strain states on the material failure. For example, given a multiaxial stress state (e.g., $\bar{\sigma}_{22} = \bar{\sigma}_{33} > 0$), the material may not fail when $\bar{\sigma}_{22} = \bar{\sigma}_{33} = Y_T$, nor when $\bar{\sigma}_{22} = \bar{\sigma}_{33} = Z_T$. Interactive failure criteria propose an equation (or equations) that include multiple stress or strain components. In this chapter, two popular interactive failure criteria, Tsai-Hill and Tsai-Wu, are presented.

All of the above discussion, which used composite laminates as the example, is equally applicable to the composite materials themselves. That is to say that, given the local, constituent level stresses or strains predicted by micromechanics, comparison can be made to constituent stress and strain allowables to assess the structure's safety under given loads. Of course, if a constituent material is isotropic, the number of independent allowables may be taken to be even lower (e.g., $X_T = Y_T = Z_T$). Using micromechanics, the failure criterion can be used in a stand-alone micromechanics analysis (with applied composite scale stresses or strains) or in a micromechanics-based laminate analysis (with applied laminate force and moment resultants).

In addition to assessing whether or not a material point can safely support its global load state, failure criteria provide a means to determine the global load state at which the failure criteria would first be exceeded. For example, one can determine the laminate loading level at which a given failure criterion will first be exceeded in one of the plies. This is done by assuming the loading remains proportional and scaling the loading to just fulfill the failure criterion. That is, for example, if one is applying $N_x = N_y = 100$ N/mm, the question becomes, at what load level N, where $N = N_x = N_y$, will the failure criterion first be fulfilled. The load level N is then the allowable load for the laminate, given the specified loading proportionality. N may be greater than the applied load (100 N/mm in this example), indicating safety, or it may be less than the applied loading, indicating failure or that the design is not good. As will be discussed in detail in Section 4.2, the margin of safety (MoS) quantifies how close the load state is to the boundary between safe and unsafe, a good design vs. failure. This boundary corresponds to MoS $= 0$, with negative MoS indicating unsafe, failed designs. A safety factor (FS) (or factor of safety) can also be included, which, in effect, multiplies the loads by some value greater than one to include addition conservatism in the design and safety evaluation.

The scaling operates the same way for any arbitrary loading. For instance, if the laminate loading is $N_x = 100$ N/mm, $N_y = -50$ N/mm a scaling factor, δ, is sought such that the failure criteria is first satisfied at a load state of $\delta \times N_x$, $\delta \times N_y$, which maintains the load component proportionality ($N_x/N_y = -2$, in this case). This procedure can be used for determining allowable loads and laminate MoS using CLT with plies represented using effective properties (as in Chapter 2) in conjunction with effective ply allowables or using micromechanics-based CLT, wherein the failure criteria are evaluated using constituent level stress and strain fields.

In the context of allowables and MoS, thermal loading (i.e., including a temperature change in the analysis) represents a special case. In the case of mechanical loading, as described above, the load components can be scaled proportionally from zero (the completely unloaded case) to determine when the failure criterion will first be fulfilled. Scaling the temperature change in this manner is not useful because MoS are not evaluated based on temperature. For instance, it is not of practical value to know that, given some temperature change, the structure could withstand twice that temperature change before local stresses exceed a failure criterion. Rather, what is useful is knowing how the stresses induced by temperature combine with those induced by the mechanical loading to alter the allowable mechanical load. This, in essence, means that the thermal loading can be treated as a "preload" that is present prior to scaling the mechanical loading to determine when the failure criterion is first fulfilled. For example, for uniaxial N_x loading, a laminate might have an allowable load of $N_x = 400$ N/mm, but with a $\Delta T = -100\ °C$, this allowable might change to 300 N/mm. This is discussed in detail in Section 4.2.2.

In this chapter, four commonly used, representative failure criteria are considered: max stress, max strain, Tsai-Hill, and Tsai-Wu. These will first be introduced, followed by the development of the equations to calculate the MoS for each criterion under mechanical loading. These equations are then expanded to consider thermomechanical loading, wherein the thermal load is considered to be a preload. The MATLAB (2018) implementation is discussed, followed by example problems considering (1) laminate failure based on ply-level allowables, (2) composite failure based on constituent scale allowables, and (3) laminate failure based on constituent scale allowables. Included is the ability to generate failure envelopes for both unidirectional composites and laminates. Again, the term "failure" in this chapter is used loosely to define the point at which allowables are exceeded at some point within the laminate or composite according to the chosen failure criteria.

4.1   Failure criteria

Here, four common failure criteria are presented. The maximum stress and maximum strain criteria are simply based on comparison of the magnitudes of the individual stress and strain components to the appropriate allowables. In contrast, the Tsai-Hill and Tsai-Wu criteria combine the stress components into a single equation, thereby including interaction of the components. According to Muzel et al. (2020), the maximum stress, Tsai-Hill, and Tsai-Wu are among the most widely used failure criteria for composites.

4.1.1   Maximum stress and strain criteria

The maximum stress criterion (e.g., Tsai and Hahn (1980), Daniel and Ishai (1994), and Herakovich (1998)) is based on any stress component exceeding a limit. The limits on the stress components (to avoid failure) can be expressed as,

$$X_C < \sigma_{11} < X_T, \quad Y_C < \sigma_{22} < Y_T, \quad Z_C < \sigma_{33} < Z_T$$
$$|\tau_{23}| < Q, \quad |\tau_{13}| < R, \quad |\tau_{12}| < S \tag{4.1}$$

where, X_T, Y_T, and Z_T are the tensile material normal strengths (or allowable stresses) in the x_1-, x_2-, and x_3-directions, respectively. X_C, Y_C, and Z_C are the compressive material normal strengths (or allowable stresses) in the three directions, and Q, R, and S are the material shear strengths (or allowable stresses). Note that compressive strengths/allowables are negative.

The maximum strain criterion (e.g., Tsai and Hahn (1980), Daniel and Ishai (1994), and Herakovich (1998)) is identical to the maximum stress criterion, but the limits are expressed in terms of strain components,

$$X_{\varepsilon C} < \varepsilon_{11} < X_{\varepsilon T}, \quad Y_{\varepsilon C} < \varepsilon_{22} < Y_{\varepsilon T}, \quad Z_{\varepsilon C} < \varepsilon_{33} < Z_{\varepsilon T}$$
$$\left|\gamma_{23}\right| < Q_\varepsilon, \quad \left|\gamma_{13}\right| < R_\varepsilon, \quad \left|\gamma_{12}\right| < S_\varepsilon \tag{4.2}$$

where $X_{\varepsilon T}$, $Y_{\varepsilon T}$, and $Z_{\varepsilon T}$ are the tensile material normal failure (or allowable) strains in the x_1-, x_2-, and x_3-directions, respectively. $X_{\varepsilon C}$, $Y_{\varepsilon C}$, and $Z_{\varepsilon C}$ are the compressive material normal failure (or allowable) strains in the three directions, and Q_ε, R_ε, and S_ε are the material shear failure (or allowable) strains. Note that compressive failure/allowables strains are negative and the shear strain allowables are taken herein to be engineering strains. The maximum strain criterion is usually implemented such that the mechanical strain components, rather than the total strain components, are used. That is, the normal strain components are replaced with $\varepsilon^{mech} = \varepsilon - \alpha \, \Delta T$, where ε^{mech} is the mechanical strain, ε is the total strain, $\alpha \, \Delta T$ is the thermal strain. In this way, a free thermal expansion will not contribute to failure.

4.1.2 Tsai-Hill criterion

The fully multiaxial Tsai-Hill criterion (Tsai, 1968) interactive equation can be expressed as,

$$TH = \frac{\sigma_{11}^2}{X^2} + \frac{\sigma_{22}^2}{Y^2} + \frac{\sigma_{33}^2}{Z^2} + \frac{\tau_{23}^2}{Q^2} + \frac{\tau_{13}^2}{R^2} + \frac{\tau_{12}^2}{S^2} - \sigma_{11}\sigma_{22}\left(\frac{1}{X^2} + \frac{1}{Y^2} - \frac{1}{Z^2}\right)$$
$$- \sigma_{11}\sigma_{33}\left(\frac{1}{X^2} - \frac{1}{Y^2} + \frac{1}{Z^2}\right) - \sigma_{22}\sigma_{33}\left(-\frac{1}{X^2} + \frac{1}{Y^2} + \frac{1}{Z^2}\right) \tag{4.3}$$

and the limit on the Tsai-Hill function (to avoid failure) is,

$$TH < 1 \tag{4.4}$$

where, in order to incorporate differing tensile and compressive strengths/allowables,

$$X = \begin{cases} X_T & \sigma_{11} \geq 0 \\ X_C & \sigma_{11} < 0 \end{cases} \qquad Y = \begin{cases} Y_T & \sigma_{22} \geq 0 \\ Y_C & \sigma_{22} < 0 \end{cases} \qquad Z = \begin{cases} Z_T & \sigma_{33} \geq 0 \\ Z_C & \sigma_{33} < 0 \end{cases} \quad (4.5)$$

Failure is indicated when the *TH* function in Eq. (4.3) is greater than 1. Note that, in the case of plane stress, Z is taken to equal Y. Beware that incorporation of differing tensile and compressive strengths/allowables, as shown in Eq. (4.5), can be problematic as a slight transition from tension to compression can cause a major instantaneous change in the *TH* function as the allowables switch from tensile to compressive (or vice versa). This is alleviated in the Tsai-Wu criterion as tensile and compressive allowables are explicitly incorporated.

4.1.3   Tsai-Wu criterion

The Tsai-Wu (1971) criterion interactive equation can be written as,

$$TW = F_1\sigma_{11} + F_2\sigma_{22} + F_3\sigma_{33} + F_{11}\sigma_{11}^2 + F_{22}\sigma_{22}^2 + F_{33}\sigma_{33}^2 + F_{44}\sigma_{23}^2 + F_{55}\sigma_{13}^2$$
$$+ F_{66}\sigma_{12}^2 + 2F_{12}\sigma_{11}\sigma_{22} + 2F_{13}\sigma_{11}\sigma_{33} + 2F_{23}\sigma_{22}\sigma_{33} \qquad (4.6)$$

and the limit on the Tsai-Wu function (to avoid failure) is,

$$TW < 1 \qquad (4.7)$$

where the compressive strengths/allowables are negative and,

$$F_1 = \left(\frac{1}{X_T} + \frac{1}{X_C}\right), \quad F_2 = \left(\frac{1}{Y_T} + \frac{1}{Y_C}\right), \quad F_3 = \left(\frac{1}{Z_T} + \frac{1}{Z_C}\right)$$
$$F_{11} = \frac{-1}{X_T X_C}, \qquad F_{22} = \frac{-1}{Y_T Y_C}, \qquad F_{33} = \frac{-1}{Z_T Z_C},$$
$$F_{44} = \frac{1}{Q^2}, \qquad F_{55} = \frac{1}{R^2}, \qquad F_{66} = \frac{1}{S^2} \qquad (4.8)$$

The interaction coefficients, F_{12}, F_{13}, and F_{23}, are often taken according to Tsai and Hahn (1980),

$$F_{12} = \frac{-1}{2\sqrt{X_T X_C Y_T Y_C}}, \qquad F_{13} = \frac{-1}{2\sqrt{X_T X_C Z_T Z_C}}, \qquad F_{23} = \frac{-1}{2\sqrt{Y_T Y_C Z_T Z_C}} \qquad (4.9)$$

These Tsai-Hahn interaction coefficients are used throughout the remainder of this book. Failure is indicated when the *TW* function in Eq. (4.6) is greater than 1.

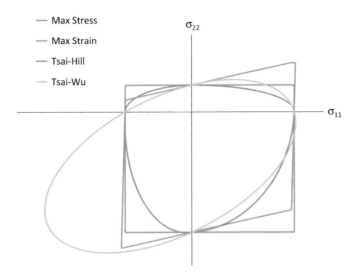

Figure 4.1 Schematic showing representative shapes of the failure envelopes for the four presented failure criteria.

4.1.4   Failure envelopes

A failure envelope quantifies the loads (or stresses) at which a given failure criterion is first satisfied in a given loading space. For example, in σ_{11}-σ_{22} stress space, Fig. 4.1 shows representative plots of the failure envelopes associated with each of the failure criteria discussed above. All stress values outside the envelope would exceed the given failure criterion, whereas inside the failure envelope, the stress state is considered "safe" (i.e., not failed). The difference between the interactive (Tsai-Hill and Tsai-Wu) and non-interactive (max stress and max strain) failure criteria is clear. It should also be noted that the points where the failure envelopes cross the stress axes in Fig. 4.1 correspond for all failure criteria. This is because these "anchor points" correspond to the uniaxial failure/allowable data (X_T, X_C, Y_T, Y_C) for the stress-based criteria. Note, for the max strain envelope to hit these points the allowable strains must be taken as the allowable stresses divided by the appropriate moduli.

4.2   Margins of safety (MoS)

The MoS quantifies how close the stress or strain state at a point in a structure is to the allowable stress or strain state. In general, the MoS is given by,

$$MoS = \frac{Allowable\ Stress\ or\ Strain}{Actual\ Stress\ or\ Strain} - 1 \qquad (4.10)$$

Thus, if the stress or strain exceeds the allowable, the MoS will be negative. A positive MoS indicates that the stress or strain state is within allowable limits (safe), while a MoS of zero means that the state is exactly at the allowable limit. As such, the MoS is commonly used in structural design. If the entire structure can be shown to have positive (or zero) MoS for all necessary failure modes and all load cases, then the design is considered closed. Note that, if the stress or strain far exceeds the allowable, the MoS approaches -1, whereas, if the stress or strain is very small, the MoS approaches infinity.

In the design process, a safety factor (SF) is often applied to the loads to ensure that the material or structure can sustain loading well above what is expected to add extra conservatism. In the presence of a SF, the MoS quantifies how close the stress or strain state is to the factored loads. That is, a MoS of zero would indicate that the current state times the SF is exactly at the allowable state. In some cases, the loading applied in the analysis is premultiplied by the SF. In such cases, Eq. (4.10) gives the MoS with respect to the factored loads. If the applied loading has not been premultiplied by the SF, then Eq. (4.10) is altered to include the SF,

$$MoS = \frac{Allowable\ Stress\ or\ Strain}{SF \times (Actual\ Stress\ or\ Strain)} - 1 \qquad (4.11)$$

For example, if a SF of 2 is employed, then the stress or strain state could only reach one-half of the allowable state to maintain a positive margin.

It is important to note that, in the presence of combined mechanical and thermal loading, the thermal loading is typically treated as a "preload", and subsequently, the MoS is calculated to quantify how much additional (or how much less) mechanical loading is required to reach the failure criterion. The thermal loading essentially shifts the starting point of the projection of the imposed stress or strain state to the failure envelope. Therefore, the thermal and mechanical load cases must be considered separately when evaluating the MoS. As such, below, the MoS calculations according to the four considered failure criteria are first presented for the case of mechanical loading only, and then in the case of combined thermo-mechanical loading.

4.2.1   Margins of safety for mechanical loading

4.2.1.1   Maximum stress and maximum strain margins of safety for mechanical loading

For the max stress criterion, the six MoS are each determined, per stress component, according to Eq. (4.10), as,

$$MoS_{11} = \frac{X}{\sigma_{11}} - 1, \qquad MoS_{22} = \frac{Y}{\sigma_{22}} - 1, \qquad MoS_{33} = \frac{Z}{\sigma_{33}} - 1$$

$$MoS_{23} = \frac{Q}{|\sigma_{23}|} - 1, \qquad MoS_{13} = \frac{R}{|\sigma_{13}|} - 1, \qquad MoS_{12} = \frac{S}{|\sigma_{12}|} - 1 \qquad (4.12)$$

Where the normal stress allowables (X, Y, Z) can be distinct in tension and compression according to Eq. (4.5).

Similarly, the max strain criterion MoS can be written as,

$$MoS_{11} = \frac{X_\varepsilon}{\varepsilon_{11}} - 1, \qquad MoS_{22} = \frac{Y_\varepsilon}{\varepsilon_{22}} - 1, \qquad MoS_{33} = \frac{Z_\varepsilon}{\varepsilon_{33}} - 1$$

$$MoS_{23} = \frac{Q_\varepsilon}{|\gamma_{23}|} - 1, \qquad MoS_{13} = \frac{R_\varepsilon}{|\gamma_{13}|} - 1, \qquad MoS_{12} = \frac{S_\varepsilon}{|\gamma_{12}|} - 1$$

(4.13)

where,

$$X_\varepsilon = \begin{cases} X_{\varepsilon T} & \varepsilon_{11} \geq 0 \\ X_{\varepsilon C} & \varepsilon_{11} < 0 \end{cases} \qquad Y_\varepsilon = \begin{cases} Y_{\varepsilon T} & \varepsilon_{22} \geq 0 \\ Y_{\varepsilon C} & \varepsilon_{22} < 0 \end{cases} \qquad Z_\varepsilon = \begin{cases} Z_{\varepsilon T} & \varepsilon_{33} \geq 0 \\ Z_{\varepsilon C} & \varepsilon_{33} < 0 \end{cases}$$

(4.14)

Note both the maximum stress and strain criteria are considered to be noninteractive, that is each direction is independent of the others. Also, the shear strain components and associated allowables are taken herein to be engineering shear strains.

4.2.1.2 Tsai-Hill margins of safety for mechanical loading

For interactive failure criteria like Tsai-Hill and Tsai-Wu, the calculation of the MoS becomes more complex. Following the methodology described in HyperSizer Structural Sizing Software on-line documentation (HyperSizer, 2019) a simple interactive failure criterion in terms of two independent stress ratios R_1 and R_2 can be written as,

$$R_1^2 + R_2^2 = 1 \qquad \text{at failure}$$

(4.15)

where the stress ratio is defined as (applied stress/allowable stress), see Tsai and Hahn (1980). That is, for example,

$$R_1 = \frac{\sigma_{11}}{X}, \qquad R_2 = \frac{\sigma_{22}}{Y}$$

(4.16)

The failure envelope (or curve) associated with an interactive failure criterion can then be plotted in terms of R_1 and R_2, as shown in Fig. 4.2, such that the curve will intersect the axes at a value of 1. Consider a point, e, in Fig. 4.2, whose coordinates are given by (R_1, R_2). It represents a positive MoS, as it is located inside the interactive failure curve. If a line from the origin (point o) is drawn through point e, point f, whose coordinates are (R_{1f}, R_{2f}), is the point at which the line intersects the interactive curve. Point f represents failure (MoS = 0) as it is located on the curve. Therefore one can

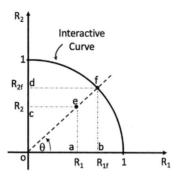

Figure 4.2 Definition of MoS for a failure criterion with two interactive terms (HyperSizer, 2019).

write the margin of safety as:

$$MoS = \frac{of}{oe} - 1 = \delta - 1 \tag{4.17}$$

where δ is the scaling factor to take the current load to the failure curve, and it is equal to of/oe in the example shown in Fig. 4.2. Given uniaxial loading, which is along the R_1 and R_2 axes, the MoS can be written as,

$$MoS = \frac{ob}{oa} - 1 = \frac{R_{1f}}{R_1} - 1 \tag{4.18}$$

$$MoS = \frac{od}{oc} - 1 = \frac{R_{2f}}{R_2} - 1 \tag{4.19}$$

It can be shown that, for any arbitrary angle θ, $of/oe = R_{1f} / R_1 = R_{2f} / R_2 = \delta$. Given Eq. (4.15), an expression for δ in terms of R_1 and R_2 can be obtained. Using Eq. (4.15),

$$R_{1f}^2 + R_{2f}^2 = 1 \tag{4.20}$$

And substituting $R_{1f} = \delta\,R_1$ and $R_{2f} = \delta\,R_2$ into Eq. (4.20) gives,

$$\left(R_1^2 + R_2^2\right)\delta^2 = 1 \tag{4.21}$$

so that,

$$\delta = \frac{1}{\sqrt{R_1^2 + R_2^2}} \tag{4.22}$$

thus providing, in conjunction with Eq. (4.17), the following interactive MoS expression for the quadratic interactive curve shown in Fig. 4.2,

$$MoS = \frac{1}{\sqrt{R_1^2 + R_2^2}} - 1 \tag{4.23}$$

When three or more loading components exist, the interaction equation, e.g., Eq. (4.15), will represent an n-dimensional surface (where n is the number of load components), and the same procedure employed previously can be utilized to arrive at the appropriate MoS expression. For example in the case of the Tsai-Hill criterion at failure, see Eqs. (4.3) and (4.4), a six dimensional extension of Eq. (4.15) can be written as,

$$R_{1f}^2 + R_{2f}^2 + R_{3f}^2 + R_{4f}^2 + R_{5f}^2 + R_{6f}^2 + C_1 R_{1f} R_{2f} + C_2 R_{1f} R_{3f} + C_3 R_{2f} R_{3f} = 1 \tag{4.24}$$

Then substituting in the universal relations, $R_{1f} = \delta R_1$, $R_{2f} = \delta R_2$, $R_{3f} = \delta R_3$, $R_{4f} = \delta R_4$, $R_{5f} = \delta R_5$, and $R_{6f} = \delta R_6$, a quadratic equation in terms of δ is obtained

$$\left(R_1^2 + R_2^2 + R_3^2 + R_4^2 + R_5^2 + R_6^2 + C_1 R_1 R_2 + C_2 R_1 R_3 + C_3 R_2 R_3\right)\delta^2 = 1 \tag{4.25}$$

where,

$$R_1 = \frac{\sigma_{11}}{X}, \qquad R_2 = \frac{\sigma_{22}}{Y}, \qquad R_3 = \frac{\sigma_{33}}{Z},$$
$$R_4 = \frac{|\sigma_{23}|}{Q}, \qquad R_5 = \frac{|\sigma_{13}|}{R}, \qquad R_6 = \frac{|\sigma_{12}|}{S} \tag{4.26}$$

and,

$$C_1 = XY\left(\frac{1}{X^2} + \frac{1}{Y^2} - \frac{1}{Z^2}\right), \qquad C_2 = XZ\left(\frac{1}{X^2} - \frac{1}{Y^2} + \frac{1}{Z^2}\right),$$
$$C_3 = YZ\left(-\frac{1}{X^2} + \frac{1}{Y^2} + \frac{1}{Z^2}\right) \tag{4.27}$$

Solving Eq. (4.25) for δ, we obtain,

$$\delta = \frac{1}{\sqrt{R_1^2 + R_2^2 + R_3^2 + R_4^2 + R_5^2 + R_6^2 + C_1 R_1 R_2 + C_2 R_1 R_3 + C_3 R_2 R_3}} \tag{4.28}$$

Consequently, the following interactive MoS expression for the Tsai-Hill criterion becomes,

$$MoS = \frac{1}{\sqrt{TH}} - 1 \tag{4.29}$$

where TH is defined in Eq. (4.3).

4.2.1.3   Tsai-Wu margins of safety for mechanical loading

Until now only interactive equations containing quadratic terms have been addressed. The Tsai-Wu criterion, however, contains both linear and quadratic terms and therefore requires a different solution for δ. To this end consider the following example of an interactive equation,

$$P_1 + P_2 = 1 \qquad \text{at failure} \tag{4.30}$$

where P_1 contains some linear combination of linear stress ratios and P_2 contains some linear combination of quadratic stress ratios. That is,

$$P_1 = \sum_{m=1}^{M} D_m R_{mf} \tag{4.31}$$

$$P_2 = \sum_{i=1}^{N} \sum_{j=1}^{L} K_{ij} R_{if} R_{jf} \tag{4.32}$$

Consequently, following the previous procedure (with, for example, $N = L = 2$, $K_{11} = K_{22} = 1$, $K_{12} = K_{21} = 0$, and $M = 4$, $D_1 = D_2 = 0$ and $D_3 = D_4 = 1$) Eq. (4.30) becomes

$$R_{1f}^2 + R_{2f}^2 + R_{3f} + R_{4f} = 1 \tag{4.33}$$

and substituting the universal relations $R_{1f} = \delta R_1$, $R_{2f} = \delta R_2$, $R_{3f} = \delta R_3$, $R_{4f} = \delta R_4$ one can obtain the following quadratic equation in terms of δ,

$$\left(R_1^2 + R_2^2\right)\delta^2 + \left(R_3 + R_4\right)\delta - 1 = 0 \tag{4.34}$$

Solving Eq. (4.34) and taking the positive root yields

$$\delta = \frac{-B + \sqrt{B^2 - 4AC}}{2A} \tag{4.35}$$

which can be written as,

$$\delta = \frac{-2C}{B + \sqrt{B^2 - 4AC}} \tag{4.36}$$

where, $A = \left(R_1^2 + R_2^2\right) = P_2$, $B = \left(R_3 + R_4\right) = P_1$, and $C = -1$. In fact, it can be shown that, in general, $A = P_2$, $B = P_1$, and $C = -1$.

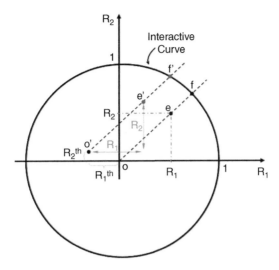

Figure 4.3 For MoS calculations in the presence of thermomechanical loading, the thermal load is treated as a preload, which shifts the origin from point o to point o', and the scaling factor is $\delta =$ o'f'/o'e'.

Consequently, the MoS for an interactive equation containing both linear and quadratic terms, e.g., Tsai-Wu, is given in general by,

$$MoS = \frac{2}{P_1 + \sqrt{(P_1)^2 + 4(P_2)}} - 1 \tag{4.37}$$

where, in the case of Tsai-Wu,

$$
\begin{aligned}
P_1 &= F_1\sigma_{11} + F_2\sigma_{22} + F_3\sigma_{33} \\
P_2 &= F_{11}\sigma_{11}^2 + F_{22}\sigma_{22}^2 + F_{33}\sigma_{33}^2 + F_{44}\sigma_{23}^2 + F_{55}\sigma_{13}^2 + F_{66}\sigma_{12}^2 \\
&\quad + 2F_{23}\sigma_{22}\sigma_{33} + 2F_{13}\sigma_{11}\sigma_{33} + 2F_{12}\sigma_{11}\sigma_{22}
\end{aligned} \tag{4.38}
$$

4.2.2 Margins of safety for combined thermomechanical loading

As mentioned, in the presence of combined thermomechanical loading, the thermal load is typically treated as a "preload". As shown in Fig. 4.3, the MoS quantifies how much more (or less) of a specified mechanical load, in addition to the thermal "preload", is required to reach the failure envelope, according to a given failure criterion. In essence, the starting point for the projection to the failure envelope is shifted from point o to point o' by the stresses (or strains) induced by the thermal loading. In Fig. 4.3, for example, this point is shifted by R_1^{th} and R_2^{th}, which are the stresses (or strains) due to the thermal loading divided by the appropriate allowable. As a result, the calculation of the MoS requires the thermal load case and the mechanical load case to be considered separately, with the ply level and microscale stresses/strains from each load case serving as the input. As such, the stresses and strains from the mechanical load case will be denoted as σ_{ij}^{mech} and ε_{ij}^{mech}, whereas, the stresses and strains from the thermal load case will be denoted as σ_{ij}^{th} and ε_{ij}^{th}. These are not the "thermal stresses" and "thermal strains" derivable from the material properties and ΔT. Rather, they are quantities arising from mismatches in the thermal expansion within the composite or laminate. For example, in a cross-ply laminate, the 0 and 90 plies typically have very different CTEs in a given direction, leading to ply-level stresses from a purely thermal load applied to the laminate (see Exercise 2.19). These ply-level stresses and strains are σ_{ij}^{th} and ε_{ij}^{th}. Furthermore, because all deformations are being considered to be linearly elastic, it should be noted that superposition is valid, and the combined thermomechanical load case stresses/strains are identical to the summation of the individual thermal and mechanical load case stresses/strains.

Fig. 4.3 shows that, with the thermal preload shifting point o to point o', a new arbitrary point e' is defined by the applied mechanical loading (R_1, R_2) such that o'e' = oe. Now, point f', on the failure curve, given by $(R_{1f'}, R_{2f'})$, represents failure (MoS = 0), and o'f' is typically not equal to of. This implies that the margin of safety in the presence of the thermal preload is different and is given by,

$$MoS = \frac{o'f'}{o'e'} - 1 = \delta - 1 \qquad (4.39)$$

4.2.2.1 Max stress and max strain margins of safety for combined thermomechanical loading

Rewriting the axial max stress criterion in terms of the stress ratio, R_1, see Eq. (4.16), at failure (MoS = 0), the following equation is satisfied,

$$\frac{\sigma_{11}}{X} = R_{1f} = 1 \qquad (4.40)$$

In the case of combined thermomechanical loading, $\sigma_{11} = \sigma_{11}^{mech} + \sigma_{11}^{th}$. Then, Eq. (4.40) can be written as,

$$\frac{\left(\sigma_{11}^{mech}\right)_f}{X} + \frac{\sigma_{11}^{th}}{X} = R_{1f}^{mech} + R_1^{th} = 1 \qquad (4.41)$$

where $\left(\sigma_{11}^{mech}\right)_f$ is the stress from a purely mechanical load case that would result in failure (MoS = 0) *when combined with the stress from the preload thermal load case*, σ_{11}^{th}. Note, the value of X is determined based on the sign of the mechanical load case stress (see Eq. 4.5). As before, the stress ratio at failure can be written in terms of the stress ratio from the arbitrary applied mechanical loading,

$$\frac{\left(\sigma_{11}^{mech}\right)_f}{X} = R_{1f}^{mech} = \delta\frac{\sigma_{11}^{mech}}{X} = \delta R_1^{mech} \qquad (4.42)$$

Substituting Eq. (4.42) into Eq. (4.41),

$$\delta R_1^{mech} + R_1^{th} = 1 \tag{4.43}$$

and solving for δ,

$$\delta = \frac{1 - R_1^{th}}{R_1^{mech}} \tag{4.44}$$

Then, from (4.39),

$$MoS_{11} = \delta - 1 = \frac{1 - R_1^{th}}{R_1^{mech}} - 1 \tag{4.45}$$

The remaining thermomechanical max stress criterion MoS can similarly be written as,

$$MoS_{22} = \frac{1 - R_2^{th}}{R_2^{mech}} - 1, \quad MoS_{33} = \frac{1 - R_3^{th}}{R_3^{mech}} - 1,$$

$$MoS_{23} = \frac{1 - R_4^{th}}{R_4^{mech}} - 1, \quad MoS_{13} = \frac{1 - R_5^{th}}{R_5^{mech}} - 1, \quad MoS_{12} = \frac{1 - R_6^{th}}{R_6^{mech}} - 1 \tag{4.46}$$

The max strain criterion MoS are identical to those of maxstress. However, the mechanical-load case R terms are now strain ratios, for example,

$$R_1^{mech} = \frac{\varepsilon_{11}^{mech}}{X_\varepsilon} \tag{4.47}$$

Furthermore, the thermal load case R terms include only the mechanical strains that arise from the thermal load case due to, for example, property mismatches between plies. That is, for example,

$$R_1^{th} = \frac{\varepsilon_{11}^{th} - \alpha_{11}\Delta T}{X_\varepsilon} \tag{4.48}$$

The numerator in Eq. (4.48) is the mechanical strain resulting from the thermal load case, where ε_{11}^{th} is the total strain in the x_1-direction arising from the thermal load case. The use of the mechanical strain rather than total strain is because free thermal expansion is typically not considered to contribute to failure. For example, for a monolithic material subjected to a temperature change from a purely thermal load case, the total strain is equal to $\alpha_{11}\Delta T$. Therefore, according to Eq. (4.48), $R_1^{th} = 0$, so such a thermal preload would have no effect on the MoS calculation.

4.2.2.2 Tsai-Hill and Tsai-Wu margins of safety for combined thermomechanical loading

In addressing the Tsai-Hill and Tsai-Wu MoS in the presence of thermomechanical loading, it is convenient to write the Tsai-Hill criterion in a similar format to the Tsai-Wu criterion, as defined in Eqs. (4.30) to (4.32). As such, for the quadratic Tsai-Hill criterion, it can be readily confirmed that Eq. (4.30) is valid with $P_1 = 0$ (requiring $D_m = 0$). In addition,

$$P_2 = \sum_{i=1}^{N} \sum_{j=1}^{L} K_{ij} R_{if} R_{jf} \tag{4.49}$$

with $L = N = 6$, and,

$$K_{11} = K_{22} = K_{33} = K_{44} = K_{55} = K_{66} = 1,$$

$$K_{23} = -YZ\left(-\frac{1}{X^2} + \frac{1}{Y^2} + \frac{1}{Z^2}\right),$$

$$K_{13} = -XZ\left(\frac{1}{X^2} - \frac{1}{Y^2} + \frac{1}{Z^2}\right), \tag{4.50}$$

$$K_{12} = -XY\left(\frac{1}{X^2} + \frac{1}{Y^2} - \frac{1}{Z^2}\right),$$

with all other $K_{ij} = 0$.

In the case of Tsai-Wu, which has both linear and quadratic terms, P_1 and P_2 are both non-zero, and $M = 3$ and $N = L = 6$. Then, from Eq. (4.38),

$$D_1 = F_1 X, \qquad D_2 = F_2 Y, \qquad D_3 = F_3 Z,$$

$$K_{11} = F_{11} X^2, \qquad K_{22} = F_{22} Y^2, \qquad K_{33} = F_{33} Z^2,$$

$$K_{44} = F_{44} Q^2, \qquad K_{55} = F_{55} R^2, \qquad K_{66} = F_{66} S^2, \tag{4.51}$$

$$K_{23} = 2F_{23} YZ, \qquad K_{13} = 2F_{13} XZ, \qquad K_{12} = 2F_{12} XY,$$

with all other $K_{ij} = 0$. This combined notation will allow the thermomechanical MoS for both Tsai-Hill and Tsai-Wu to be represented with the same equation.

Now, in the presence of combined thermomechanical loading, as before, the stress results of both the pure thermal problem and pure mechanical problem are known. To determine the MoS, the mechanical problem stresses are scaled until the failure envelope is reached, with the thermal problem stresses as the starting point (see Fig. 4.3). As such, the stress ratios at failure, R_{mf} ($m = 1, \ldots, 6$), are mechanical problem stress ratios, which have been scaled to the failure envelope, added to the preload thermal problem stress ratio,

$$R_{mf} = R_{mf}^{mech} + R_m^{th} \tag{4.52}$$

where, as before, $R_{mf}^{mech} = \delta R_m^{mech}$. Thus, using Eq. (4.31),

$$P_1 = \sum_{m=1}^{M} D_m \left(R_{mf}^{mech} + R_m^{th} \right) = \sum_{m=1}^{M} D_m R_{mf}^{mech} + \sum_{m=1}^{M} D_m R_m^{th} = P_1^{mech} + P_1^{th} \tag{4.53}$$

Similarly, from Eq. (4.32),

$$P_2 = \sum_{i=1}^{N} \sum_{j=1}^{L} K_{ij} \left(R_{if}^{mech} + R_i^{th} \right) \left(R_{jf}^{mech} + R_j^{th} \right) = P_2^{mech} + P_2^{mech-th} + P_2^{th} \tag{4.54}$$

where,

$$P_2^{mech} = \sum_{i=1}^{N} \sum_{j=1}^{L} K_{ij} R_{if}^{mech} R_{jf}^{mech}$$

$$P_2^{mech-th} = \sum_{i=1}^{N} \sum_{j=1}^{L} K_{ij} \left(R_{if}^{mech} R_j^{th} + R_i^{th} R_{jf}^{mech} \right) \tag{4.55}$$

$$P_2^{th} = \sum_{i=1}^{N} \sum_{j=1}^{L} K_{ij} R_i^{th} R_j^{th}$$

The MoS can then be determined identically as done for the mechanical-only Tsai-Wu criterion as shown in Section 4.2.1.3. That is, substituting $R_{mf}^{mech} = \delta R_m^{mech}$ into Eq. (4.30), where P_1 and P_2 are defined by Eqs. (4.53) and (4.54) respectively, a quadratic equation in δ is formed,

$$A\delta^2 + B\delta + C = 0 \tag{4.56}$$

where,

$$A = P_2^{mech}$$
$$B = P_2^{mech-th} + P_1^{mech} \tag{4.57}$$
$$C = P_2^{th} + P_1^{th} - 1$$

Solving Eq. (4.56) and taking the positive root, once again gives Eq. (4.36). Finally, substituting into Eq. (4.17),

$$MoS = \frac{-2C}{B + \sqrt{B^2 - 4AC}} - 1 \tag{4.58}$$

thus providing the MoS for the combined thermomechanical case for both the Tsai-Hill and Tsai-Wu criteria.

4.3   MATLAB implementation of margins of safety calculation

The MATLAB code provided with this text is available from:

https://github.com/nasa/Practical-Micromechanics

In this section, the provided MATLAB (2018) functions and scripts for calculating Margins of Safety (MoS), based on the four included failure criteria, are discussed. MoS can be calculated based on (1) ply-level stresses/strains in a laminate analysis wherein the plies are represented using effective properties, (2) constituent scale stresses/strains in a micromechanics-based laminate analysis, and (3) constituent scale stresses/strains in a stand-alone micromechanics problem (material point). In order to calculate and output MoS, two additional steps have been introduced into MicroAnalysis.m (MATLAB Script 3.5) and LamAnalysis.m (MATLAB Script 2.6). Updated flow charts showing these additions (steps 7 and 8) are shown in Fig. 4.4. In MicroAnalysis.m, the MATLAB function CalcMicroMargins.m checks if the problem involves combined thermomechanical loading. If so, the problem is divided into a thermal load case (involving only the temperature change, ΔT, with all global stress components equal to zero), and a mechanical load case (involving the applied mechanical loading with $\Delta T = 0$). For each of these two load cases, the local (constituent level) fields are calculated by calling the MATLAB Function MicroFields.m (see MATLAB Function 3.10). If the problem involves purely mechanical or purely thermal loading, the local fields have already been calculated in step 4 (see Fig. 4.4A). CalcMicroMargins.m then calls MicroMargins.m, which loops through the subcells (for MOC) or the fiber and matrix constituents (for Voigt, Reuss, and MT), and passes the local fields to the function FCriteria.m (shown in MATLAB Function 4.1). FCriteria.m then (for each subcell or constituent) calls the specified failure criteria to calculate the MoS and determines the controlling failure criterion (which is the one that provides the lowest MoS).

In step 8, WriteMicroMoS.m (see Fig. 4.4A) writes the MoS for each subcell or constituent to the output file (text based or MS Word based) for each specified failure criterion, and the overall controlling subcell or constituent and failure criterion (which has the lowest overall MoS) is determined and output. Finally, the composite allowable mechanical load is calculated (based on scaling the applied mechanical loads using the lowest MoS) and written to the output file.

The procedure for incorporating MoS calculations for laminates, as outlined in Fig. 4.4B, is similar. CalcLamMargins.m, in step 7, divides thermomechanical problems into separate thermal and mechanical load cases. For each, it calls CLT.m (MATLAB Function 2.8) and LamMicroFields.m (MATLAB Function 3.11) to solve

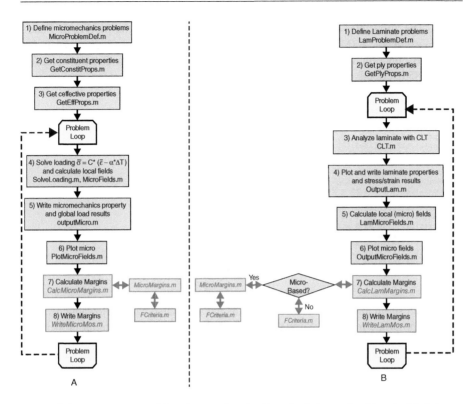

Figure 4.4 Flowcharts describing the steps involved in calculating and writing the MoS for (A) a stand-alone composite micromechanics problems in the MATLAB Script 3.5 `MicroAnalysis.m` and (B) a laminate problem, wherein the ply materials are based on effective properties or micromechanics, in the MATLAB Script 2.6 `LamAnalysis.m`.

the laminate problem and obtain ply-level and local fields (for any plies that are micromechanics-based). Then, `CalcLamMargins.m` loops through the plies and calculates the MoS for plies represented using effective properties (calling `FCriteria.m` directly) and for micromechanics-based plies (by calling `MicroMargins.m`). Then, in step 8, `WriteLamMoS.m` operates similarly to `WriteMicroMoS.m`, but it writes output for every ply, be it effective property based or micromechanics based. `WriteLamMoS.m` also determines the controlling MoS, and associated information, for the entire laminate.

For the sake of brevity, `CalcLamMargins.m`, `CalcMicroMargins.m`, `MicroMargins.m`, `WriteLamMoS.m`, and `WriteMicroMoS.m` are not shown in this text.

To calculate MoS in the MATLAB code, it is necessary to (1) specify ply-level or constituent-level material allowables and (2) set the flag `OutInfo.WriteMargins = True` (in `MicroProblemDef.m` or `LamProblemDef.m`) to obtain output. The allowables are added to the structure `getprops` in the MATLAB Function `GetPlyProps.m` (MATLAB Function 2.1) or the structure `constitprops` in the MATLAB Function `GetConstitProps.m` (MATLAB Function 3.9). For example, for the IM7/8552 effective ply level material, the property specification in `GetPlyProps.m` becomes,

```
plyprops{1}.name = "IM7/8552";
plyprops{1}.E1 = 146.8e3;
plyprops{1}.E2 = 8.69e3;
plyprops{1}.G12 = 5.16e3;
plyprops{1}.v12 = 0.32;
plyprops{1}.a1 = -0.1e-6;
plyprops{1}.a2 = 31.e-6;
% -- Material allowables introduced in Chapter 4
plyprops{1}.allowables.XT = 2323;
plyprops{1}.allowables.XC = -1531;
plyprops{1}.allowables.YT = 52.3;
plyprops{1}.allowables.YC = -235;
plyprops{1}.allowables.S = 88;
plyprops{1}.allowables.XeT = plyprops{1}.allowables.XT/plyprops{1}.E1;
plyprops{1}.allowables.XeC = plyprops{1}.allowables.XC/plyprops{1}.E1;
plyprops{1}.allowables.YeT = plyprops{1}.allowables.YT/plyprops{1}.E2;
plyprops{1}.allowables.YeC = plyprops{1}.allowables.YC/plyprops{1}.E2;
plyprops{1}.allowables.Se = plyprops{1}.allowables.S/plyprops{1}.G12;
```

where each strain allowable has been taken as the associated stress allowable divided by the appropriate modulus. Similarly, the IM7 fiber constituent material properties, specified in `GetConstitProps.m` become,

```
% -- Constituent Mat#1 - IM7 (Transversely Isotropic)
Mat = 1;
constitprops{Mat}.name = "IM7";
constitprops{Mat}.constID = Mat;
constitprops{Mat}.EL = 262.2E3;
constitprops{Mat}.ET = 11.8E3;
constitprops{Mat}.vL = 0.17;
constitprops{Mat}.vT = 0.21;
constitprops{Mat}.GL = 18.9E3;
constitprops{Mat}.aL = -0.9E-06;
constitprops{Mat}.aT = 9.E-06;
constitprops{Mat}.allowables.XT = 4335.0;
constitprops{Mat}.allowables.XC = -2608.;
constitprops{Mat}.allowables.YT = 113.;
constitprops{Mat}.allowables.YC = -354.;
constitprops{Mat}.allowables.ZT = 113.;
constitprops{Mat}.allowables.ZC = -354.;
constitprops{Mat}.allowables.Q = 128.;
constitprops{Mat}.allowables.R = 138.;
constitprops{Mat}.allowables.S = 138.;
constitprops{Mat}.allowables.XeT = constitprops{Mat}.allowables.XT/constitprops{Mat}.EL;
constitprops{Mat}.allowables.XeC = constitprops{Mat}.allowables.XC/constitprops{Mat}.EL;
constitprops{Mat}.allowables.YeT = constitprops{Mat}.allowables.YT/constitprops{Mat}.ET;
constitprops{Mat}.allowables.YeC = constitprops{Mat}.allowables.YC/constitprops{Mat}.ET;
constitprops{Mat}.allowables.ZeT = constitprops{Mat}.allowables.ZT/constitprops{Mat}.ET;
constitprops{Mat}.allowables.ZeC = constitprops{Mat}.allowables.ZC/constitprops{Mat}.ET;
GT = constitprops{Mat}.ET/(2*(1+constitprops{Mat}.vT));
constitprops{Mat}.allowables.Qe = constitprops{Mat}.allowables.Q/GT;
constitprops{Mat}.allowables.Re = constitprops{Mat}.allowables.R/constitprops{Mat}.GL;
constitprops{Mat}.allowables.Se = constitprops{Mat}.allowables.S/constitprops{Mat}.GL;
```

where, again, each strain allowable has been taken as the associated stress allowable divided by the appropriate modulus. The MATLAB code will check if allowables are specified for each material used in a problem, and use them to calculate MoS if they are present. For constituent materials without specified allowables, the allowable magnitudes are set to a very high value (i.e., 1×10^{99}) so they will never be exceeded.

The failure criteria that will be considered in a given problem can be specified in `MicroProblemDef.m` (MATLAB Function 3.6) or `LamProblemDef.m` (MATLAB Function 2.7) by adding a flag to the `Loads{NP}` cell/structure (wherein 1 indicates the criterion is on, and 0 indicates the criterion is off). For example,

```
Loads{NP}.CriteriaOn = [1, 0, 1, 0]; % [max stress, max strain, Tsai-Hill, Tsai-Wu]
```

indicates that the max stress criterion is on, the max strain criterion is off, the Tsai-Hill criterion is on, and the Tsai-Wu Criterion is off. By default, if `CriteriaOn` is not specified for a given problem, only the Tsai-Wu criterion is active.

As mentioned, the max stress, max strain, Tsai-Hill, and Tsai-Wu criteria have each been implemented to calculate MoS (with or without a thermal preload) as MATLAB functions (`MaxStress`, `MaxStrain`, `TsaiHill`, and `TsaiWu`, respectively). These functions are located within the same file as MATLAB Function 4.1– `FCriteria.m`.

MATLAB Function 4.1 – *FCriteria.m*

This function calculates the margins of safety (MoS) for the max stress, max strain, Tsai-Hill, and Tsai-Wu criteria based on the thermal preload stresses/strains and mechanical problem stresses and strains (at a point) along with the material allowables

```
function [MoS] = FCriteria(stress, strain, stressth, strainth, allowables, ...
                           CriteriaOn, micro, MoS)
% +++++++++++++++++++++++++++++++++++++++++++++++++++++++++++++++++++++++++++++
% Copyright 2020 United States Government as represented by the Administrator of the
% National Aeronautics and Space Administration. No copyright is claimed in the
% United States under Title 17, U.S. Code. All Other Rights Reserved. By downloading
% or using this software you agree to the terms of NOSA v1.3. This code was prepared
% by Drs. B.A. Bednarcyk and S.M. Arnold to complement the book "Practical
% Micromechanics of Composite Materials" during the course of their government work.
% +++++++++++++++++++++++++++++++++++++++++++++++++++++++++++++++++++++++++++++
% Purpose: Given a stress and strain state, calculate the MoS by calling functions
%          (contained herein) for each failure criterion
% Input:
% - stress: Stresses (from mechanical problem only if ThermoMech)
% - strain: Strains (from mechanical problem only if ThermoMech)
% - stressth: Stresses from thermal problem only if ThermoMech, 0 otherwise
% - strainth: Strains from thermal problem only if ThermoMech, 0 otherwise
% - allowables: Struct containing material allowables
% - CriteriaOn: Flags indicating which failure criteria are turned on
% - micro: Flag indicating if problem is micromechanics-based
```

```
% - MoS: Struct containing MoS results and info
% Output:
% - MoS: Updated struct containing MoS results and info
% ++++++++++++++++++++++++++++++++++++++++++++++++++++++++++++++++++++++++++++++++

% -- Determine if problem is combined thermomechanical
thermomech = false;
for i = 1:6
    if stressth(i) ~= 0
        thermomech = true;
    end
end

% -- Check if thermal preload causes failure
if thermomech
    zero6 = zeros(6,1);

    % -- Calculate margins from thermal preload only
    [MoSTh.Crit{1}] = MaxStress(stressth, zero6, allowables, micro);
    [MoSTh.Crit{2}] = MaxStrain(strainth, zero6, allowables, micro);
    [MoSTh.Crit{3}] = TsaiHill(stressth, zero6, allowables, micro);
    [MoSTh.Crit{4}] = TsaiWu(stressth, zero6, allowables, micro);

    for j = 1:4
        if j < 3
            num = 6;
        else
            num = 1;
        end
        for i = 1:num
            if MoSTh.Crit{j}.MoS(i) <= 0
                error(['** Negative Margin Detected Due to Thermal Preload ', ...
                    ' To see results, run with thermal load only **'])
            end
        end
    end

end

% -- Standard margin calculations, call each criterion
[MoS.Crit{1}] = MaxStress(stress, stressth, allowables, micro);
[MoS.Crit{2}] = MaxStrain(strain, strainth, allowables, micro);
[MoS.Crit{3}] = TsaiHill(stress, stressth, allowables, micro);
[MoS.Crit{4}] = TsaiWu(stress, stressth, allowables, micro);

% -- Set MoS high if criterion is turned off
for i = 1:4
    if CriteriaOn(i) == 0
        MoS.Crit{i}.MoS = [99999, 99999, 99999, 99999, 99999, 99999];
        MoS.Crit{i}.MinMoS = 99999;
    end
end

% -- Set and store controlling MoS information
MoS.Controlling = "none";
MoS.MinMoS = 99999;
MoS.ControllingNum = 0;
for i = 1:4
    if MoS.Crit{i}.MinMoS < MoS.MinMoS
        MoS.Controlling = MoS.Crit{i}.name;
        MoS.ControllingNum = i;
        MoS.MinMoS = MoS.Crit{i}.MinMoS;
    end
end

end

%%%%%%%%%%%%%%%%%%%%%%%%%%%%%%%%%%%%%%%%%%%%%%%%%%%%%%%%%%%%%%%%%%%%%%%%%%%%%%%%%%%
%%%%%%%%%%%%%%%%%%%%%%%%%%%%%%%%%%%%%%%%%%%%%%%%%%%%%%%%%%%%%%%%%%%%%%%%%%%%%%%%%%%
```

```
function [Crit] = MaxStress(stress, stressth, allowables, micro)

% -- Max stress failure criterion

Crit.name = "MaxStress";

% -- Calculate stress ratios (R_i)
[Rmech, Rth, ~] = GetR(stress, stressth, allowables, micro);

% -- Calculate MoS and determine controlling component
Crit.Controlling = 0;
Crit.MinMoS = 99999;
for i = 1:6 % -- Loop through 6 stress component-based MoS
    if abs(Rmech(i)) < 0.0000001 % -- Set MoS high if R is small
        Crit.MoS(i) = 99999;
    else
        Crit.MoS(i) = (1 - Rth(i))/Rmech(i) - 1; % -- Eqs. 4.45 & 4.46
    end
    if Crit.MoS(i) < Crit.MinMoS % -- Store controlling MoS
        Crit.MinMoS = Crit.MoS(i);
        Crit.Controlling = i;
    end
end

end

%%%%%%%%%%%%%%%%%%%%%%%%%%%%%%%%%%%%%%%%%%%%%%%%%%%%%%%%%%%%%%%%%%%%%%%%%%%
%%%%%%%%%%%%%%%%%%%%%%%%%%%%%%%%%%%%%%%%%%%%%%%%%%%%%%%%%%%%%%%%%%%%%%%%%%%

function [Crit] = MaxStrain(strain, strainth, allowables, micro)

% -- Max strain failure criterion

% -- Note:
%    - strain is the strain from the mechanical load case
%    - strainth is the mechanical strain from the thermal load case

Crit.name = "MaxStrain";

strainallowables.XT = allowables.XeT;
strainallowables.XC = allowables.XeC;
strainallowables.YT = allowables.YeT;
strainallowables.YC = allowables.YeC;
strainallowables.S = allowables.Se;
if micro
    strainallowables.ZT = allowables.ZeT;
    strainallowables.ZC = allowables.ZeC;
    strainallowables.Q = allowables.Qe;
    strainallowables.R = allowables.Re;
end

% -- Calculate strain ratios (R_i)
[Rmech, Rth, ~] = GetR(strain, strainth, strainallowables, micro);

% -- Calculate MoS and determine controlling component
Crit.Controlling = 0;
Crit.MinMoS = 99999;
for i = 1:6 % -- Loop through 6 strain component-based MoS
    if abs(Rmech(i)) < 0.0000001 % -- Set MoS high if R is small
        Crit.MoS(i) = 99999;
    else
        Crit.MoS(i) = (1-Rth(i))/Rmech(i) - 1; % -- Eqs. 4.45 - 4.47
    end
    if Crit.MoS(i) < Crit.MinMoS % -- Store controlling MoS
        Crit.MinMoS = Crit.MoS(i);
        Crit.Controlling = i;
    end
end

end
```

```matlab
%%%%%%%%%%%%%%%%%%%%%%%%%%%%%%%%%%%%%%%%%%%%%%%%%%%%%%%%%%%%%%%%%%%%%%%%%%
%%%%%%%%%%%%%%%%%%%%%%%%%%%%%%%%%%%%%%%%%%%%%%%%%%%%%%%%%%%%%%%%%%%%%%%%%%
function [Crit] = TsaiHill(stress, stressth, allowables, micro)

% -- Tsai-Hill failure criterion

% -- Calculate stress ratios (R_i) and current allowables (based on T vs C)
[Rmech, Rth, CurrentAllowables] = GetR(stress, stressth, allowables, micro);

X = CurrentAllowables.X;
Y = CurrentAllowables.Y;
Z = CurrentAllowables.Z;

% -- Determine D and K coefficients (Eq. 4.50)
D = zeros(3,1);
K = eye(6,6);

K(2,3) = -Y*Z*(-1/X^2 + 1/Y^2 + 1/Z^2);
K(1,3) = -X*Z*(1/X^2 - 1/Y^2 + 1/Z^2);
K(1,2) = -X*Y*(1/X^2 + 1/Y^2 - 1/Z^2);

% -- Calculate MoS
[Crit] = TsaiHillWuMoS(Rmech, Rth, D, K);

Crit.name = "Tsai-Hill";

end

%%%%%%%%%%%%%%%%%%%%%%%%%%%%%%%%%%%%%%%%%%%%%%%%%%%%%%%%%%%%%%%%%%%%%%%%%%
%%%%%%%%%%%%%%%%%%%%%%%%%%%%%%%%%%%%%%%%%%%%%%%%%%%%%%%%%%%%%%%%%%%%%%%%%%
function [Crit] = TsaiWu(stress, stressth, allowables, micro)

% -- Tsai-Wu failure criterion

% -- Calculate stress ratios (R_i) and current allowables (based on T vs C)
[Rmech, Rth, CurrentAllowables] = GetR(stress, stressth, allowables, micro);

X = CurrentAllowables.X;
Y = CurrentAllowables.Y;
Z = CurrentAllowables.Z;

XT = allowables.XT;
XC = allowables.XC;
YT = allowables.YT;
YC = allowables.YC;
S = allowables.S;

if micro % -- Micromechanics-based
    Q = allowables.Q;
    R = allowables.R;
    ZT = allowables.ZT;
    ZC = allowables.ZC;
else % -- Not micromechanics-based, Q,R,Z not used, but can't be 0
    Q = 1;
    R = 1;
    ZT = 1;
    ZC = 1;
end

% -- Eq 4.8
F1 = 1/XT + 1/XC;
F2 = 1/YT + 1/YC;
F3 = 1/ZT + 1/ZC;
F11 = -1/(XT*XC);
F22 = -1/(YT*YC);
F33 = -1/(ZT*ZC);
F44 = 1/Q^2;
F55 = 1/R^2;
F66 = 1/S^2;
```

```
% -- Tsai-Hahn (1980) interaction coefficients (Eq. 4.9)
F12 = -1/(2*sqrt(XT*XC*YT*YC));
F13 = -1/(2*sqrt(XT*XC*ZT*ZC));
F23 = -1/(2*sqrt(YT*YC*ZT*ZC));

% -- Determine D and K coefficients (Eq. 4.51)
D = zeros(3,1);
K = zeros(6,6);
D(1) = F1*X;
D(2) = F2*Y;
D(3) = F3*Z;
K(1,1) = F11*X^2;
K(2,2) = F22*Y^2;
K(3,3) = F33*Z^2;
K(4,4) = F44*Q^2;
K(5,5) = F55*R^2;
K(6,6) = F66*S^2;
K(2,3) = 2*F23*Y*Z;
K(1,3) = 2*F13*X*Z;
K(1,2) = 2*F12*X*Y;

% -- Calculate MoS
[Crit] = TsaiHillWuMoS(Rmech, Rth, D, K);

Crit.name = "Tsai-Wu";

end


%%%%%%%%%%%%%%%%%%%%%%%%%%%%%%%%%%%%%%%%%%%%%%%%%%%%%%%%%%%%%%%%%%%%%%%%%%%%%
%%%%%%%%%%%%%%%%%%%%%%%%%%%%%%%%%%%%%%%%%%%%%%%%%%%%%%%%%%%%%%%%%%%%%%%%%%%%%

function [Crit] = TsaiHillWuMoS(Rmech, Rth, D, K)

% -- Calculate actual MoS for Tsai-Hill and Tsai-Wu

% -- Eq. 4.53
P1mech = Rmech(1:3)*D;
P1th = Rth(1:3)*D;

% -- Eq. 4.55
P2mech = 0;
P2mechth = 0;
P2th = 0;
for i = 1:6
    for j = 1:6
        P2mech = P2mech + K(i,j)*Rmech(i)*Rmech(j);
        P2mechth = P2mechth + K(i,j)*(Rmech(i)*Rth(j) + Rth(i)*Rmech(j));
        P2th = P2th + K(i,j)*Rth(i)*Rth(j);
    end
end

% -- Eq. 4.57
A = P2mech;
B = P2mechth + P1mech;
C = P2th + P1th - 1;

% -- Check for sqrt term in 4.58 less than zero (should only be numerics)
if B^2 - 4*A*C < 0
    den = 0;
else
    den = B + sqrt(B^2 - 4*A*C);
end

if den == 0
    Crit.MoS = 99999;
else
    Crit.MoS = -2*C/den - 1; % -- Eq. 4.58
end
```

```
if Crit.MoS > 99999
   Crit.MoS = 99999;
end

% -- Set controlling to 1 - there is only one MoS for Tsai-Hill and Tsai-Wu
Crit.Controlling = 1;
Crit.MinMoS = Crit.MoS;

end
```

4.4 Examples problems

The failure criteria presented above can be applied at either the ply level or the micro level. At the micro level, the stress and strain allowables for the constituent (fiber/matrix) materials are required for use in conjunction with the constituent or subcell stresses and strains. Examples for both ply-level and microlevel MoS are presented below using the MATLAB functions given in the previous section. In all cases, the minimum MoS indicates the first ply or constituent that experiences stresses or strains that exceed the allowables (according to some criterion). It is important to understand that this may or may not be representative of composite or laminate failure/fracture. It is common practice to design composites to avoid ply failure/damage initiation (i.e., first ply failure). Therefore, the controlling minimum MoS is very useful information used in the design of composite parts (see HyperSizer, 2019). Examples involving ply level MoS are first presented followed by stand-alone micromechanics-based MoS and micromechanics-based laminate MoS analysis.

4.4.1 Laminate ply-level MoS

4.4.1.1 Unidirectional laminate

To introduce MoS calculations, a unidirectional laminate, in which all fibers are aligned in a single direction, θ, will first be considered. Unidirectional (e.g., [0] and [90]) laminates are the simplest of all laminates and provide the same response as would a single ply. In Chapter 2, Table 2.1, ply level thermoelastic properties were given for all three classes of composites, PMC, MMC, and CMC. In this chapter only PMC systems will be examined with respect to failure and MoS, with representative ply level stress allowables being given in Table 4.2 (see NCAMP (2011a,b)). The ply level strain allowables have been taken as the stress allowables divided by the appropriate moduli, see code in Section 4.3.

Table 4.2 Ply material stress allowables.

Mat	Material name	X_T (MPa)	X_C (MPa)	Y_T (MPa)	Y_C (MPa)	S (MPa)
1	IM7/8552	2323	−1531	52.3	−235	88
2	Glass/epoxy	1346	−944	50	−245	122

* Note: strain allowables have been taken as the stress allowables divided by appropriate modulus.

In the first case, consider a unidirectional IM7/8552 laminate [0], comprised of a single layer of thickness 1 mm, loaded in the fiber direction, i.e., x-direction, subjected to 1000 MPa, i.e., $N_x = 1000$ N/mm (with all other force and moment resultants equal to zero). Note that, since the [0] laminate has a thickness of 1 mm and no moments are applied, the magnitude of the force resultant is equivalent to the ply level stress magnitude. Given that the allowable in the x-direction in tension, X_T, is 2323 MPa (see Table 4.2), from Eq. (4.10) it is obvious that the margin of safety is,

$$MoS = 2323/1000 - 1 = 1.323$$

This problem can be executed (using MATLAB Script 2.6 – LamAnalysis.m) once the ply allowables (see Table 4.2 and code given in Section 4.3) are added to the ply material properties in GetPlyProps.m (see MATLAB Function 2.1). Otherwise the laminate problem is defined in LamProblemDef.m (see MATLAB Function 2.7) as in the previous chapters:

```
NP = NP + 1;
OutInfo.Name(NP) = "Section 4.4.1.1 - [0] Laminate Nx = 1000";
Geometry{NP}.Orient = [0];
nplies = length(Geometry{NP}.Orient);
Geometry{NP}.plymat(1:nplies) = 1;
Geometry{NP}.tply(1:nplies) = 1;
Loads{NP}.DT = 0.;
Loads{NP}.Type  = [  NM,  NM, NM, NM, NM, NM];
Loads{NP}.Value = [1000,   0,  0,  0,  0,  0];
Loads{NP}.CriteriaOn = [1, 1, 1, 1];
```

As mentioned previously, to write the margins to the output file, Out-Info.WriteMargins must be set equal to true and the active failure criteria can be set using Loads{NP}.CriteriaOn. Note that, if Loads{NP}.CriteriaOn is not specified, the code will default to using only the Tsai-Wu criterion. In this example all criteria are active.

The stresses and MoS results indicate that all four criteria, as expected, provide the same MoS, see Fig. 4.5. Note that the maximum strain criterion also provides a MoS in the y-direction since, due to Poisson effects, a transverse strain is induced by the applied uniaxial N_x load. If the stress or strain is close to zero (or the MoS is extremely high), a MoS of 99999 is reported in the MATLAB results.

Clearly, if the applied $N_x = 1000$ N/mm were multiplied (or "scaled") by the factor MoS + 1 (i.e., $N_x = 2323$ N/mm), the laminate would reach the failure criteria (or MoS = 0) in the x-direction. If one applied $N_y = 10.0$ N/mm to the above [0] laminate, what would the resulting MoS become? The answer is 4.23 since the tensile allowable in the y-direction, Y_T, is 52.3 MPa (see Table 4.2) and $52.3/10 - 1 = 4.23$. The stresses and MoS resulting from application of $N_y = 10.0$ N/mm to the laminate are shown in Fig. 4.6. The maximum strain criterion MoS in the fiber direction, which is again due to Poisson effects, is two orders of magnitude larger than that in the loading direction.

To illustrate the effect of off-axis plies, consider the case of a unidirectional IM7/8552 laminate whose fibers are oriented at 45 degrees, for example [45], with thickness 1 mm subjected to a uniaxial load in the x-direction, i.e., $N_x = 1000$ N/mm (with all other force and moment resultants equal to zero). This problem can be defined in LamProblemDef.m as follows:

Ply stresses through the thickness

Z-location	Angle	SigmaX	SigmaY	TauXY	Sigma11	Sigma22	Tau12
-0.5	0	1000	3.208e-15	0	1000	3.208e-15	0
0.5	0	1000	3.208e-15	0	1000	3.208e-15	0

Ply margins of safety through the thickness

Z-location	Angle	Criterion	MoS or MoS11	MoS22	MoS12
-0.5	0	Maxstress	1.323	99999	99999
		Maxstrain	1.323	11.4058	99999
		Tsai-Hill	1.323		
		Tsai-Wu	1.323		
0.5	0	Maxstress	1.323	99999	99999
		Maxstrain	1.323	11.4058	99999
		Tsai-Hill	1.323		
		Tsai-Wu	1.323		

Figure 4.5 Stresses and MoS for a unidirectional [0] laminate loaded in the x-direction, $Nx = 1000$ N/mm.

Ply stresses through the thickness

Z-location	Angle	SigmaX	SigmaY	TauXY	Sigma11	Sigma22	Tau12
-0.5	0	6.7843e-16	10	0	6.7843e-16	10	0
0.5	0	6.7843e-16	10	0	6.7843e-16	10	0

Ply margins of safety through the thickness

Z-location	Angle	Criterion	MoS or MoS11	MoS22	MoS12
-0.5	0	Maxstress	99999	4.23	99999
		Maxstrain	477.4375	4.23	99999
		Tsai-Hill	4.23		
		Tsai-Wu	4.23		
0.5	0	Maxstress	99999	4.23	99999
		Maxstrain	477.4375	4.23	99999
		Tsai-Hill	4.23		
		Tsai-Wu	4.23		

Figure 4.6 Stresses and MoS for a unidirectional [0] laminate loaded in the y-direction, $Ny = 10$ N/mm.

```
NP = NP + 1;
OutInfo.Name(NP) = "Section 4.4.1.1 - [45] Laminate Nx = 1000";
Geometry{NP}.Orient = [45];
nplies = length(Geometry{NP}.Orient);
Geometry{NP}.plymat(1:nplies) = 1;
Geometry{NP}.tply(1:nplies) = 1;
Loads{NP}.DT = 0.;
Loads{NP}.Type  = [  NM,  NM, NM, NM, NM, NM];
Loads{NP}.Value = [1000,   0,  0,  0,  0,  0];
Loads{NP}.CriteriaOn = [1, 1, 1, 1];
```

In this case the stress in the fiber direction (σ_{11}), as well as transverse to the fiber (σ_{22}), is 500 MPa, as shown in Fig. 4.7. Additionally, in the material coordinate system, a

Ply stresses through the thickness

Z-location	Angle	SigmaX	SigmaY	TauXY		Sigma11	Sigma22	Tau12
-0.5	45	1000	8.241e-14	2.7439e-13		500	500	-500
0.5	45	1000	8.241e-14	2.7439e-13		500	500	-500

Ply margins of safety through the thickness

Z-location	Angle	Criterion	MoS or MoS11	MoS22	MoS12
-0.5	45	Maxstress	3.646	-0.8954	-0.824
		Maxstrain	5.8324	-0.89338	-0.824
		Tsai-Hill	-0.91008		
		Tsai-Wu	-0.91465		
0.5	45	Maxstress	3.646	-0.8954	-0.824
		Maxstrain	5.8324	-0.89338	-0.824
		Tsai-Hill	-0.91008		
		Tsai-Wu	-0.91465		

Figure 4.7 Stresses and MoS for a unidirectional [45] laminate loaded in the x-direction, $N_x = 1000$ N/mm.

negative shear stress (σ_{12}) of 500 MPa is present. Due to the high magnitude multiaxial state, now both positive and negative margins of safety are calculated depending upon the criterion used and the direction considered, see results in Fig. 4.7.

Recall that a negative MoS indicates that the laminate is not acceptable due to the stresses caused by the applied loading exceeding the allowables. Considering the maximum stress criterion, it is clear that, in the fiber direction (MoS11 = 3.646) the ply has a positive MoS, while in the transverse direction (MoS22 = -0.8954) and for in-plane shear (MoS12 = -0.824) the ply has negative MoS. The transverse component is the controlling max stress MoS as it is the margin with the lowest value for this criterion. The two interactive criteria, Tsai-Hill and Tsai-Wu, provide negative MoS as well, with Tsai-Wu being the lowest and thus controlling MoS (see highlighted value in Fig. 4.7). The off-axis laminate illustrates nicely the advantage of using a criterion that has interacting stress components; that is, only a single MoS value need be considered to assess failure of a given ply. In contrast, in the case of the component-dependent (non-interactive) criteria (max stress and max strain), three different ply level MoS values must be considered, with the minimum taken as controlling.

As shown previously, the "failure load" for a laminate (according to a given failure criterion) can be calculated by scaling the controlling MoS. Therefore, assuming max stress, the failure load of this laminate is,

$$N_x^f = N_x(MoS + 1) = 1000\,\text{N/mm}(-0.8954 + 1) = 104.6\,\text{N/mm}$$

To confirm this, a load of $N_x = 104.6$ N/mm has been applied to the same unidirectional [45] IM7/8552 laminate. The resulting MoS are shown in Fig. 4.8, where the controlling max stress MoS (MoS22) is now essentially zero (see highlighted value).

Alternatively, based on the Tsai-Wu criterion (MoS = -0.91465, see Fig. 4.7) and applying the laminate failure load, $N_x^f = (1000\,\text{N/mm})(-0.91465 + 1) = 85.35\,\text{N/mm}$, the resulting ply margins are given in Fig. 4.9. Consequently, there is

Ply margins of safety through the thickness

Z-location	Angle	Criterion	MoS or MoS11	MoS22	MoS12
-0.5	45	Maxstress	43.4168	1.1102e-15	0.6826
		Maxstrain	64.3189	0.019309	0.6826
		Tsai-Hill	-0.14036		
		Tsai-Wu	-0.18404		
0.5	45	Maxstress	43.4168	1.1102e-15	0.6826
		Maxstrain	64.3189	0.019309	0.6826
		Tsai-Hill	-0.14036		
		Tsai-Wu	-0.18404		

Figure 4.8 Stresses and MoS for a unidirectional [45] laminate loaded in the x-direction, $N_x = 104.6$ N/mm.

Ply margins of safety through the thickness

Z-location	Angle	Criterion	MoS or MoS11	MoS22	MoS12
-0.5	45	Maxstress	53.4347	0.22554	1.0621
		Maxstrain	79.051	0.24921	1.0621
		Tsai-Hill	0.053525		
		Tsai-Wu	-9.9846e-06		
0.5	45	Maxstress	53.4347	0.22554	1.0621
		Maxstrain	79.051	0.24921	1.0621
		Tsai-Hill	0.053525		
		Tsai-Wu	-9.9846e-06		

Figure 4.9 Stresses and MoS for a unidirectional [45] laminate loaded in the x-direction, $N_x = 85.35$ N/mm.

a significant effect on the "failure load" of a given laminate depending upon loading, stacking sequence (which impacts the level of multiaxial loading in a given ply), material properties, and the specific criterion used.

4.4.1.2   Quasi-isotropic laminate $[60/-60/0]_s$

Consider the case of a 6-ply, quasi-isotropic, symmetric IM7/8552 laminate, $[60/-60/0]_s$, with typical 0.15 mm thick plies, loaded only in the x-direction, $N_x = 300$ N/mm. This laminate problem has been defined in LamProblemDef.m as follows:

```
NP = NP + 1;
OutInfo.Name(NP) = "Section 4.4.1.2 - [60,-60,0]s Laminate Nx = 300";
Geometry{NP}.Orient = [60,-60,0,0,-60,60];
nplies = length(Geometry{NP}.Orient);
Geometry{NP}.plymat(1:nplies) = 1;
Geometry{NP}.tply(1:nplies) = 0.15;
Loads{NP}.DT = 0.;
Loads{NP}.Type  = [ NM,  NM, NM, NM, NM, NM];
Loads{NP}.Value = [300,   0,  0,  0,  0,  0];
Loads{NP}.CriteriaOn = [1, 1, 1, 1];
```

Ply stresses through the thickness for top half of laminate

Z-location	Angle	SigmaX	SigmaY	TauXY	Sigma11	Sigma22	Tau12
-0.45	60	63.5479	-0.3142	14.1848	27.9357	35.298	-34.7455
-0.3	60	63.5479	-0.3142	14.1848	27.9357	35.298	-34.7455
-0.3	-60	63.5479	-0.3142	-14.1848	27.9357	35.298	34.7455
-0.15	-60	63.5479	-0.3142	-14.1848	27.9357	35.298	34.7455
-0.15	0	872.9042	0.62839	1.4121e-15	872.9042	0.62839	1.4121e-15
0	0	872.9042	0.62839	5.5559e-16	872.9042	0.62839	5.5559e-16
Symmetric							

Ply margins of safety through the thickness for top half of laminate

Z-location	Angle	Criterion	MoS or MoS11	MoS22	MoS12
-0.45	60	Maxstress	82.1552	0.48167	1.5327
		Maxstrain	138.6003	0.50422	1.5327
		Tsai-Hill	0.27894		
		Tsai-Wu	0.21225		
-0.3	60	Maxstress	82.1552	0.48167	1.5327
		Maxstrain	138.6003	0.50422	1.5327
		Tsai-Hill	0.27894		
		Tsai-Wu	0.21225		
-0.3	-60	Maxstress	82.1552	0.48167	1.5327
		Maxstrain	138.6003	0.50422	1.5327
		Tsai-Hill	0.27894		
		Tsai-Wu	0.21225		
-0.15	-60	Maxstress	82.1552	0.48167	1.5327
		Maxstrain	138.6003	0.50422	1.5327
		Tsai-Hill	0.27894		
		Tsai-Wu	0.21225		
-0.15	0	Maxstress	1.6612	82.2279	99999
		Maxstrain	1.6618	13.7735	99999
		Tsai-Hill	1.6608		
		Tsai-Wu	1.6543		
0	0	Maxstress	1.6612	82.2279	99999
		Maxstrain	1.6618	13.7735	99999
		Tsai-Hill	1.6608		
		Tsai-Wu	1.6543		
Symmetric					

Figure 4.10 Stresses and MoS for a quasi-isotropic laminate $[60/-60/0]_s$, loaded in the x-direction, $N_x = 300$ N/mm.

The resulting stress state in the laminate and material coordinate systems as well as ply MoS are shown in Fig. 4.10. Note that results from only the top half of the laminate are shown because the results are symmetric.

The material coordinate stresses in the 60 and −60 plies are the same, except for the sign of the shear stress component, and the magnitudes of the various components are comparable. Also, as expected, the [0] ply, has the highest axial stress value since the applied loading is along the x-axis. The controlling failure criterion is Tsai-Wu, with a MoS = 0.21225, and it indicates that failure within the ±60 degree plies would occur at an applied N_x load of 363.67 N/mm. However, significant margin remains in the [0] ply (i.e., a load of $N_x = 796.29$ N/mm would meet the 0 ply allowable according to

Ply stresses through the thickness

Z-location	Angle	SigmaX	SigmaY	TauXY	Sigma11	Sigma22	Tau12
-0.45	60	126.8602	476.5727	233.3689	591.248	12.1849	34.7455
-0.3	60	126.8602	476.5727	233.3689	591.248	12.1849	34.7455
-0.3	-60	126.8602	476.5727	-233.3689	591.248	12.1849	-34.7455
-0.15	-60	126.8602	476.5727	-233.3689	591.248	12.1849	-34.7455
-0.15	0	-253.7204	46.8545	3.3595e-15	-253.7204	46.8545	3.3595e-15
0	0	-253.7204	46.8545	1.5346e-15	-253.7204	46.8545	1.5346e-15
Symmetric							

Ply margins of safety through the thickness

Z-location	Angle	Criterion	MoS or MoS11	MoS22	MoS12
-0.45	60	Maxstress	2.929	3.2922	1.5327
		Maxstrain	2.9551	52.094	1.5327
		Tsai-Hill	0.91173		
		Tsai-Wu	0.97304		
-0.3	60	Maxstress	2.929	3.2922	1.5327
		Maxstrain	2.9551	52.094	1.5327
		Tsai-Hill	0.91173		
		Tsai-Wu	0.97304		
-0.3	-60	Maxstress	2.929	3.2922	1.5327
		Maxstrain	2.9551	52.094	1.5327
		Tsai-Hill	0.91173		
		Tsai-Wu	0.97304		
-0.15	-60	Maxstress	2.929	3.2922	1.5327
		Maxstrain	2.9551	52.094	1.5327
		Tsai-Hill	0.91173		
		Tsai-Wu	0.97304		
-0.15	0	Maxstress	5.0342	0.11622	99999
		Maxstrain	4.6975	0.012375	99999
		Tsai-Hill	0.094262		
		Tsai-Wu	-0.0052288		
0	0	Maxstress	5.0342	0.11622	99999
		Maxstrain	4.6975	0.012375	99999
		Tsai-Hill	0.094262		
		Tsai-Wu	-0.0052288		
Symmetric					

Figure 4.11 Stresses and MoS for a quasi-isotropic laminate [60/−60/0]$_s$, loaded in the y-direction, $N_y = 300$ N/mm.

Tsai-Wu). This is a clear indication that, under purely uniaxial loading, the presence of off-axis plies is detrimental; the allowable N_x load would be much higher if all fibers were aligned with the load direction. Recall that exceeding an allowable does not necessarily indicate complete composite failure, and the laminate could potentially still carry additional load.

Given that this composite laminate is quasi-isotropic, an interesting question arises when evaluating the MoS in response to identical loading applied in the y-direction, i.e., $N_y = 300$ N/mm. Applying $N_y = 300$ N/mm one can obtain the laminate and material coordinate system stress state as well as ply MoS, see Fig. 4.11. Under this same loading but now in the y-direction, it is clear that the controlling criterion is

Tsai-Wu, with a MoS = -0.0052288, with failure indicated within the [0] ply instead of the [±60] ply under N_x loading. Also, the scaled failure load would now occur at an applied N_y load level of 298.43 N/mm (as opposed to 363.67 N/mm under N_x loading). Consequently, one can conclude that "failure" of this quasi-isotropic laminate is not isotropic, even though the effective elastic properties are. This is because the effective properties are related to the average of all stresses throughout the laminate, whereas the failure criteria are applied per ply and, as is clear from Figs. 4.10 and 4.11, the ply stresses are not the same when the [60/–60/0]$_s$ laminate is loaded in the two different directions. Note that this is not the case for all quasi-isotropic laminates, e.g., a [0/±45/90]$_s$ laminate.

4.4.1.3   Thermomechanical loading on a quasi-isotropic laminate [60/–60/0]$_s$

Consider the same 6-ply, quasi-isotropic, symmetric IM7/8552 laminate, [60/–60/0]$_s$, with 0.15 mm thick plies, loaded with $N_x = 300$ N/mm, but now with a superimposed temperature change, $\Delta T = 100$ °C or –100 °C. For the positive temperature change, this laminate problem can be defined in LamProblemDef.m as follows:

```
NP = NP + 1;
OutInfo.Name(NP) = "Section 4.4.1.3 - [60,-60,0]s Laminate Nx = 300, DT = 100";
Geometry{NP}.Orient = [60,-60,0,0,-60,60];
nplies = length(Geometry{NP}.Orient);
Geometry{NP}.plymat(1:nplies) = 1;
Geometry{NP}.tply(1:nplies) = 0.15;
Loads{NP}.DT = 100.;
Loads{NP}.Type  = [ NM,  NM, NM, NM, NM, NM];
Loads{NP}.Value = [300,   0,  0,  0,  0,  0];
Loads{NP}.CriteriaOn = [1, 1, 1, 1];
```

For both temperature change values, the results are given in Figs. 4.12 and 4.13. In the case of $\Delta T = 100$ °C, compared to the case without a temperature change (Section 4.4.1.2), the allowable load has increased from 363.67 N/mm to 497.43 N/mm. This is mainly because the addition of the thermal load significantly decreased the transverse stress in the 60 degree plies. Conversely, in the case of $\Delta T = -100$°C, the allowable load has decreased from 363.67 N/mm to 206.94 N/mm. This is mainly because the addition of the negative thermal load significantly increased the transverse stress in the 60 degree plies. Note that the allowable loads are calculated based on scaling the applied mechanical loading, while the applied thermal load is not scaled (see Section 4.2.2).

In practice, when a thermal load increases MoS, as in the case of $\Delta T = 100$°C, it is common to ignore (or knockdown) the increase provided by the thermal load. This condition is referred to as "thermal help". In contrast, when the thermal load decreases the MoS, (the case of "thermal hurt") as in the case of $\Delta T = -100$°C above, the full effect of the thermal load is utilized. Of course, positive temperature change is *not* always helpful; this is dependent on the material properties, laminate layup, and loading/boundary conditions.

Ply stresses through the thickness

Z-location	Angle	SigmaX	SigmaY	TauXY		Sigma11	Sigma22	Tau12
-0.45	60	51.2307	12.003	35.5188		52.5701	10.6636	-34.7455
-0.3	60	51.2307	12.003	35.5188		52.5701	10.6636	-34.7455
-0.3	-60	51.2307	12.003	-35.5188		52.5701	10.6636	34.7455
-0.15	-60	51.2307	12.003	-35.5188		52.5701	10.6636	34.7455
-0.15	0	897.5385	-24.006	1.8832e-15		897.5385	-24.006	1.8832e-15
0	0	897.5385	-24.006	8.4914e-16		897.5385	-24.006	8.4914e-16
Symmetric								

Ply margins of safety through the thickness

Z-location	Angle	Criterion	MoS or MoS11	MoS22	MoS12
-0.45	60	Maxstress	81.2734	1.1796	1.5327
		Maxstrain	136.6462	1.2262	1.5327
		Tsai-Hill	0.76182		
		Tsai-Wu	0.65809		
-0.3	60	Maxstress	81.2734	1.1796	1.5327
		Maxstrain	136.6462	1.2262	1.5327
		Tsai-Hill	0.76182		
		Tsai-Wu	0.65809		
-0.3	-60	Maxstress	81.2734	1.1796	1.5327
		Maxstrain	136.6462	1.2262	1.5327
		Tsai-Hill	0.76182		
		Tsai-Wu	0.65809		
-0.15	-60	Maxstress	81.2734	1.1796	1.5327
		Maxstrain	136.6462	1.2262	1.5327
		Tsai-Hill	0.76182		
		Tsai-Wu	0.65809		
-0.15	0	Maxstress	1.633	121.43	99999
		Maxstrain	1.6246	12.1955	99999
		Tsai-Hill	1.3446		
		Tsai-Wu	1.6742		
0	0	Maxstress	1.633	121.43	99999
		Maxstrain	1.6246	12.1955	99999
		Tsai-Hill	1.3446		
		Tsai-Wu	1.6742		
Symmetric					

Figure 4.12 Stresses and MoS for a quasi-isotropic laminate [60/−60/0]$_s$, loaded in the x-direction, $N_x = 300$ N/mm and $\Delta T = 100$ °C.

4.4.1.4   Ply-level failure envelopes

As discussed in Section 4.1.4, failure envelope quantifies the loads at which "failure" is first expected to occur in a given loading space (e.g., in the N_x-N_y loading plane). Points inside the failure envelope represent loading values that would not cause failure, whereas, outside the failure envelope, the loading value would cause failure. The MATLAB code can be enhanced to enable generation of failure envelopes by defining a number of problems in LamProblemDef.m that probe the desired loading space. For example, for failure envelopes in the N_x-N_y loading plane, the angle in this plane is defined as,

$$ang = \tan^{-1}\left(\frac{N_y}{N_x}\right)$$

(4.59)

Ply stresses through the thickness

Z-location	Angle	SigmaX	SigmaY	TauXY		Sigma11	Sigma22	Tau12
-0.45	60	75.8651	-12.6314	-7.1492		3.3014	59.9323	-34.7455
-0.3	60	75.8651	-12.6314	-7.1492		3.3014	59.9323	-34.7455
-0.3	-60	75.8651	-12.6314	7.1492		3.3014	59.9323	34.7455
-0.15	-60	75.8651	-12.6314	7.1492		3.3014	59.9323	34.7455
-0.15	0	848.2698	25.2627	9.4099e-16		848.2698	25.2627	9.4099e-16
0	0	848.2698	25.2627	2.6203e-16		848.2698	25.2627	2.6203e-16
Symmetric								

Ply margins of safety through the thickness

Z-location	Angle	Criterion	MoS or MoS11	MoS22	MoS12
-0.45	60	Maxstress	83.037	-0.21623	1.5327
		Maxstrain	140.5544	-0.21772	1.5327
		Tsai-Hill	-0.27778		
		Tsai-Wu	-0.31019		
-0.3	60	Maxstress	83.037	-0.21623	1.5327
		Maxstrain	140.5544	-0.21772	1.5327
		Tsai-Hill	-0.27778		
		Tsai-Wu	-0.31019		
-0.3	-60	Maxstress	83.037	-0.21623	1.5327
		Maxstrain	140.5544	-0.21772	1.5327
		Tsai-Hill	-0.27778		
		Tsai-Wu	-0.31019		
-0.15	-60	Maxstress	83.037	-0.21623	1.5327
		Maxstrain	140.5544	-0.21772	1.5327
		Tsai-Hill	-0.27778		
		Tsai-Wu	-0.31019		
-0.15	0	Maxstress	1.6895	43.0259	99999
		Maxstrain	1.6991	15.3515	99999
		Tsai-Hill	1.3491		
		Tsai-Wu	1.4954		
-5.5511e-17	0	Maxstress	1.6895	43.0259	99999
		Maxstrain	1.6991	15.3515	99999
		Tsai-Hill	1.3491		
		Tsai-Wu	1.4954		
Symmetric					

Figure 4.13 Stresses and MoS for a quasi-isotropic laminate $[60/-60/0]_s$, loaded in the x-direction, $N_x = 300$ N/mm and $\Delta T = -100\ °C$.

then the loading at a given angle can be expressed as,

$$N_x = \cos(ang) \qquad N_y = \sin(ang) \qquad\qquad (4.60)$$

By defining a number of problems in LamProblemDef.m with the identical laminate, but with the angle (*ang*) changing incrementally from 0° to 360°, the failure envelope is defined by plotting the allowable loads based on the scaled prescribed loads.

The failure envelope data is written to a file with extension ".EnvData", the name of which is placed in the variable OutInfo.EnvFile. A standard output file for each angle (problem) is written, and if OutInfo.EnvFile exists, then the MATLAB function WriteLamMoS.m writes the laminate allowable loads, controlling ply angle, controlling failure criterion, and, if applicable, controlling load component, for each angle [Eq. (4.59)] in the ".EnvData" file. Then this file can be used to plot the failure envelops. It is recommended to use text output only and turn off the plotting (both controlled in LamProblemDef.m) when generating failure envelops. Note that, if more than one failure envelope is generated, all data will be written sequentially to the same ".EnvData" file.

Considering a single [0] layer, IM7/8552 laminate, of thickness 1 mm, the N_x-N_y loading plane failure envelope data for the max stress, max strain, Tsai-Hill, and Tsai-Wu failure criteria can be generated using the following code in LamProblemDef.m:

```
% -- Laminate Failure Envelopes (Section 4.4.1.4) --
%    Comment and uncomment to execute this section examples
OutInfo.MakePlots = false; % -- Ensure plotting is off for envelope runs
tttt = datetime(datetime,'Format','yyyy-MMM-dd HH.mm.ss');
OutInfo.EnvFile = ['Output/', 'Composite ENVELOPE - ', char(tttt), '.EnvData'];
ang = 0;
anginc = 5;
anginc1 = anginc;
while (ang <= 360.1)

    % -- Extra angles for highly anisotropic cases
    if ang < 9.9 || (ang >= 170 && ang <= 190) || ang >= 350
      anginc = 0.1;
    else
      anginc = anginc1;
    end

    NP = NP + 1;
    OutInfo.Name(NP) = string(['Ang = ', num2str(ang)]);
    Geometry{NP}.Orient = [0];
    %Geometry{NP}.Orient = [60,-60,0,0,-60,60];
    nplies = length(Geometry{NP}.Orient);
    for k = 1:nplies
        Geometry{NP}.plymat(k) = 1;
        Geometry{NP}.tply(k) = 1.;
        %Geometry{NP}.tply(k) = 0.15;
    end

    NY = sin(ang*pi/180);
    NX = cos(ang*pi/180);
    Loads{NP}.Type  = [NM, NM, NM, NM, NM, NM];
    Loads{NP}.Value = [NX, NY,  0,  0,  0,  0];
    Loads{NP}.DT = 0.0;
    %Loads{NP}.DT = 100.0;
    %Loads{NP}.DT = -100.0;
    % -- Turn criteria on/off to generate all envelopes in Fig. 4.14
    Loads{NP}.CriteriaOn = [1,1,1,1];
    %Loads{NP}.CriteriaOn = [1,0,0,0];
    Loads{NP}.ang = ang;
    ang = ang + anginc;
end
```

The failure envelopes are shown in Fig. 4.14. The envelope for each criterion is generated by turning off all other failure criteria while running the LamAnalysis.m

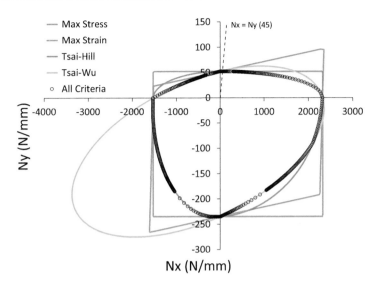

Figure 4.14 N_x-N_y failure envelopes for a single layer, [0], IM7/8552 laminate, of thickness 1 mm.

MATLAB script. Note that the axis scales in Fig. 4.14 are vastly different due to the large difference in the ply properties and allowables in the fiber and transverse directions. In the first quadrant, the line representing $N_x = N_y$ has been plotted, which corresponds to an angle, *ang* $= 45°$. Clearly, most of the envelopes (as plotted) represent angles that are close to the horizontal N_x-axis. To better define the envelopes, extra angles have been added when the angle is close to the horizontal axis, as shown in the MATLAB code above.

As shown in Fig. 4.14, if all failure criteria are active, the resulting failure envelope is the inner intersection of each individual criterion failure envelope, as denoted by the open circle symbols. Similarly, Fig. 4.15 shows the failure envelopes for the 6-ply quasi-isotropic laminate considered previously. These envelopes can be generated by uncommenting/commenting the appropriate lines in the MATLAB code given above. For the quasi-isotropic case, the axis scales are similar, so no extra angles are needed (and the appropriate section of the MATLAB code given above can be commented).

Next, still considering the case of the 6-ply, quasi-isotropic, symmetric IM7/8552 laminate, [60/–60/0]$_s$, with typical 0.15 mm thick plies, three failure envelopes are shown in Fig. 4.16 wherein only the max stress criterion is employed. Three cases are depicted: (1) with no thermal preload, (2) with a thermal preload of 100°C, and (3) with a thermal preload of -100°C. Clearly, a thermal preload results in changes to the failure envelopes, with a negative ΔT preload resulting in a smaller failure envelope and a positive ΔT preload resulting in a larger failure envelope. The failure envelope has not simply been shifted by the thermal preload. Rather, each portion of the failure envelope has been translated, with the magnitude and direction of the translation dependent on the mode (axial, transverse, or shear) of the failure initiation. For the

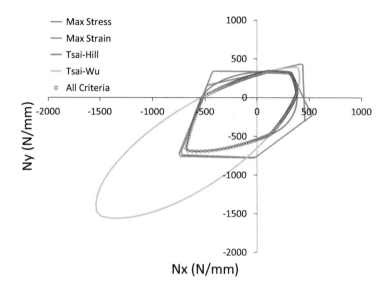

Figure 4.15 N_x-N_y failure envelopes for a 6-ply, quasi-isotropic, symmetric IM7/8552 laminate, [60/–60/0]$_s$, with typical 0.15 mm thick plies.

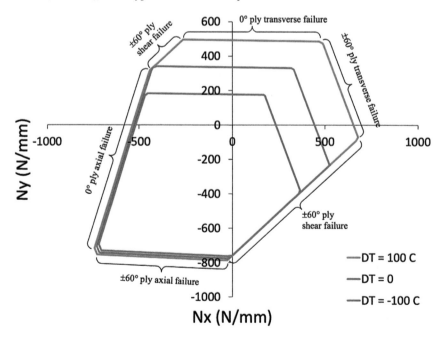

Figure 4.16 N_x-N_y max stress criterion failure envelopes for a 6-ply, quasi-isotropic, symmetric IM7/8552 laminate, [60/–60/0]$_s$, with typical 0.15 mm thick plies with and without thermal preloads (DT $= \Delta T$).

$\Delta T = 100\,°C$ case, Fig. 4.16 also includes the ply responsible for the failure, as well as the mode associated with the ply's failure. Clearly, when the mode is transverse failure (mainly quadrants I and II), the thermal preload results in a significant translation of the failure envelope in the $+N_x$ and $+N_y$ directions. In contrast, when the failure mode is axial or shear, little or no change in the failure envelope is evident from the thermal preload. This is because the thermal preload results in stresses that are quite significant compared to the ply transverse tensile stress allowable (52.3 MPa). The ply axial stress allowables are much greater (2323 MPa and -1531 MPa) so the preload stresses are much smaller in comparison and only translate the failure envelope slightly when ply axial failure is controlling. The thermal preload results in zero ply-level shear stresses (in ply material coordinates) for this $[60/{-}60/0]_s$ laminate, and thus has no effect on the failure envelopes when shear failure is controlling (see Quadrant 4 in Fig. 4.16). It is also noteworthy that, when axial failure is controlling, the failure envelope translation is in the opposite direction (the $-N_x$ and $-N_y$ directions) compared to when transverse failure is controlling. This is because the thermal preload results in tension in one direction and compression in the other direction in each ply.

4.4.2 Constituent level MoS in stand-alone micromechanics analyses

As shown in Chapter 3, micromechanics theories provide the local constituent level stresses and strains within a composite. Just as was done with the ply-level stresses in the previous section, the local stresses within a composite can be used to evaluate failure criteria, given allowables for the constituent materials.

4.4.2.1 MoS results for given constituent allowables

To illustrate the localization of MoS calculations to the constituent (i.e., fiber and matrix) level, consider the laminate case from Section 4.4.1.1: the single layer, [0] IM7/8552 laminate, of thickness 1 mm loaded in the fiber direction, that is, the x-direction, subjected to a 1000 MPa load, that is, $N_x = 1000$ N/mm. For the micromechanics calculations, a $v_f = 0.55$ has been assumed. This same unidirectional composite can be analyzed using MATLAB Script 3.5 – MicroAnalysis.m with modifications as outlined in Fig. 4.4A. That is, the constituent allowables (see Table 4.3 and the example code given in Section 4.3) are added to the constituent material properties in GetConstitProps.m (see MATLAB Function 3.9). Otherwise, the problem is defined in MicroProblemDef.m (see MATLAB Function 3.6):

```
NP = NP + 1;
OutInfo.Name(NP) = "Section 4.4.2.1 - MT IM7-8552 Vf=0.55 - S11";
MicroMat{NP} = 102;
Loads{NP}.DT = 0;
Loads{NP}.Type  = [   S, S,   S,   S,   S,   S];
Loads{NP}.Value = [1000, 0,   0,   0,   0,   0];
Loads{NP}.CriteriaOn = [1, 1, 1, 1];
```

Table 4.3 Representative constituent-stress allowables.

Mat	Material name	XT (MPa)	XC (MPa)	YT (MPa)	YC (MPa)	Q (MPa)	R (MPa)	S (MPa)
1	IM7	4335	−2608	113	−354	128	138	138
2	Glass	TBD – see below						
3	8552	59.4	−259	59.4	−259	112	112	112
4	Epoxy	TBD – see below						

*Note all constituents are assumed transversely isotropic so ZT = YT and ZC = YC.

As mentioned previously, to write the margins to the output file, Out-Info.WriteMargins must be set equal to true near the top of MicroProblemDef.m, and the active failure criteria are specified using Loads{NP}.CriteriaOn. The micromechanics-based composite material 102 is defined in GetEffProps.m as:

```
Mat = 102;
name = "MT IM7-8552 Vf = 0.55";
Theory = 'MT';
Vf = 0.55;
Constits.Fiber = constitprops{1};
Constits.Matrix = constitprops{3};
[props{Mat}] = RunMicro(Theory, name, Vf, Constits, props{Mat});
```

As mentioned, to apply the above failure theories to the constituent materials above (#1 – IM7 fiber, #3 – 8552 matrix), constituent material allowables are required, see Table 4.3. It should be noted that, in practice, obtaining these allowables is often difficult as the constituent materials sometimes fail differently within a composite than they do when tested in monolithic form. Consequently, often times they are backed out using a given micromechanics model by comparing the simulation to a measured composite response for a variety of composite layups and test conditions.

Utilizing the Mori-Tanaka (MT) micromechanics theory, the resulting MoS for the four failure criteria are given in Fig. 4.17 for the unidirectional IM7/8552 problem defined above. MoS values are output for each phase (or in the case of MOC – each subcell), as is a table of MoS associated with each criterion. Examining these tables, the controlling criterion is Tsai-Wu, MoS = 0.59275 (highlighted), in the matrix constituent. Consequently, based on the rounded minimum Tsai-Wu MoS of 0.593, the scaled load, $\bar{\sigma}_{11} = 1000$ MPa $\times$ 1.593 = 1593 MPa, would cause the local constituent stresses to meet the allowable stresses of the matrix constituent according to the Tsai-Wu criterion. It should be noted that the MT method provides the average stress in the matrix, thus it is this quantity that is being evaluated in the failure criteria at the micro scale.

The Method of Cells (MOC) micromechanics theory can be used to analyze the same composite by specifying (in MicroProblemDef.m) the micromechanics-based composite material 103, which is defined in GetEffProps.m as:

Micro Scale Margins of Safety

Maxstress

Subcell or Mat	MoS11	MoS22	MoS33	MoS23	MoS13	MoS12
F	1.4219	146.1051	146.1051	99999	99999	99999
M	0.71966	19.1958	19.1958	99999	99999	99999

Maxstrain

Subcell or Mat	MoS11	MoS22	MoS33	MoS23	MoS13	MoS12
F	1.4208	21.6994	21.6994	99999	99999	99999
M	0.86239	17.5982	17.5982	99999	99999	99999

Tsai-Hill

Subcell or Mat	MoS
F	1.4186
M	0.87972

Tsai-Wu

Subcell or Mat	MoS
F	1.4168
M	0.59275

Figure 4.17 MoS for a unidirectional 0.55 volume fraction IM7/8552 composite loaded in the x_1-direction, $\bar{\sigma}_{11} = 1000$ MPa, using the MT micromechanics theory.

```
Mat = 103;
name = "MOC IM7-8552 Vf = 0.55";
Theory = 'MOC';
Vf = 0.55;
Constits.Fiber = constitprops{1};
Constits.Matrix = constitprops{3};
[props{Mat}] = RunMicro(Theory, name, Vf, Constits, props{Mat});
```

The resulting MoS for the four failure criteria are given in Fig. 4.18. Now margins are provided for each subcell ($\beta\gamma$), where subcell 11 is associated with the fiber (F) and the remaining three subcells (12, 21, and 22) are associated with the matrix phase (M). Again, Tsai-Wu is the controlling criterion, with failure occurring in the matrix subcell 22, and the associated minimum MoS being 0.38009. The scaled failure stress, $\bar{\sigma}_{11} = 1380$ MPa, is 13% lower than that determined using MT (1593 MPa). The important distinction between the MOC and MT theories is that the MOC provides multiple points within the matrix, capturing some of the stress and strain gradients that exist. This enables assessment of the failure criteria at various points in the matrix. Subcell (22) had the lowest MoS and thus would be expected to "fail" before the other two matrix subcells. The ability to capture such gradient features is important in accurately predicting nonlinear material response resulting from both inelastic and damage, see Aboudi et al. (2013) and Chapter 7.

Of course, exceeding the allowable stress in the matrix due to axial composite loading might not be treated as failure in practice, as the strong and stiff fibers would

Microscale Margins of Safety

Maxstress

Subcell or Mat	MoS11	MoS22	MoS33	MoS23	MoS13	MoS12
11 – F	1.4215	170.5621	170.5621	99999	99999	99999
12 – M	0.76604	12.284	421.9003	99999	99999	99999
21 – M	0.76604	421.9003	12.284	99999	99999	99999
22 – M	0.59557	9.0296	9.0296	99999	99999	99999

Maxstrain

Subcell or Mat	MoS11	MoS22	MoS33	MoS23	MoS13	MoS12
11 – F	1.4205	22.0973	22.0973	99999	99999	99999
12 – M	0.86219	23.9317	13.5832	99999	99999	99999
21 – M	0.86219	13.5832	23.9317	99999	99999	99999
22 – M	0.86219	18.1919	18.1919	99999	99999	99999

Tsai-Hill

Subcell or Mat	MoS
11 – F	1.4187
12 – M	1.026
21 – M	1.026
22 – M	0.89742

Tsai-Wu

Subcell or Mat	MoS
11 – F	1.4172
12 – M	0.67269
21 – M	0.67269
22 – M	0.38009

Figure 4.18 MoS for a unidirectional 0.55 volume fraction IM7/8552 composite loaded in the x_1- direction, $\bar{\sigma}_{11} = 1000$ MPa, using the MOC micromechanics theory.

remain intact. Rather, axial failure of the composite might be defined as the point at which the fibers fail, regardless of the matrix stresses. It is possible, essentially, to turn off the matrix failure in the axial direction by using an artificially high value for the matrix axial allowable stresses, XT and XC. For example, if the values of these allowables are changed from 59.4 MPa and -259 MPa to 1 million MPa and -1 million MPa, respectively, (in GetConstitProps.m – see commented lines) the resulting Tsai-Wu MoS are shown in Fig. 4.19. Clearly, by ignoring any possibility of matrix failure in the x-direction, the minimum MoS is now in the fiber for both MT and MOC without changing the controlling failure criterion (Tsai-Wu). Now the minimum MoS values are in close agreement between MT and MOC, and also in reasonable agreement with the ply level MoS (1.417 vs. 1.323) determined in Section 4.4.1.1.

The difference between MT and MOC is even more vivid when the loading is transverse to the fiber direction, i.e., in the x_2-direction. Consider an applied loading $\bar{\sigma}_{22} = 10.0$ MPa to the same unidirectional composite considered above. The MoS results associated with the Tsai-Wu criterion for both MT and MOC are shown in

Tsai-Wu MoS – MT

Subcell or Mat	MoS
F	1.4168
M	11.0383

Tsai-Wu MoS – MOC

Subcell or Mat	MoS
11	1.4172
12	13.0527
21	13.0527
22	4.9778

Figure 4.19 Tsai-Wu criterion MoS for a unidirectional 0.55 volume fraction IM7/8552 composite loaded in the x_1-direction, $\bar{\sigma}_{11} = 1000$ MPa, using the MT and MOC micromechanics theory with the matrix x_1-direction allowable stresses (XT and XC) assigned very high magnitudes.

Tsai-Wu MoS – MT

Subcell or Mat	MoS
F	9.8361
M	3.7187

Tsai-Wu MoS – MOC

Subcell or Mat	MoS
11	9.7647
12	5.9896
21	2.5462
22	3.688

Figure 4.20 Tsai-Wu MoS results for a unidirectional 0.55 volume fraction IM7/8552 composite subjected to $\bar{\sigma}_{22} = 10$ MPa, using the MT and MOC micromechanics theory.

Fig. 4.20. Clearly, while both MT and MOC indicate that the minimum Tsai-Wu MoS is in the matrix constituent, the minimum MoS values are quite different (3.72 vs. 2.55). These MoS values correspond to composite allowable stresses of $\bar{\sigma}_{22} = 47.2$ MPa and 35.5 MPa for MT and MOC, respectively. This difference is due to the fact that MOC captures the stress concentration that occurs between fibers when the composite is loaded in the transverse direction. This difference is evident when the MT- and MOC-stress fields were compared in Fig. 3.9. Note that, unidirectional composites, when subjected to transverse tensile loading (as in this case), tend to exhibit brittle failures, with the failure initiation and final failure loads being very close to each other.

4.4.2.2 Backed-out constituent allowables

Allowable stresses are usually measured at the composite level. As such, constituent allowables can be backed out that provide composite allowable loads equal to the measured composite level allowables. Consider, for example, the known IM7/8552 ply level transverse tensile allowable stress, 52.3 MPa, given in Table 4.2. One can match this allowable by altering the values of XT, YT, and ZT of the 8552 constituent material in GetConstitProps.m (see MATLAB Function 3.9), while applying global transverse tension, $\bar{\sigma}_{22}$. However, this backed out matrix tensile allowable is micromechanics

MT – XT = 65 MPa

S11	S22	S33	S23	S13	S12
0	52.5794	0	0	0	0

MOC – XT = 80 MPa

S11	S22	S33	S23	S13	S12
0	51.9531	0	0	0	0

Figure 4.21 Allowable stresses based on scaling the prescribed loads applied to a unidirectional 0.55 volume fraction IM7/8552 composite subjected to $\bar{\sigma}_{22} = 10$ MPa, using the MT and MOC micromechanics theory.

Table 4.4 Results for a unidirectional IM7/8552 composite subjected to $\bar{\sigma}_{22} = 10$ N/mm and ΔT of 0 and $\pm 100°C$, using the MT and MOC micromechanics theories, with adjusted matrix tensile stress allowables.

	MT			MOC		
	Min. MoS	Controlling	Allowable	Min. MoS	Controlling	Allowable
ΔT (°C)	(Tsai-Wu)	Mat	$\bar{\sigma}_{22}$ (MPa)	(Tsai-Wu)	Subcell	$\bar{\sigma}_{22}$ (MPa)
100	6.44	M	74.4	4.99	2,1	59.9
0	4.26	M	52.6	4.20	2,1	52.0
–100	1.67	M	26.7	1.94	2,2	29.4

theory dependent, since the different micromechanics theories predict different local stresses.

To the nearest MPa, the backed out values of the matrix tensile allowable stress are 65 MPa and 80 MPa for the MT and MOC theories, respectively. The resulting allowable loads for the unidirectional composite are shown in Fig. 4.21 when these values are used for XT = YT = ZT in GetConstitProps.m (see commented lines). Again, because the MOC theory is able to approximate the stress concentrations in the matrix, the matrix tensile stress allowable required to match the ply level stress allowable is higher than that required for the MT method, which provides a single average stress value for the matrix. It should be noted that, in this example, only the ply-level transverse tensile allowable has been matched by backing out a new matrix tensile allowable. The other ply-level allowables have not been addressed.

4.4.2.3 Influence of thermal loading on MoS

Next consider the effect of an applied temperature change on the calculated MoS using micromechanics. Keeping the above method-specific backed out values for the matrix tensile allowable stress and applying a ΔT of 100 °C and -100 °C (in addition to the applied $\bar{\sigma}_{22} = 10$ MPa), results are compared to the case with $\Delta T = 0$ in Table 4.4. As in the quasi-isotropic laminate results (see Section 4.4.1.3), the positive ΔT results in an increase in the allowable load, whereas the negative ΔT results in a decrease. It should be noted that, although the same magnitude of ΔT has been applied (± 100 °C), the shift in the allowable $\bar{\sigma}_{22}$ is not the same (e.g., for MT, $\Delta T=100$ °C: $74.4 - 52.6 = 21.8$ MPa; $\Delta T= -100$ °C: $52.6 - 26.7$ MPa $= 25.9$ MPa). This is because the shift caused by the thermal preload is not aligned with the applied mechanical load path (see Fig. 4.3). In addition, for the MOC, the controlling subcell changes for the $\Delta T = -100$ °C case, see Table 4.4.

4.4.3 Micromechanics-based failure envelopes using stand-alone micromechanics analyses

As done at the laminate level in Section 4.4.1.4, micromechanics-based failure envelopes can be generated using the MATLAB code. Here, MT and MOC failure envelopes are compared for the max stress, max strain, and Tsai-Wu failure criteria. Note that the Tsai-Hill failure criterion has been purposely omitted due to the known issue of discontinuities associated with transitioning between tension and compression (see Eq. (4.5)). A 0.55 fiber volume fraction unidirectional IM7/8552 composite is considered, where the constituent allowables given in Table 4.3 have been employed.

The following MATLAB code (which resides in `MicroProblemDef.m`) generates the micromechanics-based failure envelopes for this example:

```
% -- Example micro scale failure envelope - Section 4.4.3
tttt = datetime(datetime,'Format','yyyy-MMM-dd HH.mm.ss');
OutInfo.EnvFile = ['Output/', 'Composite ENVELOPE - ', char(tttt), '.EnvData'];
ang = 0;
anginc = 5;
anginc1 = anginc;

while (ang <= 360.1)

    anginc = anginc1;

    % -- Extra angles for highly anisotropic cases
%   if ang < 15 || (ang >= 165 && ang <= 190) || ang >= 345
%       anginc = 0.2;
%   end

    NP = NP + 1;
    OutInfo.Name(NP) = string(['Ang = ', num2str(ang)]);
    MicroMat{NP} = 102; % -- MT
%    MicroMat{NP} = 103; % --- MOC

    S33 = sin(ang*pi/180);
    S22 = cos(ang*pi/180);
    % Define laminate applied loading and temperature change
    Loads{NP}.S = S;
    Loads{NP}.E = E;
    Loads{NP}.Type  = [S,   S,   S,   S,   S, S];
    Loads{NP}.Value = [0, S22, S33,   0,   0,  0];

    Loads{NP}.DT = 0.0;

    % Turn off individual criteria
    Loads{NP}.CriteriaOn = [0,0,1,0];
    Loads{NP}.ang = ang;
    ang = ang + anginc;
end
```

Fig. 4.22 compares the $\bar{\sigma}_{22}$-$\bar{\sigma}_{33}$ failure envelopes predicted by MT and MOC. Note that the $\bar{\sigma}_{22}$-$\bar{\sigma}_{33}$ envelope involves only transverse normal loading. The MT and MOC failure envelopes coincided whenever both theories predict that the fiber is the controlling constituent. When the matrix is controlling, as expected, the MOC envelope is inside the MT envelope because the MOC approximates stress gradients in the matrix (due

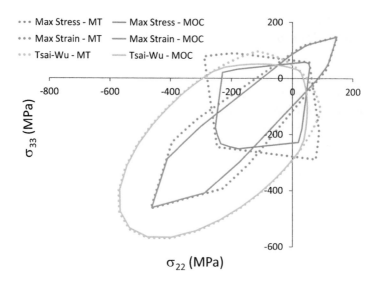

Figure 4.22 Micromechanics-based failure envelopes for a 0.55 fiber volume fraction unidirectional IM7/8552 composite in the $\bar{\sigma}_{22}$–$\bar{\sigma}_{33}$ plane, using the MT and MOC methods.

to its three matrix subcells). The largest difference between MT and MOC is exhibited when using the max stress criterion because MOC predicts that the matrix constituent is controlling for the entire envelope. It is interesting that the Tsai-Wu criterion is the most conservative in Quadrant I (tension-tension) and the least conservative in Quadrant III (compression-compression). Of course, these observations are highly dependent on the constituent allowables and the loading plane being examined.

4.4.4 Laminate micromechanics-based MoS for a quasi-isotropic laminate [60/–60/0]ₛ

Consider the case of a 6-ply, quasi-isotropic, symmetric 0.55 fiber volume fraction IM7/8552 laminate, [60/–60/0]ₛ, with 0.15 mm thick plies, loaded only in the x-direction, $N_x = 300$ N/mm, but now assessing the MoS at the micro scale. Note that *the constituent allowables from* Table 4.3 *have been utilized* rather than the backed out matrix allowables discussed in the previous section. The resulting Tsai-Wu MoS are shown in Fig. 4.23. Note that this laminate was previously modeled at the ply level, with results given in Fig. 4.10. As was the case when modeling the laminate at the ply level, for both MT and MOC, the ±60° plies are controlling, and the controlling failure criterion is Tsai-Wu. From the ply level analysis, the allowable N_x load was 363.67 N/mm. Here, at the micro scale, based on MT, the allowable N_x load is 371.08 N/mm, whereas, based on MOC, the allowable N_x load is 275.53 N/mm. As shown previously, the MOC allowable load (and controlling MoS) is lower than that of MT because MOC approximates stress concentrations that are present in the matrix

Tsai-Wu – MT

Z-location	Subcell or Mat	MoS
-0.45	F	1.2093
	M	0.23692
-0.3	F	1.2093
	M	0.23692
-0.3	F	1.2093
	M	0.23692
-0.15	F	1.2093
	M	0.23692
-0.15	F	1.7373
	M	0.83838
-5.5511e-17	F	1.7373
	M	0.83838
Symmetric		

Tsai-Wu – MOC

Z-location	Subcell or Mat	MoS
-0.45	11	1.2855
	12	0.80873
	21	-0.081564
	22	0.26195
-0.3	11	1.2855
	12	0.80873
	21	-0.081564
	22	0.26195
-0.3	11	1.2855
	12	0.80873
	21	-0.081564
	22	0.26195
-0.15	11	1.2855
	12	0.80873
	21	-0.081564
	22	0.26195
-0.15	11	1.7267
	12	0.91816
	21	0.95361
	22	0.59472
-5.5511e-17	11	1.7267
	12	0.91816
	21	0.95361
	22	0.59472
Symmetric		

Figure 4.23 MoS for a quasi-isotropic 0.55 fiber volume fraction IM7/8552 laminate $[60/-60/0]_s$, loaded in the x-direction, $N_x = 300$ N/mm, with plies modeled using the MT and MOC methods.

while MT does not. Of course, the matrix allowable stresses could be adjusted to match the ply level analysis results for the laminate for MT and MOC.

4.5   Summary

This chapter addresses failure of unidirectional composites and laminates by deter-mining the load at which allowables are first exceeded at some point in the composite. Stress and strain allowables (for either effective ply materials or constituent materials) quantify limits on the stress and strain components, below which the material can be expected to safely operate (with no loss of structural integrity). The max stress, max strain, Tsai-Hill, and Tsai-Wu failure criteria have been employed to assess failure, at the ply level in laminates, at the constituent scale in unidirectional composites, and at the constituent scale within laminates. The equations enabling calculation of Margins of Safety (MoS) for each of these criteria were presented, both in the case of pure mechanical loading and in the case of thermomechanical loading, wherein the thermal loading is treated as a preload. The MATLAB implementation was then presented enabling calculation of the laminate or composite allowable loads by proportionally scaling the applied loading to the point at which failure would first occur somewhere within the laminate or composite. Finally, example problems were presented that illustrate the ability to assess failure of laminates and unidirectional composites based on effective ply material allowables and constituent allowables.

4.6 Exercises

Exercise 4.1. Determine P_1^{mech}, P_1^{th}, P_2^{mech}, $P_2^{mech-th}$, and P_2^{th} in Eqs. (4.52) to (4.54) for the max stress criterion in the x_1-direction. For the max stress criterion in the x_1-direction, show that Eq. (4.58) reduces to Eq. (4.45).

Exercise 4.2. The safety factor, SF, discussed in Section 4.2, can be included in the MoS calculation by multiplying the actual stress (or strain) by SF [see Eq. (4.11)]. The safety factor is only applicable to the mechanical loading, not the thermal preload. Derive an equation analogous to Eq. (4.58) including the factor of safety, SF.

Exercise 4.3. Consider a regular cross-ply IM7/8552 laminate $[0/90]_{6s}$ (with ply-level allowables given in Table 4.2) with ply thicknesses of 0.15 mm and applied loading of $N_x = 1000$ N/mm (and all other force and moment resultants $= 0$). Determine the minimum MoS for the max stress, max strain, Tsai-Hill, and Tsai-Wu failure criteria (based on ply-level stresses), along with the ply orientation with the minimum MoS. For each criterion's minimum MoS, determine the laminate scaled allowable load.

Exercise 4.4. Consider a regular cross-ply glass/epoxy laminate $[0/90]_{6s}$ (with ply-level allowables given in Table 4.2) with ply thicknesses of 0.15 mm and applied loading of $N_x = 1000$ N/mm (and all other force and moment resultants $= 0$). Determine the minimum MoS for the max stress, max strain, Tsai-Hill, and Tsai-Wu failure criteria (based on ply-level stresses), along with ply orientation with the minimum MoS. For each criterion's minimum MoS, determine the scaled allowable load. Why are the MoS and scaled allowable loads lower than in the previous problem?

Exercise 4.5. Consider a regular cross-ply laminate $[0/90]_{6s}$ with ply thicknesses of 0.15 mm and applied loading of $N_{xy} = 400$ N/mm (and all other force and moment resultants $= 0$). Determine the allowable load for both IM7/8552 and glass/epoxy laminates (with ply-level allowables given in Table 4.2) such that all four failure criteria discussed in this chapter provide all positive margins (based on ply-level stresses). Why is the glass/epoxy laminate allowable load higher than that of the IM7/8552?

Exercise 4.6. Consider a quasi-isotropic IM7/8552 laminate $[-45/0/45/90]_s$ (with ply-level allowables given in Table 4.2) with ply thicknesses of 0.15 mm and applied loading of $M_x = 100$ N-mm/mm (and all other force and moment resultants $= 0$), determine the scaled allowable load such that all four failure criteria discussed in this chapter provide all positive margins (based on ply-level stresses). Repeat this problem for a layup of $[-45/90/45/0]_s$. Which laminate has a higher allowable load and why?

Exercise 4.7. Consider a regular cross-ply IM7/8552 laminate $[0/90]_{Ns}$ with ply thicknesses of 0.15 mm (with ply-level allowables given in Table 4.2). Determine the minimum number of plies such that all MoS are positive (according to all four failure criteria based on ply-level stresses) in response to loading of $N_x = 1000$ N/mm, $N_y = 1000$ N/mm (and all other force and moment resultants $= 0$). Plies should be added/subtracted four at a time to maintain symmetry and 0/90 ply pairs. What is the minimum number of plies if the loading is changed to $N_x = -1000$ N/mm, $N_y = -1000$ N/mm? Why is there such a large difference?

Exercise 4.8. Consider a regular cross-ply IM7/8552 laminate $[0/90]_{Ns}$ with ply thicknesses of 0.15 mm (with ply-level allowables given in Table 4.2). Determine the minimum number of plies such that all MoS are positive (according to all four failure criteria based on ply-level stresses) in response to loading of $N_x = 1000$ N/mm, $N_y = 1000$ N/mm, $N_{xy} = 500$ N/mm (and all other force and moment resultants $= 0$). Plies should be added/subtracted four at a time to maintain symmetry and 0/90 ply pairs.

Exercise 4.9. Consider a regular cross-ply IM7/8552 laminate $[0/90]_{5s}$ with ply thicknesses of 0.15 mm (with ply-level allowables given in Table 4.2). Determine the minimum number of $[-45/45]$ and $[45/-45]$ ply pairs that should be added to the outside of the laminate (while maintaining symmetry) such that all MoS are positive (according to all four failure criteria based on ply-level stresses) in response to loading of $N_x = 1000$ N/mm, $N_y = 1000$ N/mm, $N_{xy} = 500$ N/mm (and all other force and moment resultants $= 0$). Compare total number of plies to the Exercise 4.8 solution. Why are so many fewer plies required?

Exercise 4.10. Consider a regular cross-ply glass/epoxy laminate $[0/90]_{Ns}$ with ply thicknesses of 0.15 mm (with ply-level allowables given in Table 4.2). Determine the minimum number of plies such that all MoS are positive (according to all four failure criteria based on ply-level stresses) in response to loading of $N_x = 1000$ N/mm, $N_y = 1000$ N/mm (and all other force and moment resultants $= 0$). Plies should be added/subtracted four at a time to maintain symmetry and 0/90 ply pairs. What is the minimum number of plies if the loading is changed to $N_x = 1000$ N/mm, $N_y = 1000$ N/mm, $N_{xy} = 500$ N/mm? Compare to the IM7/8552 laminate results of Exercises 4.7 and 4.8. Why did the number of plies for the glass/epoxy vs. the IM7/8552 increase by a larger percentage for the case with $N_{xy} = 0$?

Exercise 4.11. Consider a unidirectional IM7/8552 composite subjected to an applied global transverse strain of 0.005. Using all four failure criteria, analyze this composite using (a) CLT with ply level properties/allowables (see Table 4.2) as a single ply with a thickness of 1 mm (apply $\varepsilon_y^0 = 0.005$, all other force and moment resultants $= 0$), (b) the MT micromechanics theory (constituent allowables given in Table 4.3) using a fiber volume fraction of 0.55 (apply $\bar{\varepsilon}_{22} = 0.005$, all other global stress components $= 0$), and (c) the MOC micromechanics theory (constituent properties given in

Table 4.3) using a fiber volume fraction of 0.55 (apply $\bar{\varepsilon}_{22} = 0.005$, all other global stress components $= 0$). Compare the minimum MoS, scaled allowable global strain, and controlling failure criterion across all three approaches. Which analysis provides the lowest scaled allowable applied transverse strain and why?

Exercise 4.12. Consider a regular cross-ply IM7/8552 laminate $[0/90]_s$ with ply thicknesses of 0.15 mm subjected to an applied load of $N_x = 100$ N/mm (and all other force and moment resultants $= 0$). Using all four failure criteria, analyze this laminate with CLT using (a) ply level properties (see Table 4.2), (b) plies modeled with the MT micromechanics theory (with constituent allowables given in Table 4.3), and (c) plies modeled with the MOC micromechanics theory (with constituent allowables given in Table 4.3). Assume a fiber volume fraction of 0.55. Compare the orientation of the controlling ply, the controlling failure criterion, minimum MoS, and scaled allowable load across all three analyses. Which approach provides the lowest allowable load and why.

Exercise 4.13. Consider a regular cross-ply IM7/8552 laminate $[0/90]_s$ with ply thicknesses of 0.15 mm subjected to a negative temperature change, ΔT (with all force and moment resultants $= 0$). Using all four failure criteria, analyze this laminate with CLT using (a) ply level properties (see Table 4.2), (b) plies modeled with the MT micromechanics theory (with constituent allowables given in Table 4.3), and (c) plies modeled with the MOC micromechanics theory (with constituent allowables given in Table 4.3). Assume a fiber volume fraction of 0.55. For each analysis, determine the negative ΔT value that would first cause a negative MoS. Why are these ΔT magnitudes lower for the micromechanics-based cases?

Exercise 4.14. Analyze a regular cross-ply 0.55 fiber volume fraction IM7/8552 laminate $[0/90]_s$ with ply thicknesses of 0.15 mm subjected to an applied load of $N_x = 50$ N/mm, $N_y = 50$ N/mm, $N_{xy} = 50$ N/mm (and all moment resultants $= 0$) using the MT micromechanics theory and the MOC micromechanics theory (constituent allowables given in Table 4.3) to model the plies. Determine the controlling ply, the controlling constituent, the controlling failure criterion, minimum MoS, and scaled allowable load.

Exercise 4.15. Analyze a regular angle-ply 0.55 fiber volume fraction IM7/8552 laminate $[-45/45]_s$ with ply thicknesses of 0.15 mm subjected to an applied load of $N_x = 100$ N/mm (and all other force and moment resultants $= 0$) using the MT micromechanics theory and the MOC micromechanics theory (constituent allowables given in Table 4.3) to model the plies. Determine the controlling ply, the controlling failure criterion, the controlling constituent, the minimum MoS, and the scaled allowable load. Why are the controlling criteria, controlling constituents, and minimum MoS the same as those in Exercise 4.14?

Exercise 4.16. The following constituent stress allowables were backed out for a unidirectional 0.55 fiber volume fraction glass/epoxy composite (see Table 3.4 for constituent mechanical properties) using the MT micromechanics theory and ONLY the Tsai-Wu failure criterion:

MatID	Name	XT (MPa)	XC (MPa)	YT (MPa)	YC (MPa)	Q (MPa)	R (MPa)	S (MPa)
2	Glass	2358	−1653	2358	−1653	1000	1000	1000
4	Epoxy	1×10^6	-1×10^6	42	−181	81	81	81

The epoxy axial stress allowables in tension and compression have been set to high magnitudes to suppress failure of the matrix in the fiber direction so that failure initiation is dictated by the fiber in this direction. Using the MT theory and ONLY the Tsai-Wu criterion, confirm that these allowables provide composite allowable stresses that correspond closely to the ply level glass/epoxy stress allowables given in Table 4.2 (show the scaled allowable stresses). Hint: This requires solving 5 problems sequentially. Apply $+\bar{\sigma}_{11}$ to obtain XT, apply $-\bar{\sigma}_{11}$ to obtain XC, apply $+\bar{\sigma}_{22}$ to obtain YT, apply $-\bar{\sigma}_{22}$ to obtain YC, and apply $\bar{\sigma}_{12}$ to obtain S. To turn off other failure criteria, specify `Loads{NP}.CriteriaOn = [0,0,0,1]` in `MicroProblemDef.m`.

Exercise 4.17. Given the unidirectional 0.55 fiber volume fraction glass/epoxy composite in Exercise 4.16, and using the given stress allowables backed out from MT as a starting point, follow the same procedure as in Exercise 4.16 to back out new epoxy constituent stress allowables using the MOC theory and ONLY the Tsai-Wu criterion such that the allowable loads correspond to the ply level glass/epoxy stress allowables given in Table 4.2 to the nearest MPa. Keep XT $= 1 \times 10^6$ MPa and XC $= -1 \times 10^6$ MPa to preclude axial failure in the matrix. Further assume isotropy in the shear stress allowables, i.e., $Q = R = S$. Hint: Modify the appropriate constituent stress allowables in `GetConstitProps.m`.

Exercise 4.18. Consider a unidirectional glass/epoxy composite. Plot failure envelopes (see Sections 4.4.1.4 and 4.4.3), using only the Tsai-Wu criterion, for this composite in the axial-transverse stress plane using (a) CLT with ply level properties/allowables (see Table 4.2) using a single ply with a thickness of 1 mm (apply N_x and N_y with all other force and moment resultants $= 0$), (b) the MT micromechanics theory (use the constituent stress allowables given in the Exercise 4.16 table) using a fiber volume fraction of 0.55 (apply $\bar{\sigma}_{11}$ and $\bar{\sigma}_{22}$, all other global stress components $= 0$), and (c) the MOC micromechanics theory (use the constituent stress allowables backed out in Exercise 4.17) using a fiber volume fraction of 0.55 (apply $\bar{\sigma}_{11}$ and $\bar{\sigma}_{22}$, all other global stress components $= 0$). Plot all three envelopes on the same axes. Where do all three envelopes match and why? For the MT and MOC envelopes, identify the reason for the slope discontinuity in quadrant I. Hint: the failure envelopes can be easily plotted in Excel.

Exercise 4.19. Consider the same unidirectional glass/epoxy composite as in Exercise 4.18. Plot failure envelopes (see Sections 4.4.1.4 and 4.4.3), using only the max strain criterion, for this composite in the axial-transverse stress plane using (a) CLT with ply level properties/allowables (see Table 4.2) using a single ply with a thickness of 1 mm (apply N_x and N_y with all other force and moment resultants $= 0$), (b) the MT micromechanics theory (use the constituent stress allowables given in the Exercise 4.16 table) using a fiber volume fraction of 0.55 (apply $\bar{\sigma}_{11}$ and $\bar{\sigma}_{22}$, all other global stress components $= 0$), and (c) the MOC micromechanics theory (use the constituent stress allowables backed out in Exercise 4.17) using a fiber volume fraction of 0.55 (apply $\bar{\sigma}_{11}$ and $\bar{\sigma}_{22}$, all other global stress components = 0). Plot all three envelopes on the same axes. In all cases, assume that the strain allowables are proportional to the stress allowables (as done with the IM7/8552 materials). Hint: the failure envelopes can be easily plotted in Excel.

Exercise 4.20. Consider a quasi-isotropic glass/epoxy laminate $[-45/0/45/90]_s$ with ply thicknesses of 0.15 mm. Plot N_x–N_y failure envelopes (see Sections 4.4.1.4 and 4.4.3), using only the Tsai-Wu criterion, for this laminate using CLT (a) with ply level properties/allowables (see Table 4.2), (b) with plies modeled using the MT micromechanics theory (use the constituent stress allowables given in the Exercise 4.16 table) using a fiber volume fraction of 0.55, and (c) with plies modeled using the MOC micromechanics theory (use the constituent stress allowables backed out in Exercise 4.17) using a fiber volume fraction of 0.55. Plot all three envelopes on the same axes. Hint: the failure envelopes can be easily plotted in Excel.

Exercise 4.21. Consider the same quasi-isotropic glass/epoxy laminate $[-45/0/45/90]_s$ with ply thicknesses of 0.15 mm as in Exercise 4.20. Plot N_x–N_y failure envelopes (see Sections 4.4.1.4 and 4.4.3), using only the Tsai-Wu criterion, with a thermal preload of $\Delta T = 100\ °C$, for this laminate using CLT (a) with ply level properties/allowables (see Table 4.2), (b) with plies modeled using the MT micromechanics theory (use the constituent stress allowables given in the Exercise 4.16 table) using a fiber volume fraction of 0.55, and (c) with plies modeled using the MOC micromechanics theory (use the constituent stress allowables backed out in Exercise 4.17) using a fiber volume fraction of 0.55. Compare each of these three failure envelopes to the corresponding failure envelop generated in Exercise 4.20 by plotting both on the same axes. Hint: the failure envelopes can be easily plotted in Excel.

References

Aboudi, J., Arnold, S.M., Bednarcyk, B.A., 2013. Micromechanics of Composite Materials A Generalized Multiscale Analysis Approach. Elsevier, New York, United States.

Daniel, I.M., Ishai, O., 1994. Engineering Mechanics of Composite Materials. Oxford University Press.

Herakovich, C.T., 1998. Mechanics of Fibrous Composites. John Wiley & Sons, Inc., New York, NY, United States.

HyperSizer, 2019. Methods and Equations Margin of Safety Interactions, Collier Research Corp., http://hypersizer.com/download.php?type=analysis&file=AID000_Margin_of_Safety_Interaction_Equations.HME.pdf, accessed July 12, 2020.

NCAMP, 2011a. Hexcel 8852 IM7 Unidirectional Prepreg 190 gsm & 35% RC Qualification Material Property Data Report, CAM-RP-2009-015 Rev A, FAA Special Project Number SP4614WI-Q.

NCAMP, 2011b. Hexcel 8852 IM7 Unidirectional Prepreg 190 gsm & 35% RC Qualification Statistical Analysis Report, NCP-RP-2009-028 Rev N/C, FAA Special Project Number SP4614WI-Q.

MATLAB, 2018. version R2018. The MathWorks Inc, Natick, Massachusetts.

MIL-HDBK-17-1F, 2002. Composite Materials Handbook, Volume 1. Polymer Matrix Composites Guidelines for Characterization of Structural Materials. U.S. Department of Defense.

Muzel, S.D., Bonhin, E.P., Guimaraes, N.M., Guidi, E.S., 2020. Application of the finite element method in the analysis of composite materials: a review. Polymers 12, 818. doi:10.3390/polym12040818.

Tsai, S.W., 1968. Strength theories of filamentary structures. In: Schwartz, R.T., Schwartz, H.S. (Eds.), Fundamental Aspects of Fiber Reinforced Plastic Composites. Wiley Interscience, New York, United States, pp. 3–11.

Tsai, S.W., Hahn, H.T., 1980. Introduction to Composite Materials. Technomic Publishing Company, Lancaster, United Kingdom.

Tsai, S.W., Wu, E.M., 1971. A general theory of strength for anisotropic materials. J. Compos. Mater. 5, 58–80.

The generalized method of cells (GMC) micromechanics theory

Chapter outline

In Chapter 3, the fundamental concepts of micromechanics, as well as four specific micromechanics theories [Voigt, Reuss, Mori-Tanaka (MT), and the method of cells (MOC)], were presented. These models are limited in their ability to address general microstructures, local field gradients, and composite systems with more than two constituent materials. However, the MOC represents a step towards addressing the lack of local field gradients in that the matrix was divided into three subcells, each of which having distinct fields. The aptly named generalized method of cells (GMC) (Paley and Aboudi, 1992; Aboudi, 1993), presented in this chapter, further expands the geometric representation of the composite repeating unit cell to enable consideration of an arbitrary number of subcells in the two periodic directions. The framework of GMC expands significantly on the MOC capabilities. For continuous composites, instead of being limited to a single fiber subcell, a multisubcell fiber, more representative of a cylindrical volume, can be considered. Furthermore, multiple fibers, approximating different packing arrangements, fiber types, fiber sizes, or fiber shapes, can be modeled. Additional phases to represent interfaces (e.g., around the fiber) can also be included. In addition, GMC provides an efficient and nonmesh-dependent procedure (unlike the finite element method) that automatically ensures periodicity for its repeating unit cell. These features make GMC ideal for performing rapid parametric trade studies,

Practical Micromechanics of Composite Materials. DOI: 10.1016/C2019-0-04193-3

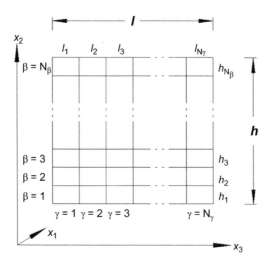

Figure 5.1 Doubly periodic GMC repeating unit cell (RUC). The RUC is composed of subcells, each of which contains a constituent material. The subcells are identified by the indices $\beta\gamma$.

especially in the context of multiscale modeling, wherein GMC can be used to represent the constitutive behavior of a material point.

The chapter begins with the development of the doubly-periodic GMC theory for the simulation of continuously reinforced composite materials. The GMC mechanical and thermal-strain concentration tensors are derived in terms of the thermomechanical properties of the constituent materials and their volume fractions/arrangements. Once the concentration tensors are known, the effective stiffness tensor and CTE of the composite can be readily determined using the closed form expressions established in Chapter 3. Examples are provided that compare GMC predictions of the effective thermoelastic properties with the four simpler micromechanics methods presented previously. In addition, local-field predictions of GMC are compared to those generated with the finite element method. Finally effective properties, local fields and failure initiation envelopes of a three phase ceramic matrix composite, in which the third phase represents a weak interface, are presented.

5.1   Geometry and basic relations

In Chapter 3, Figure 3.5 defined the repeating unit cell (RUC) for a continuously reinforced composite material in the context of the MOC. To extend this geometry to admit more than the four subcells, consider a composite material with a doubly periodic microstructure whose RUC consists of $N_\beta \times N_\gamma$ rectangular subcells, see Fig. 5.1. The volume of each subcell is $h_\beta \times l_\gamma$ (the third dimension is infinitely long, and thus can be taken as 1), where β and γ are running indices: $\beta = 1,\ldots,N_\beta$; $\gamma = 1,\ldots,N_\gamma$ in the x_2- and x_3-directions, respectively. As in the MOC, the length of all subcells in

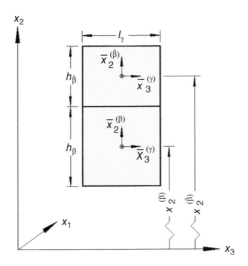

Figure 5.2 Illustration of the neighboring subcell numbering convention.

the x_1-direction is infinite, representing the fiber direction in a continuously reinforced composite. The total volume of the RUC is $h \times l$ (with the RUC dimension in the third direction taken as 1) where $h = \sum_{\beta=1}^{N_\beta} h_\beta$ and $l = \sum_{\gamma=1}^{N_\gamma} l_\gamma$. Each of the subcells can be filled, in general, by a distinct material, enabling consideration of multiphase composites with more than two constituents.

As in the MOC, the effective behavior of the composite can be established by considering the detailed interactions of the subcells within the RUC. This semi-analytical analysis requires that static equilibrium of the materials in the subcells is satisfied, and that continuity of the displacements and tractions between neighboring subcells within the RUC, as well as between neighboring RUCs, is satisfied on an average basis.

To this end, as in the MOC, local coordinates $\bar{x}_2^{(\beta)}$, $\bar{x}_3^{(\gamma)}$ are introduced whose origin is located at the center of each subcell $(\beta\gamma)$. These local coordinates are shown in Fig. 5.2 for subcell $(\beta\gamma)$ and the neighboring subcell in the x_2-direction $(\hat{\beta}\gamma)$, where $\hat{\beta}$ is defined as

$$\hat{\beta} = \begin{cases} \beta+1 & \beta < N_\beta \\ 1 & \beta = N_\beta \end{cases} \tag{5.1}$$

This definition ensures that, for $\beta < N_\beta$ the current subcell, labeled $(\beta\gamma)$, the neighboring subcell in the x_2-direction is the one labeled by $(\beta+1,\gamma)$ within the RUC, whereas for $\beta = N_\beta$ the neighboring subcell is within the next RUC whose first subcell is (1γ). Similarly, $\hat{\gamma}$ is defined by

$$\hat{\gamma} = \begin{cases} \gamma+1 & \gamma < N_\gamma \\ 1 & \gamma = N_\gamma \end{cases} \tag{5.2}$$

As in the MOC (Chapter 3), GMC is a first order theory in which the displacements $u_i^{(\beta\gamma)}$ in each subcell are expanded linearly in terms of the distances from the center of the subcell, i.e., in terms of $\bar{x}_2^{(\beta)}$ and $\bar{x}_3^{(\gamma)}$. Thus, the following first order expansion in the subcell $(\beta\gamma)$ is considered

$$u_i^{(\beta\gamma)} = w_i^{(\beta\gamma)}(\mathbf{x}) + \bar{x}_2^{(\beta)}\phi_i^{(\beta\gamma)} + \bar{x}_3^{(\gamma)}\psi_i^{(\beta\gamma)} \qquad i = 1, 2, 3;$$
$$\beta = 1, \ldots, N_\beta; \qquad \gamma = 1, \ldots, N_\gamma \qquad (5.3)$$

where $w_i^{(\beta\gamma)}(\mathbf{x})$ are the displacement components at the centroid of the subcell, and the microvariables, $\phi_i^{(\beta\gamma)}$ and $\psi_i^{(\beta\gamma)}$, characterize the linear dependence of the displacements $u_i^{(\beta\gamma)}$ on the local coordinates $\bar{x}_2^{(\beta)}$ and $\bar{x}_3^{(\gamma)}$. The vector $\mathbf{x} = (x_2, x_3)$ denotes the position of the center of the subcell with respect to the fixed global coordinate system (see Fig. 5.1). In Eq. (5.3) and the sequel, repeated Greek letters do not imply summation. Note that, due to the linearity of Eq. (5.3), as will be shown, strains and stresses are constant per subcell, and thus static equilibrium of the material within the subcell $(\beta\gamma)$ is ensured.

The components of the small strain tensor are given by

$$\varepsilon_{ij}^{(\beta\gamma)} = \frac{1}{2}\left(\partial_i u_j^{(\beta\gamma)} + \partial_j u_i^{(\beta\gamma)}\right) \quad i, j = 1, 2, 3 \qquad (5.4)$$

where $\partial_1 = \partial/\partial x_1$, $\partial_2 = \partial/\partial \bar{x}_2^{(\beta)}$ and $\partial_3 = \partial/\partial \bar{x}_3^{(\gamma)}$.

5.2   Interfacial displacement continuity conditions

The displacement components must be continuous at the various interfaces within the RUC, and at the interfaces between neighboring RUCs. This implies that for $\beta = 1, \ldots, N_\beta$ and $\gamma = 1, \ldots, N_\gamma$ the following relations hold

$$u_i^{(\beta\gamma)}\Big|_{\bar{x}_2^{(\beta)} = \frac{h_\beta}{2}} = u_i^{(\hat{\beta}\gamma)}\Big|_{\bar{x}_2^{(\beta)} = \frac{-h_\beta}{2}} \qquad (5.5)$$

$$u_i^{(\beta\gamma)}\Big|_{\bar{x}_3^{(\gamma)} = \frac{l_\gamma}{2}} = u_i^{(\beta\hat{\gamma})}\Big|_{\bar{x}_3^{(\gamma)} = \frac{-l_\gamma}{2}} \qquad (5.6)$$

Notice that with the definition of $\hat{\beta}$ and $\hat{\gamma}$, Eqs. (5.1) and (5.2), continuity of the displacement at the interfaces between neighboring RUCs is ensured, see Fig. 5.2.

The continuity conditions, Eqs. (5.5) and (5.6) are imposed at the interfaces in an average sense. For example, Eq. (5.5) is applied in the form

$$\int_{-\ell_\gamma/2}^{\ell_\gamma/2}\left[u_i^{(\beta\gamma)}\Big|_{\bar{x}_2^{(\beta)} = \frac{h_\beta}{2}}\right]d\bar{x}_3^{(\gamma)} = \int_{-\ell_\gamma/2}^{\ell_\gamma/2}\left[u_i^{(\hat{\beta}\gamma)}\Big|_{\bar{x}_2^{(\beta)} = \frac{-h_\beta}{2}}\right]d\bar{x}_3^{(\gamma)} \qquad (5.7)$$

Using Eq. (5.3) in Eq. (5.5) one obtains

$$w_i^{(\beta\gamma)} + \frac{1}{2}h_\beta\phi_i^{(\beta\gamma)} = w_i^{(\hat{\beta}\gamma)} - \frac{1}{2}h_{\hat{\beta}}\phi_i^{(\hat{\beta}\gamma)} \tag{5.8}$$

A similar equation results from the use of Eq. (5.3) in Eq. (5.6):

$$w_i^{(\beta\gamma)} + \frac{1}{2}l_\gamma\psi_i^{(\beta\gamma)} = w_i^{(\beta\hat{\gamma})} - \frac{1}{2}l_{\hat{\gamma}}\psi_i^{(\beta\hat{\gamma})} \tag{5.9}$$

For the special case in which $N_\beta = N_\gamma = 2$, Eqs. (5.8) and (5.9) are identical to the MOC Eqs. (3.69) and (3.70).

As in the MOC, a first-order Taylor series is used to relate $w_i^{(\beta\gamma)}$ of adjacent RUCs such that,

$$w_i^{(\beta\gamma)}\Big|_{\text{next}} = w_i^{(\beta\gamma)} + \Delta x \frac{\partial}{\partial x} w_i^{(\beta\gamma)} \tag{5.10}$$

where "next" refers to the next adjacent RUC in the positive coordinate direction, $\Delta x\,(= h \text{ or } l)$ is the distance between RUCs in the given direction, and $\partial x = \partial x_2$ or ∂x_3. Applying Eq. (5.8) for $\beta = 1, \ldots, N_\beta$, and utilizing the definition of $\hat{\beta}$ in Eq. (5.1),

$$w_i^{(1\gamma)} + \frac{1}{2}h_1\phi_i^{(1\gamma)} = w_i^{(2\gamma)} - \frac{1}{2}h_2\phi_i^{(2\gamma)}$$

$$w_i^{(2\gamma)} + \frac{1}{2}h_2\phi_i^{(2\gamma)} = w_i^{(3\gamma)} - \frac{1}{2}h_3\phi_i^{(3\gamma)}$$

$$\vdots$$

$$w_i^{(N_\beta-1\,\gamma)} + \frac{1}{2}h_{N_\beta-1}\phi_i^{(N_\beta-1\,\gamma)} = w_i^{(N_\beta\gamma)} - \frac{1}{2}h_{N_\beta}\phi_i^{(N_\beta\gamma)}$$

$$w_i^{(N_\beta\gamma)} + \frac{1}{2}h_{N_\beta}\phi_i^{(N_\beta\,\gamma)} = w_i^{(1\gamma)}\Big|_{\text{next}} - \frac{1}{2}h_1\phi_i^{(1\gamma)} = w_i^{(1\gamma)} + h\frac{\partial}{\partial x_2}w_i^{(1\gamma)} - \frac{1}{2}h_1\phi_i^{(1\gamma)} \tag{5.11}$$

where the first-order Taylor series, Eq. (5.10), has been used in the last equation in Eq. (5.11) to connect to the next adjacent RUC.

Summing Eqs. (5.11) eliminates the $w_i^{(\beta\gamma)}$ terms, yielding,

$$\sum_{\beta=1}^{N_\beta} h_\beta\phi_i^{(\beta\gamma)} = h\frac{\partial}{\partial x_2}w_i^{(1\gamma)} \tag{5.12}$$

Similarly, applying Eq. (5.9) for $\gamma = 1, ..., N_\gamma$, utilizing a first-order Taylor series, and summing, one arrives at,

$$\sum_{\gamma=1}^{N_\gamma} l_\gamma \psi_i^{(\beta\gamma)} = l\frac{\partial}{\partial x_3} w_i^{(\beta 1)} \tag{5.13}$$

Because the continuity equations are applicable to any RUC, Eq. (5.8) can be written for the next RUC as,

$$w_i^{(\beta\gamma)}\Big|_{\text{next}} + \frac{1}{2}h_\beta\phi_i^{(\beta\gamma)} = w_i^{(\beta\gamma)}\Big|_{\text{next}} - \frac{1}{2}h_{\hat\beta}\phi_i^{(\hat\beta\gamma)} \tag{5.14}$$

and then using the Taylor series, Eq. (5.10),

$$w_i^{(\beta\gamma)} + h\frac{\partial}{\partial x_2}w_i^{(\beta\gamma)} + \frac{1}{2}h_\beta\phi_i^{(\beta\gamma)} = w_i^{(\hat\beta\gamma)} + h\frac{\partial}{\partial x_2}w_i^{(\hat\beta\gamma)} - \frac{1}{2}h_{\hat\beta}\phi_i^{(\hat\beta\gamma)} \tag{5.15}$$

Subtracting Eq. (5.8) from Eq. (5.15), we have,

$$\frac{\partial}{\partial x_2}w_i^{(\beta\gamma)} = \frac{\partial}{\partial x_2}w_i^{(\hat\beta\gamma)} \tag{5.16}$$

and it can be similarly established that,

$$\frac{\partial}{\partial x_3}w_i^{(\beta\gamma)} = \frac{\partial}{\partial x_3}w_i^{(\beta\hat\gamma)} \tag{5.17}$$

Equations (5.16) and (5.17) are the generalizations of the MOC Eqs. (3.79) and (3.81), respectively. As in the MOC, Eqs. (5.16) and (5.17) are satisfied by assuming that common gradients of the displacement functions, w_i, exist such that,

$$\frac{\partial}{\partial x_j}w_i^{(\beta\gamma)} = \frac{\partial}{\partial x_j}w_i \tag{5.18}$$

From Eqs. (5.3), (5.4), and (5.18) it is clear that the subcell total strain components are independent of $\bar{x}_2^{(\beta)}$ and $\bar{x}_3^{(\gamma)}$ and are thus, as in the MOC, constant within a given subcell. The fact that the subcell-strain components are constant within a given subcell requires the subcell-stress components to be constant within a given subcell. These arguments show that the average fields and pointwise fields within the subcells are identical. That is,

$$\bar\sigma_{ij}^{(\beta\gamma)} = \sigma_{ij}^{(\beta\gamma)}, \quad \bar\varepsilon_{ij}^{(\beta\gamma)} = \varepsilon_{ij}^{(\beta\gamma)} \tag{5.19}$$

This also implies that the equations of equilibrium are identically satisfied in each subcell.

Using Eq. (5.18), the following set of continuum relations can then be written from Eqs. (5.12) and (5.13)

$$\sum_{\beta=1}^{N_\beta} h_\beta \phi_i^{(\beta\gamma)} = h \frac{\partial w_i}{\partial x_2}, \quad \gamma = 1, \dots, N_\gamma \tag{5.20}$$

$$\sum_{\gamma=1}^{N_\gamma} l_\gamma \psi_i^{(\beta\gamma)} = l \frac{\partial w_i}{\partial x_3}, \quad \beta = 1, \dots, N_\beta \tag{5.21}$$

The strains in the subcell, $\varepsilon_{ij}^{(\beta\gamma)}$ are given according to Eqs. (5.3) and (5.4) by

$$
\begin{aligned}
\varepsilon_{11}^{(\beta\gamma)} &= \frac{\partial w_1}{\partial x_1} \\
\varepsilon_{22}^{(\beta\gamma)} &= \phi_2^{(\beta\gamma)} \\
\varepsilon_{33}^{(\beta\gamma)} &= \psi_3^{(\beta\gamma)} \\
2\varepsilon_{23}^{(\beta\gamma)} &= \phi_3^{(\beta\gamma)} + \psi_2^{(\beta\gamma)} \\
2\varepsilon_{13}^{(\beta\gamma)} &= \psi_1^{(\beta\gamma)} + \frac{\partial w_3}{\partial x_1} \\
2\varepsilon_{12}^{(\beta\gamma)} &= \phi_1^{(\beta\gamma)} + \frac{\partial w_2}{\partial x_1}
\end{aligned}
\tag{5.22}
$$

where Eq. (5.18) has been employed for the first, fifth, and sixth equation. Note that the shear strains in these equations are tensorial quantities, and as usual the engineering shear strains are given by $\gamma_{ij} = 2\varepsilon_{ij}$, $i \neq j$. As in the MOC, perfect bonding is assumed, so the average strains in the composite can be expressed as

$$\bar{\varepsilon}_{ij} = \frac{1}{hl} \sum_{\beta=1}^{N_\beta} \sum_{\gamma=1}^{N_\gamma} h_\beta l_\gamma \, \varepsilon_{ij}^{(\beta\gamma)} \tag{5.23}$$

The constitutive equation of the elastic material that occupies subcell $(\beta\gamma)$ is expressed as,

$$\sigma_{ij}^{(\beta\gamma)} = C_{ijkl}^{(\beta\gamma)} \varepsilon_{kl}^{(\beta\gamma)} \tag{5.24}$$

where $C_{ijkl}^{(\beta\gamma)}$ is the elastic stiffness tensor of the material.

The average stress in the composite is determined from

$$\bar{\sigma}_{ij} = \frac{1}{hl} \sum_{\beta=1}^{N_\beta} \sum_{\gamma=1}^{N_\gamma} h_\beta l_\gamma \, \sigma_{ij}^{(\beta\gamma)} \tag{5.25}$$

Next, the relationship between the gradients of the subcell-centroid displacements and the average RUC strains,

$$\bar{\varepsilon}_{ij} = \frac{1}{2}\left(\frac{\partial w_i}{\partial x_j} + \frac{\partial w_j}{\partial x_i}\right) \tag{5.26}$$

which is identical to the MOC Eq. (3.88), is established below in a component by component manner, similar to that employed for the MOC in the Appendix of Chapter 3.

For $i = 1$, $j = 1$, multiplying the first of Eq. (5.22) by $h_\beta l_\gamma$ and performing a summation over β from 1 to N_β, and over γ from 1 to N_γ,

$$\sum_{\beta=1}^{N_\beta}\sum_{\gamma=1}^{N_\gamma} h_\beta l_\gamma \varepsilon_{11}^{(\beta\gamma)} = hl\frac{\partial w_1}{\partial x_1} \tag{5.27}$$

Comparing Eq. (5.27) to Eq. (5.23) gives,

$$\bar{\varepsilon}_{11} = \frac{\partial w_1}{\partial x_1} \tag{5.28}$$

Clearly, from the first of Eqs. (5.22) and (5.28),

$$\varepsilon_{11}^{(\beta\gamma)} = \bar{\varepsilon}_{11} \tag{5.29}$$

For $i = j = 2$, multiplying Eq. (5.20) by l_γ and performing a summation over γ from 1 to N_γ provides

$$\sum_{\beta=1}^{N_\beta}\sum_{\gamma=1}^{N_\gamma} h_\beta l_\gamma \phi_2^{(\beta\gamma)} = hl\frac{\partial w_2}{\partial x_2} \tag{5.30}$$

Substituting the second equality in Eq. (5.22) into Eq. (5.30) and then comparing with Eq. (5.23) gives

$$\bar{\varepsilon}_{22} = \frac{\partial w_2}{\partial x_2} \tag{5.31}$$

It can similarly be established, for $i = 3$, $j = 3$, that

$$\bar{\varepsilon}_{33} = \frac{\partial w_3}{\partial x_3} \tag{5.32}$$

For $i = 1, j = 2$, multiplying Eq. (5.20) with $i = 1$ by l_γ and performing a summation over γ from 1 to N_γ,

$$\sum_{\beta=1}^{N_\beta} \sum_{\gamma=1}^{N_\gamma} h_\beta l_\gamma \phi_1^{(\beta\gamma)} = hl \frac{\partial w_1}{\partial x_2} \tag{5.33}$$

Substituting the sixth of Eq. (5.22) into Eq. (5.33),

$$\sum_{\beta=1}^{N_\beta} \sum_{\gamma=1}^{N_\gamma} h_\beta l_\gamma \left(2\varepsilon_{12}^{(\beta\gamma)} - \frac{\partial w_2}{\partial x_1} \right) = hl \frac{\partial w_1}{\partial x_2} \tag{5.34}$$

Rearranging Eq. (5.34) gives

$$\frac{1}{hl} \sum_{\beta=1}^{N_\beta} \sum_{\gamma=1}^{N_\gamma} h_\beta l_\gamma \varepsilon_{12}^{(\beta\gamma)} = \frac{1}{2} \left(\frac{\partial w_1}{\partial x_2} + \frac{\partial w_2}{\partial x_1} \right) \tag{5.35}$$

Comparing Eq. (5.35) to Eq. (5.23) results in

$$\bar{\varepsilon}_{12} = \frac{1}{2} \left(\frac{\partial w_1}{\partial x_2} + \frac{\partial w_2}{\partial x_1} \right) \tag{5.36}$$

It can similarly be established, for $i = 1, j = 3$, that

$$\bar{\varepsilon}_{13} = \frac{1}{2} \left(\frac{\partial w_1}{\partial x_3} + \frac{\partial w_3}{\partial x_1} \right) \tag{5.37}$$

For $i = 2, j = 3$, first, multiplying Eq. (5.20) with $i = 3$ by l_γ and performing summation over γ from 1 to N_γ,

$$\sum_{\beta=1}^{N_\beta} \sum_{\gamma=1}^{N_\gamma} h_\beta l_\gamma \phi_3^{(\beta\gamma)} = hl \frac{\partial w_3}{\partial x_2} \tag{5.38}$$

and multiplying Eq. (5.21) with $i = 2$ by h_β and performing summation over β from 1 to N_β,

$$\sum_{\beta=1}^{N_\beta} \sum_{\gamma=1}^{N_\gamma} h_\beta l_\gamma \psi_2^{(\beta\gamma)} = hl \frac{\partial w_2}{\partial x_3} \tag{5.39}$$

By adding Eqs. (5.38) and (5.39), one obtains

$$\sum_{\beta=1}^{N_\beta}\sum_{\gamma=1}^{N_\gamma} h_\beta l_\gamma \left(\psi_2^{(\beta\gamma)} + \phi_3^{(\beta\gamma)} \right) = hl\left(\frac{\partial w_2}{\partial x_3} + \frac{\partial w_3}{\partial x_2} \right) \tag{5.40}$$

Substituting the fourth of Eq. (5.22) into Eq. (5.40) and rearranging results in

$$\frac{1}{hl}\sum_{\beta=1}^{N_\beta}\sum_{\gamma=1}^{N_\gamma} h_\beta l_\gamma \varepsilon_{23}^{(\beta\gamma)} = \frac{1}{2}\left(\frac{\partial w_2}{\partial x_3} + \frac{\partial w_3}{\partial x_2} \right) \tag{5.41}$$

Comparing Eq. (5.41) to Eq. (5.23),

$$\bar{\varepsilon}_{23} = \frac{1}{2}\left(\frac{\partial w_2}{\partial x_3} + \frac{\partial w_3}{\partial x_2} \right) \tag{5.42}$$

Thus all components of Eq. (5.26) have been established as Eqs. (5.28), (5.31), (5.32), (5.36), (5.37), and (5.42).

It is now possible to express the continuum equations, Eqs. (5.20) and (5.21), in terms of $\varepsilon_{ij}^{(\beta\gamma)}$ and $\bar{\varepsilon}_{ij}$. Setting $i = 2$ and 3 in Eqs. (5.20) and (5.21), respectively, one obtains,

$$\sum_{\beta=1}^{N_\beta} h_\beta \varepsilon_{22}^{(\beta\gamma)} = h\bar{\varepsilon}_{22}, \quad \gamma = 1, \ldots, N_\gamma \tag{5.43}$$

$$\sum_{\gamma=1}^{N_\gamma} l_\gamma \varepsilon_{33}^{(\beta\gamma)} = l\,\bar{\varepsilon}_{33}, \quad \beta = 1, \ldots, N_\beta \tag{5.44}$$

The addition of $h\,\partial w_2/\partial x_1$ to Eq. (5.20) with $i = 1$, and using Eq. (5.36) gives,

$$\sum_{\beta=1}^{N_\beta} h_\beta\left(\phi_1^{(\beta\gamma)} + \frac{\partial w_2}{\partial x_1} \right) = h\left(\frac{\partial w_1}{\partial x_2} + \frac{\partial w_2}{\partial x_1} \right) = 2h\bar{\varepsilon}_{12}, \quad \gamma = 1, \ldots, N_\gamma \tag{5.45}$$

and from the sixth of Eq. (5.22),

$$\sum_{\beta=1}^{N_\beta} h_\beta \varepsilon_{12}^{(\beta\gamma)} = h\bar{\varepsilon}_{12}, \quad \gamma = 1, \ldots, N_\gamma \tag{5.46}$$

Table 5.1 Summary of the GMC equations resulting from imposing displacement continuity conditions.

Equation	Equation number
$\varepsilon_{11}^{(\beta\gamma)} = \bar{\varepsilon}_{11}$	(5.29)
$\displaystyle\sum_{\beta=1}^{N_\beta} h_\beta \varepsilon_{22}^{(\beta\gamma)} = h\bar{\varepsilon}_{22}, \quad \gamma = 1, \ldots, N_\gamma$	(5.43)
$\displaystyle\sum_{\gamma=1}^{N_\gamma} l_\gamma \varepsilon_{33}^{(\beta\gamma)} = l\,\bar{\varepsilon}_{33}, \quad \beta = 1, \ldots, N_\beta$	(5.44)
$\displaystyle\sum_{\beta=1}^{N_\beta}\sum_{\gamma=1}^{N_\gamma} h_\beta l_\gamma \varepsilon_{23}^{(\beta\gamma)} = hl\bar{\varepsilon}_{23}$	(5.48)
$\displaystyle\sum_{\gamma=1}^{N_\gamma} l_\gamma \varepsilon_{13}^{(\beta\gamma)} = l\bar{\varepsilon}_{13}, \quad \beta = 1, \ldots, N_\beta$	(5.46)
$\displaystyle\sum_{\beta=1}^{N_\beta} h_\beta \varepsilon_{12}^{(\beta\gamma)} = h\bar{\varepsilon}_{12}, \quad \gamma = 1, \ldots, N_\gamma$	(5.47)

It can similarly be established that

$$\sum_{\gamma=1}^{N_\gamma} l_\gamma \varepsilon_{13}^{(\beta\gamma)} = l\bar{\varepsilon}_{13}, \quad \beta = 1, \ldots, N_\beta \tag{5.47}$$

Finally, from Eqs. (5.41) and (5.42),

$$\sum_{\beta=1}^{N_\beta}\sum_{\gamma=1}^{N_\gamma} h_\beta l_\gamma \varepsilon_{23}^{(\beta\gamma)} = hl\bar{\varepsilon}_{23} \tag{5.48}$$

Thus, we have arrived at a set of linear algebraic equations that result from imposing displacement continuity in the GMC theory. These equations are summarized in Table 5.1. They represent a system of $2(N_\beta + N_\gamma) + N_\beta N_\gamma + 1$ relations, which can be written in a matrix form as follows

$$\mathbf{A}_G \boldsymbol{\varepsilon}_s = \mathbf{J}\bar{\boldsymbol{\varepsilon}} \tag{5.49}$$

where the six-order average strain vector is defined by

$$\bar{\boldsymbol{\varepsilon}} = (\bar{\varepsilon}_{11}, \ \bar{\varepsilon}_{22}, \ \bar{\varepsilon}_{33}, \ 2\bar{\varepsilon}_{23}, \ 2\bar{\varepsilon}_{13}, \ 2\bar{\varepsilon}_{12})^{Tr} \tag{5.50}$$

and the $6N_\beta N_\gamma$-order subcell strain vector, $\boldsymbol{\varepsilon}_s$, is defined by

$$\boldsymbol{\varepsilon}_s = \left(\boldsymbol{\varepsilon}^{(11)}, \ \ldots, \ \boldsymbol{\varepsilon}^{(N_\beta N_\gamma)}\right)^{Tr} \tag{5.51}$$

where the 6 components of each vector $\boldsymbol{\varepsilon}^{(\beta\gamma)}$ are arranged as in Eq. (5.50). The order of matrix $\mathbf{A}_G$ is $2(N_\beta + N_\gamma) + N_\beta N_\gamma + 1$ by $6N_\beta N_\gamma$, while $\mathbf{J}$ is a $2(N_\beta + N_\gamma) + N_\beta N_\gamma + 1$ by 6 matrix. It should be noted that the matrix $\mathbf{A}_G$ involves the geometric dimensions of the subcells only.

5.3   Interfacial continuity of tractions

The tractions must be continuous at the interfaces between the subcells of the RUCs, and at the interfaces between neighboring RUCs. Because the stresses are constant per subcell, these traction continuity conditions, imposed in an average sense, can be expressed as

$$\sigma_{2i}^{(\beta\gamma)} = \sigma_{2i}^{(\hat{\beta}\gamma)}$$
$$\sigma_{3i}^{(\beta\gamma)} = \sigma_{3i}^{(\beta\hat{\gamma})}$$

$$(5.52)$$

where $i = 1, 2, 3$, $\beta = 1, \ldots, N_\beta$, $\gamma = 1, \ldots, N_\gamma$ and $\hat{\beta}$ and $\hat{\gamma}$ are defined in Eqs. (5.1) and (5.2). It can be easily verified that these equations involve some repetitions. It can be shown that the system of independent (non-repeated) interfacial conditions is

$$\sigma_{22}^{(\beta\gamma)} = \sigma_{22}^{(\hat{\beta}\gamma)} \quad \begin{matrix} \beta = 1, \ldots, N_\beta - 1 \\ \gamma = 1, \ldots, N_\gamma \end{matrix}$$

$$(5.53)$$

$$\sigma_{33}^{(\beta\gamma)} = \sigma_{33}^{(\beta\hat{\gamma})} \quad \begin{matrix} \beta = 1, \ldots, N_\beta \\ \gamma = 1, \ldots, N_\gamma - 1 \end{matrix}$$

$$(5.54)$$

$$\sigma_{23}^{(\beta\gamma)} = \sigma_{23}^{(\hat{\beta}\gamma)} \quad \begin{matrix} \beta = 1, \ldots, N_\beta - 1 \\ \gamma = 1, \ldots, N_\gamma \end{matrix}$$

$$(5.55)$$

$$\sigma_{32}^{(\beta\gamma)} = \sigma_{32}^{(\beta\hat{\gamma})} \quad \begin{matrix} \beta = N_\beta \\ \gamma = 1, \ldots, N_\gamma - 1 \end{matrix}$$

$$(5.56)$$

$$\sigma_{31}^{(\beta\gamma)} = \sigma_{31}^{(\beta\hat{\gamma})} \quad \begin{matrix} \beta = 1, \ldots, N_\beta \\ \gamma = 1, \ldots, N_\gamma - 1 \end{matrix}$$

$$(5.57)$$

$$\sigma_{21}^{(\beta\gamma)} = \sigma_{21}^{(\hat{\beta}\gamma)} \quad \begin{matrix} \beta = 1, \ldots, N_\beta - 1 \\ \gamma = 1, \ldots, N_\gamma \end{matrix}$$

$$(5.58)$$

Equations (5.53) through (5.58) provide $5N_\beta N_\gamma - 2(N_\beta + N_\gamma) - 1$ independent equations, thus, when combined with the equations shown in Table 5.1, a system of $6N_\beta N_\gamma$ linear algebraic equations in the unknown variables $\varepsilon_{ij}^{(\beta\gamma)}$ is formed.

By using the subcell orthotropic constitutive relations,

$$
\begin{bmatrix} \sigma_{11}^{(\beta\gamma)} \\ \sigma_{22}^{(\beta\gamma)} \\ \sigma_{33}^{(\beta\gamma)} \\ \sigma_{23}^{(\beta\gamma)} \\ \sigma_{13}^{(\beta\gamma)} \\ \sigma_{12}^{(\beta\gamma)} \end{bmatrix}
=
\begin{bmatrix}
C_{11}^{(\beta\gamma)} & C_{12}^{(\beta\gamma)} & C_{13}^{(\beta\gamma)} & 0 & 0 & 0 \\
C_{12}^{(\beta\gamma)} & C_{22}^{(\beta\gamma)} & C_{23}^{(\beta\gamma)} & 0 & 0 & 0 \\
C_{13}^{(\beta\gamma)} & C_{23}^{(\beta\gamma)} & C_{33}^{(\beta\gamma)} & 0 & 0 & 0 \\
0 & 0 & 0 & C_{44}^{(\beta\gamma)} & 0 & 0 \\
0 & 0 & 0 & 0 & C_{55}^{(\beta\gamma)} & 0 \\
0 & 0 & 0 & 0 & 0 & C_{66}^{(\beta\gamma)}
\end{bmatrix}
\begin{bmatrix} \varepsilon_{11}^{(\beta\gamma)} \\ \varepsilon_{22}^{(\beta\gamma)} \\ \varepsilon_{33}^{(\beta\gamma)} \\ 2\varepsilon_{23}^{(\beta\gamma)} \\ 2\varepsilon_{13}^{(\beta\gamma)} \\ 2\varepsilon_{12}^{(\beta\gamma)} \end{bmatrix}
$$

$$(5.59)$$

it is possible to represent Eqs. (5.53) through (5.58) in matrix form as,

$$\mathbf{A}_M \boldsymbol{\varepsilon}_s = 0 \tag{5.60}$$

where the $5N_\beta N_\gamma - 2(N_\beta + N_\gamma) - 1$ by $6N_\beta N_\gamma$ matrix $\mathbf{A}_M$ involves the elastic properties of the materials in the subcells.

5.4 Overall mechanical constitutive law

The combination of Eqs. (5.49) and (5.60) leads to a system of linear algebraic equations of identical form as the MOC mechanical Eq. (3.109),

$$\mathbf{M} \boldsymbol{\varepsilon}_s = \mathbf{R} \tag{5.61}$$

where

$$\mathbf{M} = \begin{bmatrix} \mathbf{A}_M \\ \mathbf{A}_G \end{bmatrix}, \qquad \mathbf{R} = \begin{bmatrix} \mathbf{0} \\ \mathbf{J} \end{bmatrix} \bar{\boldsymbol{\varepsilon}} \tag{5.62}$$

The size of Eq. (5.62) is $6N_\beta N_\gamma \times 6N_\beta N_\gamma$ because there are six unknown strains per subcell. Recall, that for the MOC, the size of Eq. (3.109) was 24 (see Table 3.3), which also results from six unknown strain components per the four MOC subcells.

Equation (5.61) could be solved for the unknown strain components, and the results used to determine the strain concentration tensors $\mathbf{A}^{(\beta\gamma)}$ of all subcells. Instead, for consistency, although less computationally efficent, the same solution procedure used in the MOC has been employed here. That is, seeking the GMC strain concentration tensors (see Section 5.5 for the thermal strain concentration tensors) such that

$$\boldsymbol{\varepsilon}^{(\beta\gamma)} = \mathbf{A}^{(\beta\gamma)} \bar{\boldsymbol{\varepsilon}} \tag{5.63}$$

first apply the global strain component $\bar{\varepsilon}_{11} = 1$, with all other global strain components set to zero. The solution of Eq. (5.61) then yields the components $A_{ij11}^{(\beta\gamma)}$ (which in contracted matrix notation is the first column). Next, apply $\bar{\varepsilon}_{22} = 1$, with all other global strain components set equal to zero. The solution provides the components $A_{ij22}^{(\beta\gamma)}$ (which in contracted matrix notation is the second column). By continuing the application of this procedure, all components of $\mathbf{A}^{(\beta\gamma)}$ can be established. It can be verified that the general form of the nonsymmetric $\mathbf{A}^{(\beta\gamma)}$ can be represented in contracted matrix form, identically to the form for MOC in Eq. (3.111), as follows:

$$\begin{bmatrix} \varepsilon_{11}^{(\beta\gamma)} \\ \varepsilon_{22}^{(\beta\gamma)} \\ \varepsilon_{33}^{(\beta\gamma)} \\ \varepsilon_{23}^{(\beta\gamma)} \\ \varepsilon_{13}^{(\beta\gamma)} \\ \varepsilon_{12}^{(\beta\gamma)} \end{bmatrix} = \begin{bmatrix} 1 & 0 & 0 & 0 & 0 & 0 \\ A_{21}^{(\beta\gamma)} & A_{22}^{(\beta\gamma)} & A_{23}^{(\beta\gamma)} & 0 & 0 & 0 \\ A_{31}^{(\beta\gamma)} & A_{32}^{(\beta\gamma)} & A_{33}^{(\beta\gamma)} & 0 & 0 & 0 \\ 0 & 0 & 0 & A_{44}^{(\beta\gamma)} & 0 & 0 \\ 0 & 0 & 0 & 0 & A_{55}^{(\beta\gamma)} & 0 \\ 0 & 0 & 0 & 0 & 0 & A_{66}^{(\beta\gamma)} \end{bmatrix} \begin{bmatrix} \bar{\varepsilon}_{11} \\ \bar{\varepsilon}_{22} \\ \bar{\varepsilon}_{33} \\ \bar{\varepsilon}_{23} \\ \bar{\varepsilon}_{13} \\ \bar{\varepsilon}_{12} \end{bmatrix} \tag{5.64}$$

Once the concentration tensors are known, the effective composite properties are determined in a generalized manner as compared to Chapter 3. The effective stiffness is given by the volume average of the subcell stiffness times the strain concentration tensor (see Eq. 3.115),

$$\mathbf{C}^* = \frac{1}{hl} \sum_{\beta=1}^{N_\beta} \sum_{\gamma=1}^{N_\gamma} h_\beta l_\gamma \mathbf{C}^{(\beta\gamma)} \mathbf{A}^{(\beta\gamma)} \tag{5.65}$$

The stress concentration tensor, from Eq. (3.24), is given by

$$\mathbf{B}^{(\beta\gamma)} = \mathbf{C}^{(\beta\gamma)} \mathbf{A}^{(\beta\gamma)} \mathbf{C}^{*-1} \tag{5.66}$$

The global (effective) composite constitutive equation is

$$\bar{\sigma} = \mathbf{C}^* \bar{\varepsilon} \tag{5.67}$$

Given Eqs. (5.67), the effective elastic stress-strain response of the composite can be determined, and given Eqs. (5.63) and (5.59), the local strains and stresses can also be calculated. Thus, GMC has provided the complete global and local mechanical solution for the composite. As mentioned in Section 5.8, the GMC framework has been extended to include inelastic and progressive damage material behavior, fiber-matrix interfacial debonding, as well as large deformations and triply-periodic microstructures, see Aboudi et al. (2013) for details. It is also worth mentioning that, as shown in Section 3.8.6 for the MOC, the GMC strain concentration tensor components in Eq. (5.64) could be averaged to provide transversely isotropic composite behavior (for an appropriate unidirectional RUC).

5.5 Thermomechanical effects

To include thermomechanical effects in GMC, the effective coefficient of thermal expansion (CTE) and the thermal strain concentration tensors must be established. By employing the Levin Theorem (see Section 3.9.1), see Levin (1967) and Christensen (1979), the effective CTE of the composite is given by (see Eq. 3.138),

$$\boldsymbol{\alpha}^* = \sum_{k=1}^{N} v_k [\mathbf{B}_k]^{\text{Tr}} \boldsymbol{\alpha}^{(k)} = \frac{1}{hl} \sum_{\beta=1}^{N_\beta} \sum_{\gamma=1}^{N_\gamma} h_\beta l_\gamma [\mathbf{B}^{(\beta\gamma)}]^{\text{Tr}} \boldsymbol{\alpha}^{(\beta\gamma)} \tag{5.68}$$

where $\boldsymbol{\alpha}^{(\beta\gamma)} = (\alpha_1^{(\beta\gamma)}, \alpha_2^{(\beta\gamma)}, \alpha_3^{(\beta\gamma)}, 0, 0, 0)^{Tr}$, and the global thermomechanical constitutive equation is given by,

$$\bar{\sigma} = \mathbf{C}^* (\bar{\varepsilon} - \boldsymbol{\alpha}^* \Delta T) \tag{5.69}$$

As in the MOC, to include the effects of thermal strains on the local fields in the subcells, the thermal strains are first included in the subcell orthotropic constitutive equations,

$$
\begin{bmatrix}
\sigma_{11}^{(\beta\gamma)} \\
\sigma_{22}^{(\beta\gamma)} \\
\sigma_{33}^{(\beta\gamma)} \\
\sigma_{23}^{(\beta\gamma)} \\
\sigma_{13}^{(\beta\gamma)} \\
\sigma_{12}^{(\beta\gamma)}
\end{bmatrix}
=
\begin{bmatrix}
C_{11}^{(\beta\gamma)} & C_{12}^{(\beta\gamma)} & C_{13}^{(\beta\gamma)} & 0 & 0 & 0 \\
C_{12}^{(\beta\gamma)} & C_{22}^{(\beta\gamma)} & C_{23}^{(\beta\gamma)} & 0 & 0 & 0 \\
C_{13}^{(\beta\gamma)} & C_{23}^{(\beta\gamma)} & C_{33}^{(\beta\gamma)} & 0 & 0 & 0 \\
0 & 0 & 0 & C_{44}^{(\beta\gamma)} & 0 & 0 \\
0 & 0 & 0 & 0 & C_{55}^{(\beta\gamma)} & 0 \\
0 & 0 & 0 & 0 & 0 & C_{66}^{(\beta\gamma)}
\end{bmatrix}
\begin{bmatrix}
\varepsilon_{11}^{(\beta\gamma)} - \alpha_1^{(\beta\gamma)}\Delta T \\
\varepsilon_{22}^{(\beta\gamma)} - \alpha_2^{(\beta\gamma)}\Delta T \\
\varepsilon_{33}^{(\beta\gamma)} - \alpha_3^{(\beta\gamma)}\Delta T \\
2\varepsilon_{23}^{(\beta\gamma)} \\
2\varepsilon_{13}^{(\beta\gamma)} \\
2\varepsilon_{12}^{(\beta\gamma)}
\end{bmatrix}
$$

$$(5.70)$$

As a result, Eq. (5.60) is modified to include the subcell thermal strains,

$$\mathbf{A}_M\left(\boldsymbol{\varepsilon}_s - \boldsymbol{\alpha}_s \Delta T\right) = 0 \tag{5.71}$$

where

$$\boldsymbol{\alpha}_s = \left(\boldsymbol{\alpha}^{(11)}, \ldots, \boldsymbol{\alpha}^{(N_\beta N_\gamma)}\right)^{Tr} \tag{5.72}$$

with the six components of each vector $\boldsymbol{\alpha}^{(\beta\gamma)}$ arranged as $\left(\alpha_1^{(\beta\gamma)}, \alpha_2^{(\beta\gamma)}, \alpha_3^{(\beta\gamma)}, 0, 0, 0\right)^{Tr}$. The combination of Eq. (5.71) and (5.49) leads to the new system of linear algebraic equations, which now include thermal effects,

$$\mathbf{M}\boldsymbol{\varepsilon}_s = \mathbf{R} + \mathbf{R}^T \Delta T \tag{5.73}$$

where

$$\mathbf{R}^T = \begin{bmatrix} \mathbf{A}_M \\ \mathbf{0} \end{bmatrix} \boldsymbol{\alpha}_s \tag{5.74}$$

To determine the strain concentration tensors $\mathbf{A}_T^{(\beta\gamma)}$ of all subcells, where,

$$\boldsymbol{\varepsilon}^{(\beta\gamma)} = \mathbf{A}^{(\beta\gamma)}\bar{\boldsymbol{\varepsilon}} + \mathbf{A}_T^{(\beta\gamma)}\Delta T \tag{5.75}$$

one can set $\bar{\boldsymbol{\varepsilon}} = \mathbf{0}$ and $\Delta T = 1$, to obtain $\boldsymbol{\varepsilon}^{(\beta\gamma)} = \mathbf{A}_T^{(\beta\gamma)}$. Therefore, Eq. (5.73) can be employed with $\bar{\boldsymbol{\varepsilon}} = \mathbf{0}$ and $\Delta T = 1$ to determine the subcell strains and thus $\mathbf{A}_T^{(\beta\gamma)}$. This enables the full local and global thermomechanical fields to be determined using the GMC theory.

It should be noted that GMC can calculate the effective CTEs directly, without the need for the Levin Theorem. Towards this end, substituting Eq. (5.75) into Eq. (5.70) and then volume averaging, see Eq. (5.25), yields,

$$\bar{\sigma} = \frac{1}{hl} \sum_{\beta=1}^{N_\beta} \sum_{\gamma=1}^{N_\gamma} h_\beta l_\gamma \, \mathbf{C}^{(\beta\gamma)} \left(\mathbf{A}^{(\beta\gamma)} \bar{\varepsilon} + \mathbf{A}_T^{(\beta\gamma)} \Delta T - \boldsymbol{\alpha}^{(\beta\gamma)} \Delta T \right) \tag{5.76}$$

Comparing Eq. (5.76) to the global constitutive Eq. (5.69), it is clear that,

$$\boldsymbol{\alpha}^* = \frac{-\mathbf{C}^{*-1}}{hl} \sum_{\beta=1}^{N_\beta} \sum_{\gamma=1}^{N_\gamma} h_\beta l_\gamma \, \mathbf{C}^{(\beta\gamma)} \left(\mathbf{A}_T^{(\beta\gamma)} - \boldsymbol{\alpha}^{(\beta\gamma)} \right) \tag{5.77}$$

It can be numerically verified that Eq. (5.77) gives identical predictions for the effective CTE as the Levin Theory, Eq. (5.68). Thus either equation can be used.

5.6   MATLAB implementation

The MATLAB code provided with this text is available from:

   https://github.com/nasa/Practical-Micromechanics

The MATLAB implementation of the GMC micromechanics theory is shown in MATLAB Function 5.1 – GMC.m. This function assembles the $\mathbf{A}_G$ and $\mathbf{A}_M$ matrices, associated with displacement and traction continuity, respectively, as well as the $\mathbf{M}$ and $\mathbf{R}$ matrices (see Eqs. (5.61) and (5.62)). Because the number of equations, and thus the size of these matrices, increases with the number of subcells, the matrices can become quite large. For example, the commonly used 26×26 subcell GMC RUC (see Fig. 5.3) results in an $\mathbf{M}$ matrix that is 4056 × 4056, but a large number of the components are zero. For a typical GMC problem using the 26×26 subcell RUC, of the 16,451,136 components in the $\mathbf{M}$ matrix, only 43,356 (or 0.26%) are nonzero. Such a matrix, which has a large percentage of components equal to zero is referred to as *sparse*, and the *sparsity* of the discussed example $\mathbf{M}$ matrix is 99.74%.

Computational advantage can be taken of sparse matrices by storing them in sparse format. In the MATLAB implementation of GMC, the $\mathbf{A}_G$, $\mathbf{A}_M$, and $\mathbf{M}$ matrices are assembled and stored not as standard 2nd-order real matrices, but rather as three vectors representing the row number, column number, and value of only the non-zero entries in each matrix. This represents a tremendous memory savings as, for example, in the 26×26 subcell RUC discussed above, instead of storing 16,451,136 real components of $\mathbf{M}$, in sparse format, only 86,712 integers must be stored for the row and column numbers of the non-zero components of $\mathbf{M}$, along with 43,356 real values of the nonzero component values. In addition, efficient numerical methods have been developed (and implemented within MATLAB (2018)) to take advantage of matrix sparsity when performing matrix operations. To store a matrix in MATLAB's sparse format, the sparse function has been used (see MATLAB Function 5.1),

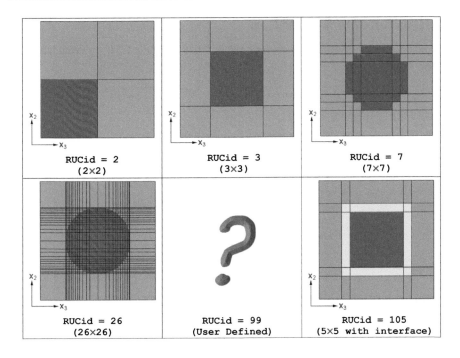

Figure 5.3 Chapter 5 GMC repeating unit cell (RUC) geometries implemented in MATLAB Function 5.2– GetRUC.m.

with arguments consisting of the matrix non-zero row and column numbers and the values.

The GMC implementation in MATLAB Function 5.1 also includes two options for solving the system of equations, Eq. (5.61) and determining the subcell strain concentration tensors (see Eq. (5.63)). In addition to the method of successively applying global strain components described in Section 5.4 (which is commented out in MATLAB Function 5.1), the direct solution of Eq. (5.61) has been implemented by first defining the matrix **K**, such that $\mathbf{R} = \mathbf{K}\bar{\varepsilon}$. This allows Eq. (5.61) to be written as,

$$\mathbf{M}\varepsilon_s = \mathbf{K}\bar{\varepsilon} \tag{5.78}$$

and, recalling that **M** is a square (invertible) matrix, the subcell strains can be directly determined in terms of the global strains,

$$\varepsilon_s = \mathbf{M}^{-1}\mathbf{K}\bar{\varepsilon} = \hat{\mathbf{A}}\bar{\varepsilon} \tag{5.79}$$

The subcell strain concentration tensors can then be extracted from the known ($\hat{\mathbf{A}}$ matrix, given the definition of ε_s in Eq. (5.51).

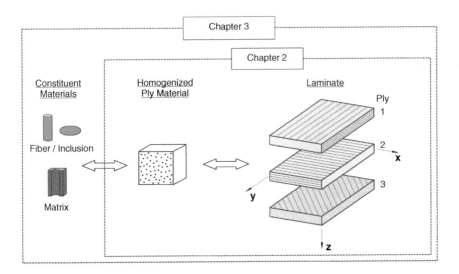

Figure 3.1 Illustration of a laminate whose plies are now represented with micromechanics, thereby incorporating the detail of the fiber and matrix constituent materials at a lower length scale.

given experimental test, for example, tensile test) is already treated as homogenous at the global scale. Thus, as with polycrystalline metallic alloys, the intrinsic, history-dependent, interactive effects of the constituents (e.g., fiber, interface, and matrix) are embedded in the experimental results. Clearly, developing macro-scale constitutive models lends itself to explicit experimental characterization and alleviates any assumptions inherent to a micromechanical approach concerning in-situ constituent response. Of course, the macromechanical approach is not universally applicable for the structural analysis of multiphased materials because its validity depends on characteristic structural dimensions, the severity of gradients (stress, temperature, etc.) in the structure, and the size (cell size) of the internal texture of the material (see the previous discussion in Section 2.1 as well as [Aboudi (1993), Liu (1988), and Onat and Wright (1991)].

Table 3.1 lists the advantages and disadvantages of the macromechanics and micromechanics approaches. The most important benefit of the macromechanical approach, as indicated in Table 3.1, is computational efficiency. No calculations nor computational effort are required because all properties and parameters of the composite material are known. In contrast, even the simplest micromechanical theory will require some additional computational effort. One of the main drawbacks of the macromechanical approach is the requirement for an anisotropic constitutive theory (for most types of composites) for which all properties and parameters must be measured for each composite material. Furthermore, any variant of the composite material (e.g., different fiber volume fractions) requires recharacterization, which can be very expensive. In the linear elastic regime, the anisotropic constitutive theory

Table 3.1 Comparison between the macromechanics and micromechanics approaches to composite modeling and analysis.

Advantages	Disadvantages
Macromechanical	

1. Computational efficiency
2. Experimental testing incorporates all in-situ effects (interface, damage, residual stresses, etc.)
3. Aligned with standard ply-level design procedures
4. Tends to work well for:
   a. Fiber-dominated situations
   b. Linear, isothermal regime
	1. Anisotropic constitutive theory required
	2. Requires additional complexities to handle:
	   a. Tension/compression
	   b. Inelasticity and damage
	   c. Multiaxiality and hydrostatic interaction
	3. Needs recalibration (retesting) for changes in
	   a. Architecture/microstructure
	   b. Volume fraction
	   c. Processing history (residual stresses)
	4. Phenomenologically based: accounts for physics of all nonlinearity on the wrong scale
	5. Form of the constitutive law is usually based on monolithic materials so applicability is questionable
	6. Adversely impacted by nonisothermal loading

Micromechanical	
1. Predicts properties/behavior of any composite material 2. Enables the capture of the physics of deformation and damage at a more fundamental scale 3. Captures varying in-situ nonproportional, multiaxial stress and strain states (iso/nonisothermal) 4. Can employ simpler isotropic constituent constitutive models 5. Can employ simpler, more fundamental, failure criteria 6. Microstructural effects explicitly captured 7. Can perform "what if" scenarios for design of materials	1. Increased computational expense 2. Need for constituent material constitutive response (fiber/matrix) as well as composite testing for validation 3. Potential need to calibrate for in-situ effects 4. Interfacial behavior is a constant unknown 5. Lower Technology Readiness Level (TRL)

is just Hooke's Law, and thus little complexity is added. However, all anisotropic elastic properties of the composite material must still be measured. In the presence of material nonlinearity, this requirement for an anisotropic constitutive model adds significantly to the macromechanics approach's theoretical and characterization complexity. In contrast, the micromechanics approach can predict not only all of the elastic properties of the composite, but also the complex, multiaxial, nonlinear response of the composite based on the properties of the constituents, see Aboudi et al. (2013). Thus, any composite, with any constituents and microstructure, can be modeled using micromechanics, regardless of if the composite exists, or even if it can currently be manufactured. Micromechanics is therefore ideal for use in the development/design of new composite materials and the optimization of current composite materials.

In this chapter, the fundamental concepts needed to establish micromechanics theories are first discussed, followed by the development of strain and stress concentration tensors. Then, four closed-form micromechanics methods are presented; namely, the Voigt and Reuss approximations, Mori-Tanaka (MT) Method, and Method of Cells (MOC), along with the MATLAB implementation of each method. Next, the MATLAB script that performs standalone micromechanics analyses of unidirectional composites (MicroAnalysis.m), and the extension of the laminate analysis script (LamAnalysis.m) to enable micromechanics-based analyses of laminates, are discussed. Finally, examples and exercises are presented.

3.2 Fundamentals of the mechanics of multiphase materials

3.2.1 Representative volume elements (RVEs) and repeating unit cells (RUCs)

Continuum mechanics is based on the concept of a homogeneous continuum, which can be repeatedly subdivided into infinitesimal subvolumes, each of which retains the properties of the bulk material. However, at some scale, all real materials are heterogeneous. The purpose of micromechanics is to account explicitly for a material's heterogeneous microstructure while allowing it to be treated as an effective (pseudohomogeneous) continuum at a higher length scale (e.g., within a structure). To account for this microstructure, micromechanics relies on either a representative volume element (RVE) or a repeating unit cell (RUC). An RVE is a volume of a material whose effective behavior is representative of (and indistinguishable from) that of the material as a whole. For a fictitious perfectly homogeneous material, the RVE would be infinitely small. For all real materials, which have a microstructure, the RVE must contain a large enough volume that it captures the essence of the microstructure from a statistical standpoint. Then, if one applies equivalent displacement or traction boundary conditions to the RVE boundaries, one would obtain an identical response that truly represents the material. Hill (1963) defined an RVE based on the requirement that strain energy densities induced by applying homogeneous traction and equivalent homogeneous displacement boundary conditions must be essentially equivalent.

Equally important is the concept of a repeating unit cell (RUC). Here, the heterogeneous microstructure is approximated as periodic, where the RUC is the volume of the material that repeats itself to generate the overall microstructure. Periodicity conditions are applied to the RUC boundaries to enforce the repeating nature of the RUC. It is often difficult to ensure that a chosen analysis volume is actually an RVE, and, as a consequence, the RUC/periodicity approach is usually utilized. Drago and Pindera (2007) provide an in depth discussion of the RVE and RUC concepts in the context of micromechanics of composites. Additional, more general, discussions of the RVE/RUC concept and its application in mechanics can be found in Galvanetto and Aliabadi (2010) and Kwon et al. (2008).

The interplay between the scales of an RVE, RUC, and structure was illustrated in Fig. 2.1. As discussed in Chapter 2, if the ratios of the dimensions (d/D and D/L) are much less than one, then the RVE should include a sufficient number of microstructural units to allow (statistically) the heterogeneous nature of the material to be homogenized. That is, the RVE of the material is such that; (a) it is entirely typical of the bulk composite on average, and (b) it contains a sufficient number of material phases and is thus large compared to the scales of the microstructure, but it is still small compared to the entire body. Consequently, the RVE response would represent the response of the composite medium, and, furthermore, the response would be insensitive to the type of boundary conditions (surface values of tractions or displacements) as long as these values are macroscopically uniform. That is, they fluctuate about a mean with a wavelength that is small compared with the dimensions of the volume, and the effects of such fluctuations become insignificant within a few wavelengths of the surface.

There are situations when the concept of an RVE, and thus applicability of homogenization, can be called into question. In the presence of an external boundary, the local fields induced by the microstructure may interact with the boundary, thus making the use of homogenized fields inaccurate near the boundary. In addition, when failure occurs, macroscopic cracks are formed in the material, and the applicability of a given RVE may breakdown. Furthermore, one or two large cracks in a structure can invalidate the approximation of periodicity, meaning the RUC no longer exists. These issues make modeling of composite failure challenging. One method that has shown promise is multiscale modeling, wherein the appropriate length scales associated with the failure mechanism are incorporated. Here, it is often necessary to link explicitly the lengths of the various scales in order to assure that not only the deformation, but also the energy dissipated during failure, is consistent between length scales. Accounting for this appropriately within multiscale micromechanics failure predictions is a topic of current research (Bazant, 2007, Pineda et al., 2011). This topic is also discussed further in Chapter 7.

The reason for emphasizing the concept of an RVE is that it appears to provide a valuable discriminator between continuum (macroscopic) theories and microscopic theories. For scales larger than the RVE one can use continuum mechanics and

MATLAB Function 5.1 – `GMC.m`

This function performs micromechanics calculations using the GMC micromechanics theory

```matlab
function [Cstar, CTEs, As] = GMC(Constits, RUC)
% ++++++++++++++++++++++++++++++++++++++++++++++++++++++++++++++++++++++++++
% Copyright 2020 United States Government as represented by the Administrator of the
% National Aeronautics and Space Administration. No copyright is claimed in the
% United States under Title 17, U.S. Code. All Other Rights Reserved. By downloading
% or using this software you agree to the terms of NOSA v1.3. This code was prepared
% by Drs. B.A. Bednarcyk and S.M. Arnold to complement the book "Practical
% Micromechanics of Composite Materials" during the course of their government work.
% ++++++++++++++++++++++++++++++++++++++++++++++++++++++++++++++++++++++++++
% Purpose: Calculate the effective stiffness, effective CTES, and the mechanical and
%          thermal strain concentration tensors for the GMC micromechanics theory
% Input:
% - Constits: Struct containing constituent material information
% - RUC: Struct containing repeating unit cell (RUC) information
% Output:
% - Cstar: Effective stiffness matrix
% - CTEs: Effective CTEs
% - As: Struct containing subcell concentration tensors
% ++++++++++++++++++++++++++++++++++++++++++++++++++++++++++++++++++++++++++

% -- Number of subcells in each direction
NB = RUC.NB;
NG = RUC.NG;

% -- Obtain computational time for large RUCs
if NB*NG > 25*25
    disp(['  ** Getting Effective Properties using GMC - RUC = ', num2str(NB), ' by ',
num2str(NG)])
    tic
end

% -- Assign stiffness and CTE of each subcell
props(NB, NG).C = zeros(6); % -- Preallocate props
for IB = 1:NB
    for IG = 1:NG
        switch(RUC.matsCh(IB, IG))
            case 'F'
                constit = Constits.Fiber;
            case '1'
                constit = Constits.FiberDam;
            case 'M'
                constit = Constits.Matrix;
            case '2'
                constit = Constits.MatrixDam;
            case 'I'
                constit = Constits.Interface;
            case '3'
                constit = Constits.InterfaceDam;
            otherwise
                error(['Invalid RUC.matsch in GMC - RUC.matsch = ', RUC.matsch])
        end
        props(IB, IG).C = GetCFromProps(constit);
        props(IB, IG).alpha = [constit.aL; constit.aT; constit.aT; 0; 0; 0];
    end
end

% -- Subcell dimensions
h = RUC.h;
l = RUC.l;
sumH = sum(h);
```

```
sumL = sum(l);

% -- Assemble AG and J (Eq. 4.59) as SPARSE matrices
NNZ = 60*NB*NG; % -- Estimate of upper bound of non-zeros in AG
AG = zeros(NNZ,1);
J = zeros(2*(NB + NG)+ NB*NG + 1, 6);
row = 0;
ic = 0;

% -- ES11(B,G) = EG11 (Eq. 5.29)
for IB = 1:NB
    for IG = 1:NG
        row = row + 1;
        J(row, 1) = 1;
        NS = NG*(IB - 1) + IG;
        ic = ic + 1;
        NR(ic) = row; % -- Row number in AG
        NC(ic) = 6*(NS - 1) + 1; % -- Column number number in AG
        AG(ic) = 1; % --Value number in AG
    end
end

% -- SUM [H(B)*ES22(B,G)]=H*EG22 (Eq. 5.43)
for IG = 1:NG
    row = row + 1;
    J(row, 2) = sumH;
    for IB = 1:NB
        NS = NG*(IB - 1) + IG;
        ic = ic + 1;
        NR(ic) = row;
        NC(ic) = 6*(NS - 1) + 2;
        AG(ic) = h(IB);
    end
end

% -- SUM [L(G)*ES33(B,G)]=L*EG33  (Eq. 5.44)
for IB = 1:NB
    row = row + 1;
    J(row, 3) = sumL;
    for IG = 1:NG
        NS = NG*(IB - 1) + IG;
        ic = ic + 1;
        NR(ic) = row;
        NC(ic) = 6*(NS - 1) + 3;
        AG(ic) = l(IG);
    end
end

% -- SUM SUM [H(B)*L(G)*ES23(B,G)]=H*L*EG23 (Eq. 5.48)
row = row + 1;
J(row, 4) = sumH*sumL;
for IB = 1:NB
    for IG = 1:NG
        NS = NG*(IB - 1) + IG;
        ic = ic + 1;
        NR(ic) = row;
        NC(ic) = 6*(NS - 1) + 4;
        AG(ic) = h(IB)*l(IG);
    end
end

% -- SUM [L(G)*ES13(B,G)]=L*EG13 (5.46)
for IB = 1:NB
    row = row + 1;
    J(row, 5) = sumL;
    for IG = 1:NG
        NS = NG*(IB - 1) + IG;
        ic = ic + 1;
        NR(ic) = row;
        NC(ic) = 6*(NS - 1) + 5;
```

```
        AG(ic) = l(IG);
    end
end

% -- SUM [H(B)*ES12(B,G)]=H*EG12 (Eq. 5.47)
for IG = 1:NG
    row = row + 1;
    J(row, 6) = sumH;
    for IB = 1:NB
        NS = NG*(IB - 1) + IG;
        ic = ic + 1;
        NR(ic) = row;
        NC(ic) = 6*(NS - 1) + 6;
        AG(ic) = h(IB);
    end
end

% -- Define AG in matlab sparse format
AGsparse = sparse( NR(1:ic), NC(1:ic), AG(1:ic) );

% -- Assemble Am (Eq. 5.60)
NNZ = 6*NB*NG; % -- Estimate of upper bound of non-zeros in AM
AM = zeros(NNZ,1);

% -- Reset counters
row = 0;
ic = 0;

% -- sig22(B,G) = sig22(B+1,G) (Eq. 5.53)
for IG = 1:NG
    for IB = 1:NB - 1
        row = row + 1;
        for i = 1:6
            NS = NG*(IB - 1) + IG;
            ic = ic + 1;
            NR(ic) = row;
            NC(ic) = 6*(NS - 1) + i;
            AM(ic) = props(IB, IG).C(2, i);
            NS = NG*(IB) + IG;
            ic = ic + 1;
            NR(ic) = row;
            NC(ic)=6*(NS - 1) + i;
            AM(ic)= -props(IB + 1, IG).C(2, i);
        end
    end
end

% -- sig33(B,G) = sig33(B,G+1) (Eq. 5.54)
for IB = 1:NB
    for IG = 1:NG - 1
        row = row + 1;
        for i = 1:6
            NS = NG*(IB - 1) + IG;
            ic = ic + 1;
            NR(ic) = row;
            NC(ic)=6*(NS - 1) + i;
            AM(ic) = props(IB, IG).C(3, i);
            NS = NG*(IB - 1) + IG + 1;
            ic = ic + 1;
            NR(ic) = row;
            NC(ic)=6*(NS - 1) + i;
            AM(ic) = -props(IB, IG + 1).C(3, i);
        end
    end
end

% -- sig23(B,G) = sig23(B+1,G) (Eq. 5.55)
for IG = 1:NG
    for IB = 1:NB - 1
```

```
        row = row + 1;
        for i = 1:6
            NS = NG*(IB - 1) + IG;
            ic = ic + 1;
            NR(ic) = row;
            NC(ic) = 6*(NS - 1) + i;
            AM(ic) = props(IB, IG).C(4, i);
            NS = NG*(IB) + IG;
            ic = ic + 1;
            NR(ic) = row;
            NC(ic)=6*(NS - 1) + i;
            AM(ic)= -props(IB + 1, IG).C(4, i);
        end
    end
end

% -- sig32(B,G) = sig32(B,G+1)  (Eq. 5.56)
IB = NB;
for IG = 1:NG - 1
    row = row + 1;
    for i = 1:6
        NS = NG*(IB - 1) + IG;
        ic = ic + 1;
        NR(ic) = row;
        NC(ic)=6*(NS - 1) + i;
        AM(ic) = props(IB, IG).C(4, i);
        NS = NG*(IB - 1) + IG + 1;
        ic = ic + 1;
        NR(ic) = row;
        NC(ic)=6*(NS - 1) + i;
        AM(ic) = -props(IB, IG + 1).C(4, i);
    end
end

% -- sig31(B,G) = sig31(B,G+1)  (eq. 5.57)
for IB = 1:NB
    for IG = 1:NG - 1
        row = row + 1;
        for i = 1:6
            NS = NG*(IB - 1) + IG;
            ic = ic + 1;
            NR(ic) = row;
            NC(ic)=6*(NS - 1) + i;
            AM(ic) = props(IB, IG).C(5, i);
            NS = NG*(IB - 1) + IG + 1;
            ic = ic + 1;
            NR(ic) = row;
            NC(ic)=6*(NS - 1) + i;
            AM(ic) = -props(IB, IG + 1).C(5, i);
        end
    end
end

% -- sig21(B,G) = sig21(B+1,G)  (Eq. 5.58)
for IB = 1:NB - 1
    for IG = 1:NG
        row = row + 1;
        for i = 1:6
            NS = NG*(IB - 1) + IG;
            ic = ic + 1;
            NR(ic) = row;
            NC(ic)=6*(NS - 1) + i;
            AM(ic) = props(IB, IG).C(6, i);
            NS = NG*(IB) + IG;
            ic = ic + 1;
            NR(ic) = row;
            NC(ic)=6*(NS - 1) + i;
            AM(ic) = -props(IB + 1, IG).C(6, i);
        end
    end
```

```
      end
end

% -- Define AM in matlab sparse format
AMsparse = sparse( NR(1:ic), NC(1:ic), AM(1:ic) );

% -- Assemble M and K (where R = K*epsb) (Eq. 5.62)
M = [AMsparse; AGsparse];
K = [zeros(5*NB*NG - 2*(NB + NG) - 1, 6); J];

% -------------------------------------------------------------------------
% -- Equations can be solved directly or by applying 6 strain components
%    See Chapter 5, Section 5.6
% -------------------------------------------------------------------------

% -- Direct Invert Solution (Faster)
% -- Solve for A
A = M\K;

% -- Extract 6x6 subcell strain concentration matrix from A
As(NB, NG).A = 0.;
for IG = 1:NG
    for IB = 1:NB
        astart=6*((IB - 1)*NG + IG - 1) + 1;
        aend = astart + 5;
        As(IB, IG).A = A(astart:aend, :);
    end
end
% -------------------------------------------------------------------------
% -- Apply 6 Components Solution (Slower)
%      for i = 1:6
%
%          % -- Apply epsb(i) = 1, all other epsb = 0
%          dtemp = 0;
%          epsb = zeros(6,1);
%          epsb(i) = 1;
%
%          R = K*epsb; % -- (see Eq. 5.62)
%          X=M\R;       % -- Subcell strains solution
%
%          % -- Extract appropriate strain components from X
%          for IG = 1:NG
%              for IB = 1:NB
%                  xstart=6*((IB - 1)*NG + IG - 1) + 1;
%                  xend = xstart + 5;
%                  As(IB, IG).A(1:6, i) = X(xstart:xend);
%              end
%          end
%
%      end
% -------------------------------------------------------------------------

% -- Calculate effective stiffness (Eq. 5.65)
Cstar = zeros(6);
for IG = 1:NG
    for IB = 1:NB
        Cstar = Cstar + h(IB)*l(IG)*props(IB,IG).C*As(IB, IG).A/(sumH*sumL);
    end
end

% ------------ THERMAL --------------
% -- Effective compliance matrix
SS = inv(Cstar);

% -- Stress concentration matrix B
B(NB, NG).B = 0;
for IG = 1:NG
    for IB = 1:NB
        B(IB, IG).B = props(IB, IG).C*As(IB, IG).A*SS; % -- Eq. 5.66
```

```
            B(IB, IG).BT = transpose(B(IB, IG).B);
        end
    end

    % -- Effective CTEs (Eq. 5.68)
    CTEs = zeros(6,1);
    for IG = 1:NG
        for IB = 1:NB
            CTEs = CTEs + h(IB)*l(IG)*B(IB, IG).BT*props(IB, IG).alpha/(sumH*sumL);
        end
    end

    % -- Subcell strains
    alpha_s = zeros(6*NB*NG,1);
    for IG = 1:NG
        for IB = 1:NB
            astart=6*((IB - 1)*NG + IG - 1) + 1;
            alpha_s(astart) = props(IB, IG).alpha(1);
            alpha_s(astart+1) = props(IB, IG).alpha(2);
            alpha_s(astart+2) = props(IB, IG).alpha(3);
        end
    end

    % -- Assemble RT (Eq. 5.74)
    RT = [AMsparse; zeros(2*(NB+NG)+NB*NG+1, 6*NB*NG)]*alpha_s;

    % -- Solve for subcell strains with epsb = 0 and DT = 1 (Eq. 5.73)
    eps_s = M\RT;

    % -- Extract 6x1 subcell thermal strain concentration matrix from eps_s
    for IG = 1:NG
        for IB = 1:NB
            astart=6*((IB - 1)*NG + IG - 1) + 1;
            aend = astart + 5;
            As(IB, IG).AT = eps_s(astart:aend);
            % -- Store subcell stiffness and CTE
            As(IB, IG).C = props(IB, IG).C;
            As(IB, IG).alpha = props(IB, IG).alpha;
        end
    end

    % -- Get computational time for large RUCs
    if NB*NG > 25*25
        ElapsedTime = toc;
        disp(['    GMC homogenization time = ', char(num2str(ElapsedTime)), ' seconds']);
    end

end
```

The `GMC.m` function can be used in stand-alone micromechanics analyses, as well as micromechanics-based CLT problems. However, because of the greater generality afforded by the GMC theory, additional input information is needed in the composite material definitions specified in `GetEffProps.m` (see MATLAB Function 3.7). In general the GMC theory admits any arbitrary number of constituent materials in any geometric arrangement within the doubly periodic RUC. In the current MATLAB implementation, only three constituents can be specified: a fiber, a matrix, and an interface material. Furthermore, the desired GMC RUC must be specified (discussed in more detail below). The following is an example GMC composite material specification from `GetEffProps.m`.

```
Mat = 134;
Theory = 'GMC';
name = "5x5 SiC-SiC Interface GMC Table 5.4";
Vf = 0.28;% Fiber Volume Fraction
Vi = 0.13;% Interface Volume Fraction
Constits.Fiber = constitprops{10};
Constits.Matrix = constitprops{11};
Constits.Interface = constitprops{12};
RUCid = 105;
[props{Mat}] = RunMicro(Theory,name,struct('Vf',Vf,'Vi',Vi),Constits,props{Mat},RUCid);
```

This is similar to the composite material specifications for MOC from Chapter 3, but now a third (interface) constituent material has been specified as Constits.Interface, with its overall volume fraction in the composite specified as vi. The desired RUC has also been indicated with an identification number (RUCid), and the call to the function RunMicro.m (see MATLAB Function 3.8) has been altered to pass the interface volume fraction and RUCid as arguments. RUCid is an optional argument to RunMicro.m that is added to the end of the argument list, while, when an interface material is specified, RunMicro.m expects the third argument to be a data structure (struct) containing both the fiber volume fraction and the interface volume fraction. The structure can be specified directly as an argument as shown above (highlighted), or, alternatively, the structure can be defined (Vol) and then passed as an argument, for example,

```
Mat = 134;
Theory = 'GMC';
name = "5x5 SiC-SiC Interface GMC Table 5.4";
Vol.Vf  = 0.28;% Fiber Volume Fraction
Vol.Vi  = 0.13;% Interface Volume Fraction
Constits.Fiber = constitprops{10};
Constits.Matrix = constitprops{11};
Constits.Interface = constitprops{12};
RUCid = 105;
[props{Mat}] = RunMicro(Theory,name,Vol,Constits,props{Mat},RUCid);
```

It can be confirmed that both composite material definitions above result in identical GMC predications. If no interface material is present in the RUC, then the composite material specification in GetEffProps.m is even more similar to Chapter 3 (with the third argument as the fiber volume fraction), where the only difference is the addition of the RUCid. For example,

```
Mat = 200;
Theory = 'GMC';
name = "GMC 7x7 IM7-8552";
Vf  = 0.28;
Constits.Fiber = constitprops{1};
Constits.Matrix = constitprops{3};
RUCid = 7;
[props{Mat}] = RunMicro(Theory, name, Vf, Constits, props{Mat}, RUCid);
```

The RUCid is a number specifying which GMC RUC will be used in the analysis. To translate this number into all the details of a GMC RUC (subcell dimensions and material positions of the subcells in the RUC), MATLAB Function 5.2 – GetRUC.m, has been implemented. As shown in the function, each RUCid number is associated with a distinct RUC, and there is an elseif block for each RUCid (so additional RUCs can easily be added). The data associated with the RUC is stored in a struct called RUC. The RUCs implemented for this chapter in GetRUC.m are shown schematically in Fig. 5.3. RUCid = 99 is user-defined and reserves space in GetRUC.m to implement any RUC definition. The remaining RUCs are implemented such that the subcell dimensions are calculated to achieve the specified fiber (and interface) volume fraction. That is, for example, in RUCid = 7, the fiber shape is constant, with the length of the 'steps' in the fiber shape equal to one quarter of the fiber flat top, bottom, and side length. The desired fiber volume fraction is then achieved by altering the size of the matrix subcells surrounding the fiber. Of course, there is also an upper limit on the fiber volume fraction, corresponding to the point at which the fiber boundaries intersect the RUC boundaries. This maximum fiber volume fraction is calculated and stored as `RUC.Vf_max`. It should also be noted that, for `RUCid = 26`, the variable `Extra` can be set to a value greater than 1 to add additional matrix subcells between the fibers (see Fig. 5.3). This will increase the total number of subcells, but not alter the fiber shape, thus having no effect on the GMC predictions. It will, however, be used for the High-Fidelity Generalized Method of Cells (HFGMC) in Chapter 6.

MATLAB Function 5.2 – `GetRUC.m`

This function sets up and stores the (RUC) information for GMC and HFGMC

```
function [RUC] = GetRUC(Constits, Vf, Vi, RUCid)
% ++++++++++++++++++++++++++++++++++++++++++++++++++++++++++++++++++++++++++++++++++
% Copyright 2020 United States Government as represented by the Administrator of the
% National Aeronautics and Space Administration. No copyright is claimed in the
% United States under Title 17, U.S. Code. All Other Rights Reserved. By downloading
% or using this software you agree to the terms of NOSA v1.3. This code was prepared
% by Drs. B.A. Bednarcyk and S.M. Arnold to complement the book "Practical
% Micromechanics of Composite Materials" during the course of their government work.
% ++++++++++++++++++++++++++++++++++++++++++++++++++++++++++++++++++++++++++++++++++
% Purpose: Obtain the repeating unit cell (RUC) information for GMC and HFGMC
% Input:
% - Constits: Struct containing constituent material information
% - Vf: Fiber volume fraction
% - Vi: Interface material volume fraction
% - RUCid: Repeating unit cell (RUC) id number
% Output:
% - RUC: Struct containing repeating unit cell (RUC) information
% ++++++++++++++++++++++++++++++++++++++++++++++++++++++++++++++++++++++++++++++++++

RUC.id = RUCid; % -- Store RUCid in the struct

F = Constits.Fiber.constID;
M = Constits.Matrix.constID;


% -- If Vf = 0, set to all matrix
if Vf == 0
    F = M;
    Vf = 0.25;
    if Vi ~= 0
        error('Vf = 0, Vi ~= 0 inadmissible (GetRUC.m)');
    end
end

% -- Store Vf and Vi in struct
RUC.Vf = Vf;
if Vi > 0
    I = Constits.Interface.constID;
    RUC.Vi = Vi;
end

% -- 2x2 RUC
if RUCid == 2
    RUC.NB = 2;
    RUC.NG = 2;
    RUC.Vf_max = 0.991;
    if (Vf >= RUC.Vf_max)
        disp([' Vf too large, Vf_max = ', num2str(RUC.Vf_max)]);
    else
        h1=sqrt(Vf);
        h2=1-h1;
        l1=h1;
        l2=h2;
        RUC.h = [h1, h2];
        RUC.l = [l1, l2];
        RUC.mats = [F, M; ...
                    M, M];
    end

% -- 3x3 RUC
elseif RUCid == 3
    RUC.NB = 3;
    RUC.NG = 3;
    RUC.Vf_max = 0.991;
```

```
    if (Vf >= RUC.Vf_max)
        disp([' Vf too large, Vf_max = ', num2str(RUC.Vf_max)]);
    else
        h2=sqrt(Vf);
        h1=(1-h2)/2;
        h3=h1;
        RUC.h = [h1, h2, h3];
        RUC.l = RUC.h;
        RUC.mats = [M, M, M; ...
                    M, F, M;
                    M, M, M];
    end

% -- 7x7 RUC with step length = 1/4 of straight length,
%    modified edge subcell dimensions to obtain desired Vf
elseif RUCid == 7
    RUC.NB = 7;
    RUC.NG = 7;
    RUC.Vf_max = 0.8125;
    if (Vf >= RUC.Vf_max)
        disp([' Vf too large, Vf_max = ', num2str(RUC.Vf_max)]);
    else
        h1 = (sqrt(13)*sqrt(Vf) - 4*Vf)/Vf;
        RUC.h = [h1, 1, 1, 4, 1, 1, h1];
        RUC.l = RUC.h;
        RUC.mats = [M, M, M, M, M, M, M; ...
                    M, M, M, F, M, M, M; ...
                    M, M, F, F, F, M, M; ...
                    M, F, F, F, F, F, M; ...
                    M, M, F, F, F, M, M; ...
                    M, M, M, F, M, M, M; ...
                    M, M, M, M, M, M, M];
    end

% -- 26x26 RUC with option to add Extra subcells between fibers,
%    modifies edge subcell dimensions to obtain desired Vf
elseif RUCid == 26
    % -- Extra = # of subcells between fiber and edge of RUC (for HFGMC)
    Extra = 1;
    %Extra = 4;

    RUC.NB = 26 + 2*(Extra - 1);
    RUC.NG = 26 + 2*(Extra - 1);
    RUC.Vf_max = 0.80613; % -- Max Vf before fibers touch
    if (Vf >= RUC.Vf_max)
        disp([' Vf too large, Vf_max = ', num2str(RUC.Vf_max)]);
    else

        XH1_TOT = sqrt(3838.D0 / Vf) - 69.D0;
        XXH1 = XH1_TOT/Extra;

        XL1_TOT = XH1_TOT;
        XXL1 = XL1_TOT/Extra;

        for II = 1:Extra
            RUC.h(II) = XXH1;
            RUC.l(II) = XXL1;
        end

        shift = Extra - 1;

        RUC.h(2 + shift)= 2.;
        RUC.h(3 + shift)= 3.;
        RUC.h(4 + shift)= 4.;
        RUC.h(5 + shift)= 3.;
        RUC.h(6 + shift)= 4.;
        RUC.h(7 + shift)= 5.;
        RUC.h(8 + shift)= 5.;
```

```
        RUC.h(9 + shift)= 4.;
        RUC.h(10 + shift)= 6.;
        RUC.h(11 + shift)= 9.;
        RUC.h(12 + shift)= 7.;
        RUC.h(13 + shift)= 17.;

        for I = 1: 12 + Extra - 1
           RUC.h(13 + Extra - 1 + I) = RUC.h(13 + Extra - I);
           RUC.l(13 + Extra - 1 + I) = RUC.h(13 + Extra - 1 + I);
           RUC.l(13 + Extra - I) = RUC.h(13 + Extra - I);
        end
        RUC.l(26 + 2*(Extra - 1)) = RUC.l(1);
        RUC.h(26 + 2*(Extra - 1)) = RUC.h(1);

        for IB = 1: RUC.NB
           for IG = 1: RUC.NG

              RUC.mats(IB, IG) = M;

              if ((IB == 2 + shift) || (IB == 25 + shift))
                 if ((IG >= 13 + shift) && (IG <= 14 + shift))
                    RUC.mats(IB, IG) = F;
                 end
              elseif ((IB == 3 + shift) || (IB == 24 + shift))
                 if ((IG >= 12 + shift) && (IG <= 15 + shift))
                    RUC.mats(IB, IG) = F;
                 end
              elseif ((IB == 4 + shift) || (IB == 23 + shift))
                 if ((IG >= 11 + shift) && (IG <= 16 + shift))
                    RUC.mats(IB, IG) = F;
                 end
              elseif ((IB == 5 + shift) || (IB == 22 + shift))
                 if ((IG >= 10 + shift) && (IG <= 17 + shift))
                    RUC.mats(IB, IG) = F;
                 end
              elseif ((IB == 6 + shift) || (IB == 21 + shift))
                 if ((IG >= 9 + shift) && (IG <= 18 + shift))
                    RUC.mats(IB, IG) = F;
                 end
              elseif ((IB == 7 + shift) || (IB == 20 + shift))
                 if ((IG >= 8 + shift) && (IG <= 19 + shift))
                    RUC.mats(IB, IG) = F;
                 end
              elseif ((IB == 8 + shift) || (IB == 19 + shift))
                 if ((IG >= 7 + shift) && (IG <= 20 + shift))
                    RUC.mats(IB, IG) = F;
                 end
              elseif ((IB == 9 + shift) || (IB == 18 + shift))
                 if ((IG >= 6 + shift) && (IG <= 21 + shift))
                    RUC.mats(IB, IG) = F;
                    RUC.matsCh(IB, IG) = 'F';
                 end
              elseif ((IB == 10 + shift) || (IB == 17 + shift))
                 if ((IG >= 5 + shift) && (IG <= 22 + shift))
                    RUC.mats(IB, IG) = F;
                 end
              elseif ((IB == 11 + shift) || (IB == 16 + shift))
                 if ((IG >= 4 + shift) && (IG <= 23 + shift))
                    RUC.mats(IB, IG) = F;
                 end
              elseif ((IB == 12 + shift) || (IB == 15 + shift))
                 if ((IG >= 3 + shift) && (IG <= 24 + shift))
                    RUC.mats(IB, IG) = F;
                 end
              elseif ((IB == 13 + shift) || (IB == 14 + shift))
                 if ((IG >= 2 + shift) && (IG <= 25 + shift))
                    RUC.mats(IB, IG) = F;
                 end
              end
           end
```

```
            end
        end

    end

% -- Spot for user to put custom RUC
elseif RUCid == 99


% -- 5x5 RUC with interface
elseif RUCid == 105
    RUC.NB = 5;
    RUC.NG = 5;
    RUC.Vf_max = 0.991 - Vi;
    if (Vf >= RUC.Vf_max)
        disp([' Vf too large, Vf_max = ', num2str(RUC.Vf_max)]);
    else
        h3=sqrt(Vf);
        h2 = ( -h3+sqrt(h3^2 + Vi) )/2;
        h1=(1-2*h2-h3)/2;
        RUC.h = [h1, h2, h3, h2, h1];
        RUC.l = RUC.h;
        RUC.mats = [M, M, M, M, M; ...
                    M, I, I, I, M; ...
                    M, I, F, I, M; ...
                    M, I, I, I, M; ...
                    M, M, M, M, M];
    end


else
    error(['Invalid RUCid, RUCid = ', num2str(RUCid)]);

end

% -- Setup the character material specifier (matsCh)
if isfield(RUC, 'mats')
    for i = 1: RUC.NB
        for j = 1: RUC.NG
            if RUC.mats(i,j) == M
                RUC.matsCh(i,j) = 'M';
            elseif RUC.mats(i,j) == F
                RUC.matsCh(i,j) = 'F';
            elseif RUC.mats(i,j) == I
                RUC.matsCh(i,j) = 'I';
            else
                error('Problem in GetRUC.m');
            end
        end
    end
end

end
```

5.7   Example problems

Examples of the GMC functionality are presented below, starting with the prediction of effective properties followed by calculations of local fields. This is done for PMC composites first and then repeated for CMC composites with and without an interface. Finally, failure envelopes are generated for CMC composites.

5.7.1   Effective properties of PMCs

In the GMC, the level of refinement used to represent the shape of a fiber within the RUC is arbitrary. Three of the five fiber shape refinements shown in Fig. 5.3 (2×2, 7×7, 26×26) will be used in this example. The 2×2 subcell RUC, which corresponds to the MOC RUC (see Chapter 3), contains a single fiber subcell with three matrix subcells. By adding additional subcells in each direction, a fiber shape that is more

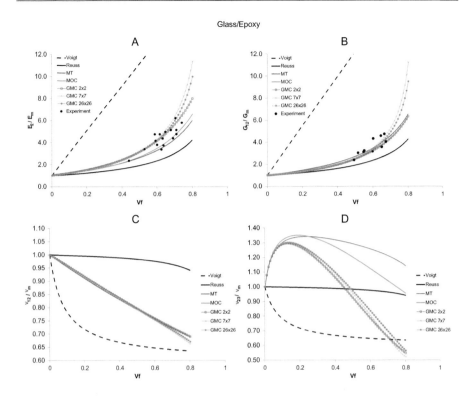

Figure 5.4 Effective property predictions for a glass/epoxy composite (Tsai and Hahn, 1980) as a function of fiber volume fraction for different micromechanics theories and GMC RUC discretizations. (A) Transverse Young's modulus normalized by matrix Young's modulus; (B) Axial shear modulus normalized by matrix shear modulus; (C) Axial Poisson's ratio normalized by matrix Poisson's ratio; (D) Transverse Poisson's ratio normalized by matrix Poisson's ratio.

representative of a circle can be achieved by "stair-stepping" the rectangular subcells. This essentially changes the local volume fraction of each constituent in a given row or column of subcells (see Liu and Ghoshal (2014) for details). It is emphasized, however, that adding subcells while keeping the fiber shape the same (i.e., simply subdividing the unit cells shown in Fig. 5.3), will have absolutely no effect on the effective properties or local stress and strain fields predicted by GMC. For this reason, GMC is "mesh independent" and there is no stress concentration associated with the rectangular subcell corners. As is the case in MOC (see Fig. 3.5), in GMC, the subcell fields are actually centroidal values.

Considering a glass/epoxy composite, which was tested by Tsai and Hahn (1980) with $E_f/E_m = 21.19$, $v_f = 0.22$, $v_m = 0.35$, the predicted normalized transverse Young's modulus, normalized-axial-shear modulus, normalized-axial-Poisson's ratio, and normalized transverse Poisson's ratio are plotted versus fiber volume fraction in Fig. 5.4. Each property is normalized by the appropriate matrix quantity. Predictions are shown for the methods presented in Chapter 3, as well as two GMC RUC discretizations approximating a circular fiber (7×7 and 26×26, Fig. 5.3). The 2×2 RUC shown in

Fig. 5.3 gives identical results for GMC and MOC unaveraged (these results are labeled as "GMC 2×2" in Fig. 5.4). Axial Young's modulus predictions are not shown as all methods other than Reuss give nearly identical results across all volume fractions (see Fig. 3.9). To obtain these results, one must introduce new constituent material properties for the Tsai-Hahn glass (Mat = 5) and Tsai-Hahn epoxy (Mat = 6) into GetConstitProps.m as shown below:

```
%Constituent Mat#5 - Tsai&Hahn Glass (Isotropic)
Mat = 5;
constitprops{Mat}.name = "Tsai&Hahn Glass";
constitprops{Mat}.constID = Mat;
constitprops{Mat}.EL = 113.4E3;
constitprops{Mat}.ET = 113.4E3;
constitprops{Mat}.vL = 0.22;
constitprops{Mat}.vT = 0.22;
constitprops{Mat}.GL = constitprops{Mat}.EL/(2*(1+constitprops{Mat}.vL)); % -- isotropic
constitprops{Mat}.aL = 5.e-06;
constitprops{Mat}.aT = 5.e-06;

%Constituent Mat#6 - Tsai&Hahn Epoxy (Isotropic)
Mat = 6;
constitprops{Mat}.name = "Tsai&Hahn Epoxy";
constitprops{Mat}.constID = Mat;
constitprops{Mat}.EL = 5.35E3;
constitprops{Mat}.ET = 5.35E3;
constitprops{Mat}.vL = 0.35;
constitprops{Mat}.vT = 0.35;
constitprops{Mat}.GL = constitprops{Mat}.EL/(2*(1+constitprops{Mat}.vL)); % -- isotropic
constitprops{Mat}.aL = 54.e-06;
constitprops{Mat}.aT = 54.e-06;
```

In addition, a new composite material (Mat = 130) must be defined within GetEff-Props.m (see MATLAB Function 3.7) as illustrated below:

```
Mat = 130;
name = "2x2-7x7-and-26x26 Glass-Epoxy Tsai-Hahn GMC - Fig 5.4";
Constits.Fiber = constitprops{5};
Constits.Matrix = constitprops{6};
% -- Call function (contained in this file) to get eff props across all Vf
RunOverVf(name, Constits, props, Mat);
props{Mat}.Quit = true; % -- Tells MicroAnalysis to quit after getting eff props
```

where the function RunOverVf within GetEffProps.m has been expanded to include the analysis of GMC using the three different RUC discretizations. Note RunOverVf is called to run all micromechanics theories over all fiber volume fractions and write the results to a text file ("Props vs. Vf ... .txt"). Plots comparing each property among the theories can then be made using any desired software. The analyses are initiated by executing the MicroAnalysis.m script (see MATLAB Script 3.5) with the problem definition in MicroProblemDef.m as follows:

```
NP = NP + 1;
OutInfo.Name(NP) = "Tsai-Hahn Glass/Epoxy Chap 5 Example 1";
MicroMat{NP} = 130;
```

Note that loading need not be defined because of the inclusion of the props{Mat}.Quit = true statement in the composite material definition shown earlier.

Fig. 5.4 shows that, for this glass/epoxy system, for all properties aside from axial Poisson's ratio, there is a significant difference between the simplified methods from Chapter 3 and the GMC results, particularly at higher fiber volume fractions. For GMC, the predicted effective properties tend to converge as the representation of the circular fiber shape is refined. For the Young's and shear moduli, below a fiber volume fraction of 0.6, all three RUCs give very similar results, with the differences among the curves becoming greater as the volume fraction rises. Note that high volume fractions require the size of the fiber in Fig. 5.3 to grow with respect to the matrix, which results in thinner and thinner matrix regions between adjacent fibers. The Poisson's ratio predictions in Fig. 5.4C and D show that the three RUCs give very similar results for the axial Poisson's ratio and also agree quite well for the transverse Poisson's ratio.

It is noted that, for the Tsai and Hahn (1980) glass/epoxy composite, the mismatch in the transverse Young's moduli of the constituents is a factor of 21.19. In contrast, for carbon/epoxy systems, this transverse mismatch is much less. For example, considering a carbon/epoxy composite tested by Dean and Turner (1973) and Kriz and Stinchcomb (1979) (fiber properties: $E_1 = 232$ GPa, $E_2 = 15$ GPa, $G_{12} = 24$ GPa, $\nu_{12} = 0.279$, $\nu_{23} = 0.49$ (see `Mat = 7` in `GetConstitProps.m`); isotropic matrix properties: $E = 5.35$ GPa, $\nu = 0.35$, see `Mat = 6`) this factor is 2.8. Due to this lower transverse property mismatch, the effect of micromechanics theory and RUC refinement on the predicted effective properties (see `Mat = 131` within `GetEffProps.m`), as shown in Fig. 5.5, is considerably less than in the case of glass/epoxy. For the transverse Young's modulus and the axial Poisson's ratio, the predictions of all three RUCs are nearly coincident. The influence of the RUC refinement on the axial shear modulus (Fig. 5.5B) and transverse Poisson's ratio (Fig. 5.5D) is noticeable. For both material systems in Figs. 5.4 and 5.5, the MT, MOC, and GMC predictions are within the scatter of the plotted experimental data.

5.7.2   Local fields in PMCs

In this section, GMC local field predictions are compared to the finite element (FE) micromechanics approach. The FE micromechanics approach, in which the fiber and matrix are explicitly meshed within the finite element model (see Fig. 5.6A), can be used to determine the local fields within a composite RUC. This approach is usually considered to be the "gold standard" that provides the "correct" local stress fields, and thus other micromechanics model results are often compared to FE results. However, care must be taken to employ proper mesh discretization and boundary conditions within the FE analyses to ensure that the FE solution represents a truly periodic material as intended. Sun and Vaidya (1996) address this issue, stating that, for normal loading, it is sufficient to analyze an RUC like that shown in Fig. 5.6A (or a quarter of this geometry) by ensuring that plane boundaries remain plane. That is, for normal loading in the x_3-direction in Fig. 5.6A, it is sufficient to apply a uniform displacement in the x_3-direction on the $+x_3$ boundary, enforce symmetry on the $-x_3$ and $-x_2$ boundaries, and require that the $+x_2$ boundary remains horizontal through constraint conditions. For transverse shear and axial shear loading, on the other hand, Sun and Vaidya (1996) show that an analogous approach, where distortion of the

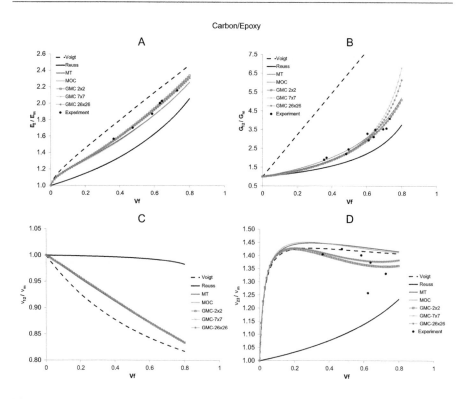

Figure 5.5 Effective property predictions for a carbon/epoxy composite (Dean and Turner, 1973; Kriz and Stinchcomb, 1979) as a function of fiber volume fraction for different micromechanics theories and GMC RUC discretizations. (A) Transverse Young's modulus normalized by matrix Young's modulus; (B) Axial shear modulus normalized by matrix shear modulus; (C) Axial Poisson's ratio normalized by matrix Poisson's ratio; (D) Transverse Poisson's ratio normalized by matrix Poisson's ratio.

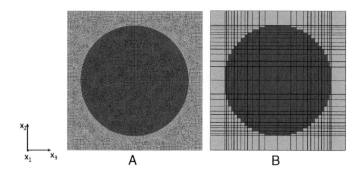

Figure 5.6 (A) Example finite element mesh of a fiber/matrix unit cell. (B) 26×26 GMC RUC.

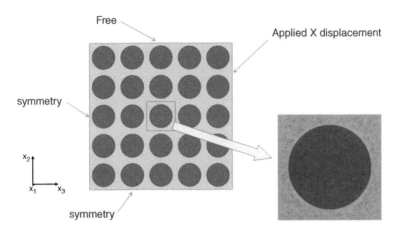

Figure 5.7 Finite element approach using a 25-fiber domain wherein the middle fiber RUC is examined.

boundaries is constrained (i.e., for transverse shear, the RUC is forced to remain a parallelogram) is incorrect. Rather, a representative volume element including multiple fibers must be employed, or truly periodic boundary conditions must be imposed in the FE analysis.

For transverse normal loading of an FE RUC mesh, three approaches are compared below. First, the often used method of requiring plane boundaries to remain plane is employed with a single fiber RUC. Second, a domain consisting of 25 fibers is employed, where the $+x_2$ boundary is left free (see Fig. 5.7). Third, a single fiber RUC is employed where true periodic boundary conditions are employed by linking the deformation of the nodes on opposite faces, see Totry et al. (2008) and Stier et al. (2015). In this case, care must be taken to ensure that the boundary nodes are perfectly aligned (e.g., every node on the $-x_2$ boundary must have a node on the $+x_2$ boundary with the identical x_3-coordinate). These three approaches are compared for a glass/epoxy composite with a fiber volume fraction of 0.5. The properties of the isotropic glass fiber (Mat = 8 in GetConstitProps.m) are: $E = 69$ GPa, $v = 0.2$; while the isotropic epoxy matrix (Mat = 9 in GetConstitProps.m) properties are: $E = 3.42$ GPa, $v = 0.34$. Transverse (x_3-direction) displacement loading was applied that is equivalent to a global strain of 0.02. Generalized plane strain elements (approximately 11,000 elements in each model) were used in the Abaqus (Simulia, 2011) commercial FE software. Note that a less dense mesh could have been employed, but, for the comparisons presented, smooth stress fields were desirable. von Mises stress field results for the second (25 fiber) approach are shown in Fig. 5.8, where it is clear that the effects of the top boundary are limited to the first row of fibers. The middle fiber RUC, as indicated, is used to examine the local fields. Fig. 5.9 compares results for all three FE approaches. The von Mises stress fields shown in Fig. 5.9 indicate that the three approaches produce nearly identical results, and the first approach, which is the simplest, is sufficient for this load case, as stated (but not explicitly shown) by Sun and Vaidya (1996).

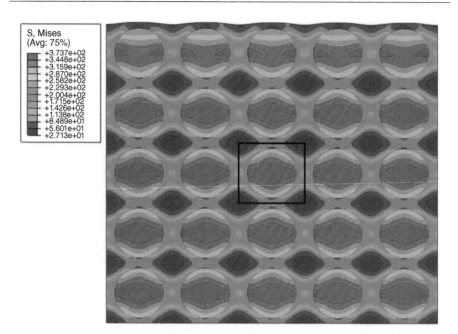

Figure 5.8 Deformed plot (deformation factor = 5) of the von Mises stress (MPa) predicted using the finite element approach using a 25-fiber domain for a 0.5 fiber volume fraction glass/epoxy composite with an applied horizontal strain of 0.02.

Fig. 5.10 compares the FE stress fields for the 0.5 fiber volume fraction glass/epoxy composite discussed above with results predicted by GMC using a 26×26 RUC (Fig. 5.6B). The GMC results can be generated using the MATLAB script 3.5 – MicroAnalysis.m and applying a transverse strain of 0.02 in the x_3-direction. The problem is defined in MicroProblemDef.m as follows:

```
NP = NP + 1;
OutInfo.Name(NP) = "Contour Glass-Epoxy Chap 5 Fig 5.10";
MicroMat{NP} = 132;
Loads{NP}.DT = 0;
Loads{NP}.Type  = [ S,    S,    E,    S,   S,   S];
Loads{NP}.Value = [ 0,  0.0,  0.02,   0,   0,   0];
```

while the composite material (Mat = 132) within GetEffProps.m (see MATLAB Function 3.7) is defined as:

```
Mat = 132;
Theory = 'GMC';
name = "26x26 Glass-Epoxy Contour GMC Fig 5.10";
Vf  = 0.50;
Constits.Fiber = constitprops{8};
Constits.Matrix = constitprops{9};
RUCid = 26;
[props{Mat}] = RunMicro(Theory, name, Vf, Constits, props{Mat}, RUCid);
```

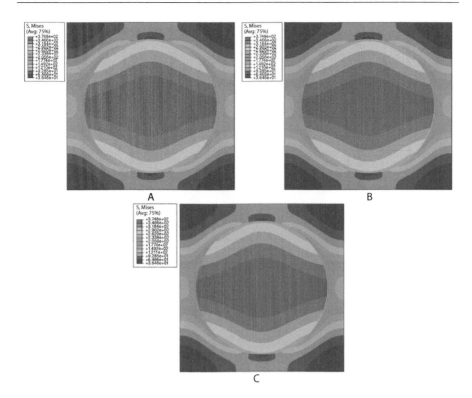

Figure 5.9 Von Mises stress fields (MPa) predicted by the FE approach for (A) a single fiber RUC subjected to planes remain plane conditions, (B) the middle fiber/matrix RUC within a 25-fiber domain, and (C) a single fiber RUC subjected to true periodic boundary conditions. The composite is 0.5-fiber volume fraction glass/epoxy with an applied horizontal strain of 0.02.

The MATLAB code will produce pseudocolor plots of the local stress and strain fields provided the flag `OutInfo.MakePlots` is set equal to true in `MicroProblemDef.m` (MATLAB Function 3.6). The limits of the colorbar scales for each component between the FE and GMC results are identical. However, although care has been taken to make the colorbars similar between MATLAB and Abaqus, they are not identical. Note that, to obtain the von Mises stress and pressure plots, the flag `PlotSvm = true` must be set in `PlotMicroFields.m`.

The most obvious feature within Fig. 5.10 is the lack of shear coupling within GMC. When transverse normal loading is applied, as shown, GMC predicts zero transverse shear stress (σ_{23}) throughout the RUC. In the FE results, transverse shear stresses arise at the fiber-matrix interface whose maxima are approximately 15% of the maximum transverse normal stress in the loading direction. Note that the average of the FE transverse shear stresses is zero. Further, both the FE and GMC simulations are based on generalized plane strain conditions and isotropic constituents, so the axial shear stresses are zero (and thus are not plotted). A further manifestation of the lack of shear

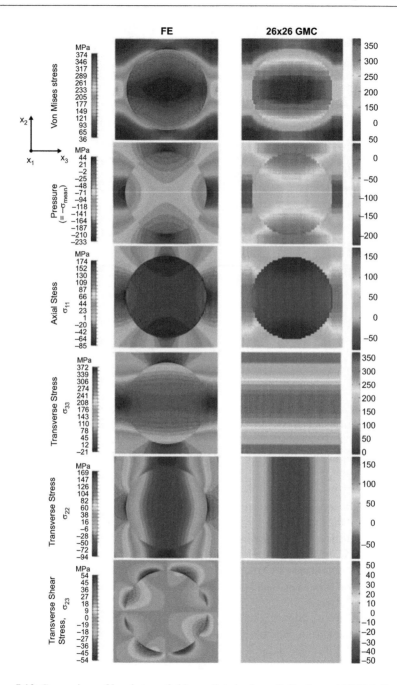

Figure 5.10 Comparison of local stress fields predicted using a finite element RUC (left column) and a 26×26 GMC RUC (right column) for a glass/epoxy composite subjected to 0.02 global transverse strain in the x_3-direction.

coupling in the GMC results is indicated by the transverse normal stress results. The GMC results for σ_{33} are constant in rows of subcells along the x_3-direction, while the GMC results for σ_{22} are constant in columns of subcells along the x_2-direction. This is because the subcells are unable to transfer stress via shear while traction continuity conditions require continuity of the normal stresses in the appropriate direction. This character of the transverse stress fields is precisely what enables GMC to be so efficient (see Aboudi et al., 2013). Comparing these transverse normal stress fields to those of the FE model, it is clear that GMC does not capture the maximum and minimum values evident in the FE results that occur due to local variations (in GMC, stresses do not vary within a subcell). However, the general character of the FE results is captured by GMC. In fact, the appearances of the GMC transverse stress fields are as if the FE fields were smeared (or averaged) in the appropriate directions. This is natural as GMC is an approximate method based on imposing averaged continuity conditions between the subcells.

Examining the axial (fiber direction) stress fields in Fig. 5.10, it is clear that GMC does quite a good job of reproducing the local fields from the FE analysis, although the maximum and minimum axial stresses in the matrix are under predicted. The von Mises stress and pressure (pressure $= -\sigma_{\mathrm{mean}} = -(\sigma_{11} + \sigma_{22} + \sigma_{33})/3$) fields are also well represented, although more average in nature, within the GMC results. The absolute maximum and minimum magnitudes are again not captured by GMC and, in the case of the von Mises stress field, there is some discrepancy in the fiber near the top and bottom. The level of correlation between the FE and GMC von Mises stress fields in the matrix is noteworthy given that GMC predicts zero shear stresses throughout the composite. It is emphasized that GMC is an efficient approximate method that provides all effective properties in one step and can simulate normal, shear, or fully multiaxial loading on the composite without the need to apply boundary conditions. It is thus significantly more efficient, both in terms of execution and pre- and postprocessing, than the FE approach. Consequently, GMC is ideal to be embedded within a structural-scale model (e.g., CLT) to represent the local composite material behavior. However, unlike most other efficient micromechanics theories (e.g., mean field theories like Mori-Tanaka), it can still provide a good representation of the spatial gradients in the local fields. Chapter 6 will demonstrate the HFGMC, which provides local fields that match very well with FE results but is less computationally efficient than GMC.

In summary, compared to the MOC (see Chapter 3), the GMC enables an arbitrary number of rectangular subcells within the composite unit cell. This allows the circular nature of the fiber cross-section to be better captured and, by nature of the fact that the local fields can vary from subcell to subcell, allows greater variation in the local fields to be captured. In addition, GMC can be used to model composites with more than two phases and much more general microstructures. The generality is obviously far greater than the simpler analytical micromechanics approaches, most of which provide predictions of only the mean fields in the constituents and thus provide no variation of the stresses and strains in the constituents. Further, GMC's lack of shear coupling is shared by simpler analytical micromechanics theories (e.g., the Mori-Tanaka method). GMC stands apart from these other theories in that, as mentioned, it admits an arbitrary number of subvolumes (subcells) and provides a good approximation of the local

Table 5.2 SiC/SiC composite constituent properties (Arnold et al., 2016).

Mat	Material name	E (GPa)	ν	$\alpha \times 10^{-6}/°C$	$X_T = -X_C = Y_T = -Y_C$ (MPa)	$Q = R = S$ (MPa)
10	SiC Fiber	385	0.17	3.2	1800	700
11	SiC Matrix	327	0.22	3.1	600	350
12	BN Coating	10	0.23	4.0	70	45

fields within continuous fiber composites. The lack of shear coupling renders these fields more average in nature compared to the local fields that can be predicted with the FE micromechanics procedure. However, GMC has an advantage over FE in that it extremely computationally efficient, and it provides semiclosed form constitutive equations for a composite material. Thus, it is ideal for use as an anisotropic constitutive model in higher scale structural analyses.

5.7.3   Effective properties of CMCs

In ceramic matrix composites (CMCs), unlike PMCs, the mismatch in properties between the fiber and matrix constituents is small. For example, in SiC/SiC composites, the ratio of axial fiber to matrix stiffness is approximately 1.2, whereas in PMCs, this ratio is on the order of 20 to 60. The purpose of the fibers in a CMC is not to add stiffness or strength, but rather to add toughness. This is accomplished by coating the SiC fibers with a weak and compliant interface material whose purpose is to deflect cracks and prevent them from growing into the fiber. Consequently, SiC/SiC CMCs consist of three phases: a fiber, a matrix, and an interface. The micromechanics theories, as presented in the previous chapters, are limited to two-phase composite materials and thus are unable to predict accurately SiC/SiC effective properties. To illustrate this, consider a 0.28 fiber volume fraction SiC/SiC CMC with no interface whose constituent properties are given in Table 5.2 and specified in `GetConstitProps.m` as `Mat` numbers 10 and 11. GMC using a 2×2 RUC (which is equivalent to MOC unaveraged) predicts the effective properties shown in Table 5.3. The composite material is defined in `GetEffProps.m` as `Mat = 133`. The predicted Young's moduli in the axial and transverse directions are nearly the same (343 GPa), whereas an actual SiC/SiC composite, see Table 2.1, has $E_1 = 300$ GPa and $E_2 = 152.4$ GPa. Thus, with no interface, the micromechanics predictions are erroneous in terms of magnitude and character.

In contrast, if a boron nitride (BN) coating with volume faction 0.13 is included in the GMC RUC, predictions of the SiC/SiC composite properties are much more realistic. To this end, the following problem has been defined in `MicroProblemDef.m`:

```
NP = NP + 1;
OutInfo.Name(NP) = "SiC-SiC BN Interface Chap 5 Fig 5.13";
MicroMat{NP} = 134;
Loads{NP}.DT = 0;
Loads{NP}.Type  = [  S,       E,    S,    S,  S,  S];
Loads{NP}.Value = [  0, 0.0006,   0.,    0,  0,  0];
```

Table 5.3 SiC/SiC composite effective properties predicted by GMC with no interface, as generated by the MATLAB code.

E1	E2	Nu12	G12	Nu23	G23	Alpha_1	Alpha_2
343360.7262	342605.9109	0.2059	141732.9253	0.20767	141356.837	3.1306e-06	3.1267e-06

Table 5.4 SiC/SiC composite effective properties predicted by GMC with interface, as generated by the MATLAB code.

E1	E2	Nu12	G12	Nu23	G23	Alpha_1	Alpha_2
302069.4622	164804.2841	0.21354	64814.1649	0.13322	26257.5636	3.1385e-06	3.1575e-06

with the composite material defined in `GetEffProps.m` as:

```
Mat = 134;
Theory = 'GMC';
name = "5x5 SiC-SiC Interface GMC Table 5.4";
Vf = 0.28;% Fiber Volume Fraction
Vi = 0.13;% Interface Volume Fraction
Constits.Fiber = constitprops{10};
Constits.Matrix = constitprops{11};
Constits.Interface = constitprops{12};
RUCid = 105;
[props{Mat}] = RunMicro(Theory,name,struct('Vf',Vf,'Vi',Vi),Constits,props{Mat},RUCid);
```

This material utilizes the 5×5 RUC shown in Fig. 5.3, wherein the SiC/SiC composite's compliant BN interface (see Table 5.2) is included as a third constituent material around the fiber, see `RUCid = 105` in MATLAB Function 5.2 - `GetRUC.m`. The effective composite properties predicted by GMC for this case are given in Table 5.4. Including the interface in the RUC reduced both the Young's and shear moduli. Further, there is now a significant difference between E_1 and E_2, and also between G_{12} and G_{23}, which is much more representative of an actual SiC/SiC composite.

5.7.4 Local fields in CMCs

The local stress fields in the CMC considered in the previous section (5×5 RUC, subjected to a transverse strain in the x_2-direction equal to 0.0006) with the BN interface included, are shown in Fig. 5.11. The location of the BN interface is evident in the plot of σ_{11} as the highly compliant interface material leads to lower stress magnitudes. The σ_{22} plot shows that all subcells located in columns that contain interface subcells have very low σ_{22} stress as GMC stresses in the x_2-direction are continuous in each column of subcells (as shown earlier in Fig. 5.10). Thus the vast majority of the σ_{22} stress is carried by the continuous columns of matrix in the x_2-direction. The same stress continuity is evident in the x_3-direction in the plot of σ_{33}.

5.7.5 CMC failure envelopes

Considering the same SiC/SiC CMC from the previous section, failure envelopes can be predicted as described in Chapter 4, but now using GMC. As done in Chapter 4 (see Section 4.4.3) the failure envelopes can be generated using the following code in GetEffProps.m:

```
% -- Micro scale failure envelopes - Example Fig 5.14
tttt = datetime(datetime,'Format','yyyy-MMM-dd HH.mm.ss');
OutInfo.EnvFile = ['Output/','SiC-SiC-BN Composite ENVELOPE - ',char(tttt),'.EnvData'];
ang = 0;
anginc = 2;
anginc1 = anginc;

while (ang <= 360.1)
    anginc = anginc1;
    % -- Extra angles for highly anisotropic cases
    if ang < 15 || (ang >= 165 && ang <= 190) || ang >= 345
      anginc = 0.2;
    end

    NP = NP + 1;
    OutInfo.Name(NP) = string(['Ang = ', num2str(ang)]);
    %MicroMat{NP} = 133; % -- 2x2 GMC - NO interface
    MicroMat{NP} = 134; % -- 5x5 GMC - RUCid=105
    S22 = sin(ang*pi/180);
    S11 = cos(ang*pi/180);
    Loads{NP}.Type  = [S,   S,   S,   S,   S, S];
    Loads{NP}.Value = [S11, S22, 0,   0,   0,  0];
    Loads{NP}.CriteriaOn = [1,1,1,1];
    Loads{NP}.ang = ang;
    ang = ang + anginc;
end
```

The CMC isotropic constituent stress allowables are given in Table 5.2. Note that all four implemented failure criteria, max stress, max strain, Tsai-Hill, and Tsai-Wu are being evaluated, and, as described in Chapter 4, the failure envelope quantifies the load at which any failure criterion is first exceeded.

The GMC failure envelopes for the SiC/SiC composite are shown in Fig. 5.12. The largest envelope represents the case where the BN interface is not included in the micromechanics analysis (i.e., Mat = 133, commented out in the code above). The remaining two failure envelopes represent cases where the interface is included (i.e., Mat = 134). For the innermost envelope, the interface is permitted to control failure (using BN properties in Table 5.2) while, for the middle failure envelope, the interface is included in the analysis, but not permitted to control failure (accomplished by omitting the allowables for the interface material or setting them to a large value in GetConstitProps.m). Including the interface reduces the size of the failure envelope, which is expected as the purpose of the compliant interface material is to stop cracks from growing into the fibers. This gives the composite greater toughness (a property not considered here) compared to monolithic ceramics, which typically fail in a brittle manner as soon as a crack initiates. However, the compliant interface reduces the composite stiffness and lowers the global stress at which the local stresses exceed the allowables. Fig. 5.12 also indicates the controlling failure criterion and material (per each section of each envelope), which is output to the .EnvData file. It is clear that when there is appreciable σ_{22} loading, the interface will initiate failure when permitted. This example demonstrates one of the advantages of micromechanics in that it enables one to identify the underlying modes such as those shown in Fig. 5.12.

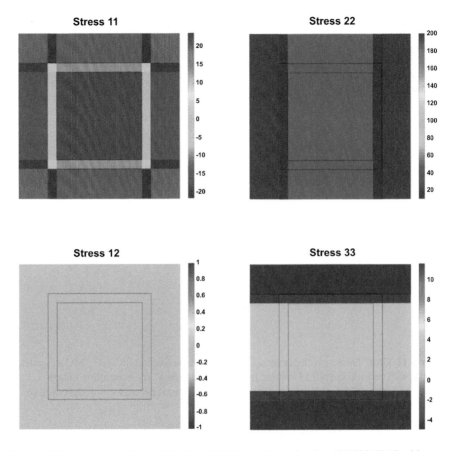

Figure 5.11 Local stress fields (MPa) in a 0.28 fiber volume fraction SiC/SiC CMC with a 0.13 volume fraction BN interface as modeled by a 5×5 GMC RUC subjected to a transverse strain in the x_2-direction equal to 0.0006.

5.8 Advanced topics

The GMC micromechanics theory has been further generalized, extended, and applied to analyze a wide variety of important composite problems. The following extensions are discussed in detail by Aboudi et al. (2013).

- Three-dimension (triply-periodic) microstructures (e.g., for particulate, short fiber, and woven composites)
- Inelastic composites – plastic and viscoplastic constituents
- Fiber/matrix interfacial debonding
- Finite strain composites – constituents undergoing large deformation
- Smart composites – piezo-electro-magnetic behavior of the constituents
- Effective thermal conductivity calculations

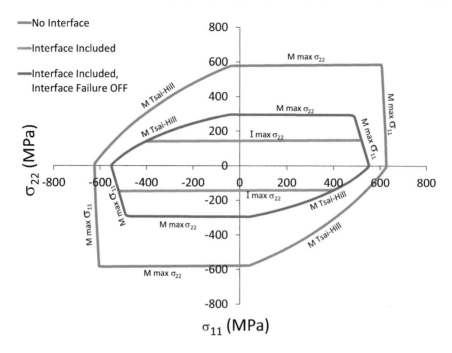

Figure 5.12 GMC failure envelopes for a 0.28 fiber volume fraction SiC/SiC CMC as modeled by GMC with and without a 0.13 volume fraction BN interface. The controlling failure criterion and material (M = matrix, I = interface) are indicated per segment of the envelope.

- Reformulation of GMC to minimize number of unknowns and maximize computational efficiency
- Multiscale GMC – recursive micromechanics wherein each subcell is represented by a composite RUC
- Progressive damage analysis (see also Chapter 7)
- Multiscale finite element implementation – GMC called from within Abaqus as a UMAT

5.9 Summary

This chapter presented the GMC micromechanics theory. The theory is a generalization of the MOC in the sense that it allows an arbitrary number of subcells in the RUC, rather than the MOC's four subcell limitation. Consequently, this generalization enables more than two constituent materials and different microstructures to be considered. After the theory and associated MATLAB code were presented, example problems were given focusing on PMCs, as well as CMCs, which include an explicit interfacial material. GMC's capabilities were illustrated by examining the predicted properties, local fields, and failure envelopes. A key advantage of using micromechanics is in providing insight into local mechanisms of failure.

5.10 Exercises

Exercise 5.1. Starting with Eq. (5.9), establish Eq. (5.13).

Exercise 5.2. Starting with Eq. (5.9) establish Eq. (5.17).

Exercise 5.3. Starting with Eq. (5.21), establish Eq. (5.32).

Exercise 5.4. Using Eqs. (5.21), (5.22), and (5.37) establish Eq. (5.47).

Exercise 5.5. Consider the unidirectional glass/epoxy composite used in Section 5.7.2 to examine the local fields in PMCs (glass fiber properties: $E = 69$ GPa, $v = 0.2$ (Mat $= 8$); epoxy matrix properties: $E = 3.42$ GPa, $v = 0.34$ (Mat $= 9$)) with a fiber volume fraction of 0.5 and the 26×26 RUC. Determine execution times in MATLAB for the direct solution approach and the component by component solution approach (see Sections 5.4 and 5.6) for the determination of the strain concentration tensors, $\mathbf{A}^{(\beta\gamma)}$. What is the ratio of these execution times (slower/faster) and which approach is faster? Show the composite effective mechanical properties calculated using each approach. Are the properties the same? Hint: Comment and uncomment the appropriate sections of the function GMC.m and use the tic and toc MATLAB commands around the sections to time the execution.

Exercise 5.6. Add to the MATLAB Function GetRUC.m a new 13×13 RUC to represent a composite material with an interface, as shown below, with subcell dimensions,

```
RUC.h = RUC.l = [0.15555, 0.050931, 0.022449, 0.050931, 0.022449, ...
                 0.050931, 0.29352, 0.050931, 0.022449, 0.050931, ...
                 0.022449, 0.050931, 0.15555];
```

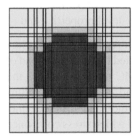

Also, within GetRUC.m, calculate the fiber volume fraction and the interface volume fraction and store them in RUC.Vf and RUC.Vi.

Exercise 5.7. Using the new 13×13 RUC from Exercise 5.6, determine the fiber and interface volume fractions and the effective properties of a SiC/SiC composite with BN interface, given the constituent properties in Table 5.2. Compare to the effective properties for the 5×5 RUC given in Table 5.4 by calculating a % difference for each property. Which three properties in the table are affected the most by the change in RUC from 5×5 to 13×13?

Exercise 5.8. Add plotting of the von Mises stress ($\sqrt{J_2}$) and pressure ($-\sigma_{\text{mean}}$) to PlotMicroFields.m by setting the flag PlotSvm $=$ true in the MATLAB Function PlotMicroFields.m. Make plots of the von Mises stress and pressure for the

26×26 RUC GMC representation of the glass/epoxy composite presented in Section 5.7.2 and compare to plots for the same composite for the MT method subjected to an applied strain in the x_2-direction, $\bar{\varepsilon}_{22} = 0.02$. The glass and epoxy constituent properties are as follows. Fiber: E = 69 GPa, $v = 0.2$ (Mat = 8), Epoxy: E = 3.42 GPa, $v = 0.34$ (Mat = 9).

Exercise 5.9. Add the capability to calculate the volume-averaged von Mises stress and pressure for each constituent material within the RUC to the MATLAB Function PlotMicroFields.m. Determine the volume-averaged von Mises stress and volume-averaged pressure for the composite described in Exercise 5.8 using the 26×26 GMC RUC. Then compare the volume-averaged von Mises stress and volume-averaged pressure for the same composite using both the 2×2 GMC RUC and the MT method. All cases should be subjected to $\bar{\sigma}_{22} = 207.6153$ MPa.

Exercise 5.10. Calculate the effective laminate properties of a regular $[0/90]_s$ SiC/SiC laminate using the *ply properties* given in Table 2.1. Next, using the 5×5 GMC RUC from Fig. 5.3 and Exercise 5.7, calculate the effective laminate properties for a $[0/90]_s$ SiC/SiC laminate with ply properties determined from micromechanics with a fiber volume fraction of 0.28 and a BN interface volume fraction of 0.13. The constituent properties can be found in Table 5.2. Compare the laminate effective property predictions between the ply-based and micromechanics-based approaches.

Exercise 5.11. Using GMC and the 13×13 RUC from Exercise 5.6, calculate the effective laminate properties for a regular $[0/90]_s$ SiC/SiC laminate with ply properties determined from micromechanics with a fiber volume fraction of 0.28 and a BN interface volume fraction of 0.13. Compare these effective property predictions with those calculated using ply-level properties and the 5×5 RUC in Exercise 5.10.

Exercise 5.12. Analyze a regular cross-ply SiC/SiC laminate $[0/90]_s$ with ply thicknesses of 0.25 mm subjected to an applied load of $N_y = 100$ N/mm with ply properties determined using the GMC micromechanics theory, the 5×5 RUC, and with a fiber volume fraction of 0.28 and a BN interface volume fraction of 0.13. Plot the local stress component fields for both the 0 and 90 degree plies. For each of the two ply orientations, which constituent has and which stress component has the highest magnitude? Explain why this is the case.

Exercise 5.13. Analyze a regular cross-ply SiC/SiC laminate $[0/90]_s$ with ply thicknesses of 0.25 mm subjected to an applied load of $N_y = 100$ N/mm with ply properties determined using the GMC micromechanics theory, the 13×13 RUC (from Exercise 5.6), and with a fiber volume fraction of 0.28 and a BN interface volume fraction of 0.13. Plot the local strain fields for both the 0 and 90 degree plies. What is the value of ε_{11} in each ply?

Exercise 5.14. Analyze a regular cross-ply SiC/SiC laminate $[0/90]_s$ with ply thicknesses of 0.25 mm subjected to an applied load of $N_y = 200$ N/mm with ply properties determined using the GMC micromechanics theory and with a fiber volume fraction of 0.28 and a BN interface volume fraction of 0.13. Use a) a 5×5 RUC and b) a 13×13 RUC (from Exercise 5.6) representation and properties in Table 5.2. Based on the calculated margins of safety (MoS) using all four failure criteria, compare the orientation of the controlling ply, controlling subcell, controlling failure criterion, minimum MoS, and scaled allowable load between the RUCs.

Exercise 5.15. Analyze a regular cross-ply SiC/SiC laminate $[0/90]_s$ with ply thicknesses of 0.25 mm subjected to an applied load of $N_x = 200$ N/mm with ply properties determined using the GMC micromechanics theory and with a fiber volume fraction of 0.28 and a BN interface volume fraction of 0.13. Use a) a 5×5 RUC and b) a 13×13 RUC (from Exercise 5.6) representation and properties in Table 5.2. Based on the calculated margins of safety (MoS) using all four failure criteria, compare the orientation of the controlling ply, controlling subcell, controlling failure criterion, minimum MoS, and scaled allowable load between the RUCs.

Exercise 5.16. Consider a 0.28 fiber volume fraction, 0.13 BN interface volume fraction, cross ply SiC/SiC laminate, $[0/90]_s$, with ply thicknesses of 0.25 mm. Using the MATLAB code developed in Section 4.4.1.4, generate two N_x-N_y failure envelopes using the GMC micromechanics theory and all four failure criteria with a) a 5×5 RUC and b) a 13×13 RUC (from Exercise 5.6) representation. Plot both failure envelopes on the same axes. What is the controlling criterion and controlling constituent for each RUC? What causes the abrupt changes of slope in the envelopes? Why are the 5×5 RUC laminate envelope and the unidirectional 5×5 RUC envelope shown in Fig. 5.12 different.

Exercise 5.17. Consider a 0.55 fiber volume fraction glass/epoxy composite whose constituent properties are given in Table 3.4. Determine and plot the stress components, von Mises stress, and pressure resulting from a temperature change $\Delta T = 100°C$ using (a) the 2×2 GMC RUC and (b) the 26×26 GMC RUC.

Exercise 5.18. Consider the same glass/epoxy composite from Exercise 5.17. Plot $\bar{\sigma}_{11} - \bar{\sigma}_{22}$ failure envelopes using only the Tsai-Wu criterion and the constituent allowables given below (taken from Exercise 4.16) for both 2×2 and 26×26 GMC RUCs and both with and without a temperature change of $\Delta T = 100°C$.

MatID	Name	XT (MPa)	XC (MPa)	YT (MPa)	YC (MPa)	Q (MPa)	R (MPa)	S (MPa)
2	Glass	2358	−1653	2358	−1653	1000	1000	1000
4	Epoxy	1×10^6	-1×10^6	42	−181	81	81	81

References

Aboudi, J., Arnold, S.M., Bednarcyk, B.A., 2013. Micromechanics of Composite Materials A Generalized Multiscale Analysis Approach. Elsevier, New York, United States.

Arnold, S.M., Mital, S.K., Murthy, P.L., Bednarcyk, B.A., 2016. Multiscale Modeling of Random Microstructures in SiC/SiC Ceramic Matrix Composites within MAC/GMC Framework. In: American Society for Composites 31st Technical Conference, September 19-21.

MATLAB, 2018. version R2018. The MathWorks Inc, Natick, Massachusetts, United States.

Christensen, R.M., 1979. Mechanics of Composite Materials. John Wiley and Sons, New York, United States.

Dean, G.D., Turner, P., 1973. The elastic properties of carbon fibres and their composites. Composites 4, 174–180.

Kriz, R.D., Stinchcomb, W.W., 1979. Elastic moduli of transversely isotropic graphite fibers and their composites. Exp. Mech. 19, 41–49.

Levin, V.M., 1967. On the coefficients of thermal expansion of heterogeneous materials. Mech. Solids 2, 58–61.

Liu, K.C., Ghoshal, A., 2014. Inherent symmetry and microstructure ambiguity in micromechanics. Compos. Struct. 108, 311–318.

Paley, M., Aboudi, J., 1992. Micromechanical analysis of composites by the generalized cell model. Mech. Mater. 14, 127–139.

Simulia, Inc., 2011. Abaqus Finite Element Software Providence, RI.

Stier, B., Bednarcyk, B.A., Simon, J.-W., Reese, S, 2015. Investigation of Micro-Scale Effects on Damage of Composites. NASA Technical Memorandum, 218740.

Sun, C.T., Vaidya, R.S., 1996. Prediction of composite properties from a representative volume element. Comp. Sci. Tech. 56, 171–179.

Totry, E., Gonlzalez, C., Llorca, J., 2008. Influence of the loading path on the strength of fiber-reinforced composites subjected to transverse compression and shear. Int. J. Solids Struct. 45, 1663–1675.

Tsai, S.W., Hahn, H.T., 1980. Introduction to Composite Materials. Technomic, Westport.

The high-fidelity generalized method of cells (HFGMC) micromechanics theory

Chapter outline

A major shortcoming with mean-field micromechanics theories (e.g., Mori-Tanaka (MT)) is that they provide only a single mean value for the stress (and strain) tensor in each constituent. In reality, and in finite element (FE) micromechanics models, the local fields can vary significantly in the constituents. The method of cells (MOC), presented in Chapter 3, took a step at overcoming this limitation by subdividing the matrix into three regions, each with distinct local fields. In Chapter 5, the generalized method of cells (GMC) was presented, enabling further subdivision of the composite into an arbitrary number of regions, or subcells, allowing more variations in the local fields to be captured (as well as consideration of more than two constituent materials). The GMC is more closely related to analytical micromechanics methods, and for this reason, it benefits from a high level of computational efficiency. However, it still shares the lack of shear coupling that is the hallmark of these more efficient methods. In reality, when global normal stresses or strains are applied to a composite, local shear stresses and strains arise within the composite constituents. The efficient analytical micromechanics models like the MT Method, MOC, and GMC do not capture shear coupling, and thus predict zero-shear stresses and strains throughout the composite when only normal stresses and strains are applied (see Figure 5.10). Of course, the actual local shear stresses and strains must sum to zero to reflect the zero applied global shear state, and this is correctly captured by all of the aforementioned methods. Similarly, these methods predict zero local normal stresses when only global shear stresses or strains are applied.

The lack of shear coupling manifested by the GMC is often not much of a problem, and it actually enables the GMC equations to be reformulated to reduce the number of unknowns significantly. Although not implemented in the provided MATLAB code, this reformulation can increase the efficiency of GMC by orders of magnitude (see Aboudi et al., 2013). Figure 5.10 indicates that, for a continuously reinforced composite with constituent property mismatches typical of PMCs, aside from the transverse shear stress, GMC's local stress fields are reasonable approximations of the fields predicted by detailed finite element analysis. However, under certain conditions, such as small matrix gaps between fibers, random microstructures, and woven composites, shear coupling can become a first-order effect that renders non-shear-coupled models ineffective. For this reason, the high-fidelity generalize method of cells (HFGMC) was developed.

The HFGMC addresses the GMC's lack of shear coupling by employing a second-order displacement field, as opposed to the GMC's first-order displacement field. This results in stress and strain fields that now vary linearly (rather than being constant) within the HFGMC subcells, and local shear stresses and strains can arise from global normal loading (and vice versa). As one might expect, the higher fidelity afforded to HFGMC by incorporation of shear coupling comes at a computational cost in that the second-order displacement field necessarily introduces more unknown variables compared to GMC (particularly in the case of the reformulated version not presented herein). The linearly varying elastic fields also cause the HFGMC model to have a dependence on the subcell grid discretization that is absent in the GMC. In essence, the HFGMC is a step further away from the fully analytical micromechanics models and a step closer toward fully numerical approaches like FE micromechanics analyses. The HFGMC is still semi-closed form in that it can be implemented as an anisotropic constitutive model, so it is still well-suited for multiscale analysis (e.g., for modeling the plies within a laminate).

This chapter begins with a derivation of the HFGMC theory with double-periodicity for continuously reinforced thermoelastic composites. As in the GMC approach, closed-form expressions for the effective stiffness tensor and CTE in terms of the HFGMC strain concentration tensors are established. These strain concentration tensors are derived in terms of the thermomechanical properties of the constituents and their microstructure, with enhanced accuracy compared to GMC. Examples are provided that compare HFGMC predictions of the effective thermoelastic properties with the GMC micromechanics method. In addition, local field predictions of HFGMC are compared to those generated with the FE method. Finally, the effect of fiber-packing arrangement is examined through detailed comparisons of GMC and HFGMC, including random microstructures with stochastic variation.

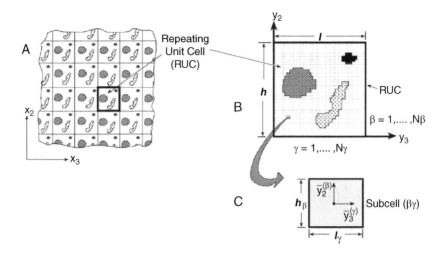

Figure 6.1 (A) A multiphase composite with doubly-periodic microstructure defined with respect to global coordinates (x_2, x_3). (B) The RUC, divided into N_β and N_γ subcells, in the y_2 and y_3 directions, respectively. (C) A characteristic subcell $(\beta\gamma)$ with local coordinates $(\bar{y}_2^{(\beta)}, \bar{y}_3^{(\gamma)})$ whose origin is located at the center of the subcell.

6.1 Geometry and second-order displacement expansion

Consider a periodic multiphase composite with global coordinates (x_2, x_3), see Fig. 6.1A. Fig. 6.1B shows the repeating unit cell (RUC) of the periodic composite microstructure. The RUC is represented with respect to the coordinates (y_2, y_3). Like GMC, it is divided into N_β and N_γ subcells, in the y_2 and y_3 directions, respectively. A characteristic subcell $(\beta\gamma)$ can be represented, with local coordinates $(\bar{y}_2^{(\beta)}, \bar{y}_3^{(\gamma)})$, whose origin is located at the subcell center. Note that this is different than the $(\bar{x}_2^{(\beta)}, \bar{x}_3^{(\gamma)})$ coordinates, used in GMC.

Let the composite be subjected to homogeneous far-field loading,

$$\bar{u}_i = \bar{\varepsilon}_{ij} x_j, \quad i, j = 1, 2, 3 \tag{6.1}$$

where $\bar{\varepsilon}_{ij}$ are the externally applied average strain components. Due to the heterogeneity of the composite, the *perturbed* displacement field in subcell $(\beta\gamma)$ is given by the following second-order displacement expansion:

$$u_i^{(\beta\gamma)} = W_{i(00)}^{(\beta\gamma)} + \bar{y}_2^{(\beta)} W_{i(10)}^{(\beta\gamma)} + \bar{y}_3^{(\gamma)} W_{i(01)}^{(\beta\gamma)}$$
$$+ \frac{1}{2}\left(3\bar{y}_2^{(\beta)2} - \frac{h_\beta^2}{4}\right) W_{i(20)}^{(\beta\gamma)} + \frac{1}{2}\left(3\bar{y}_3^{(\gamma)2} - \frac{l_\gamma^2}{4}\right) W_{i(02)}^{(\beta\gamma)} \tag{6.2}$$

where $W^{(\beta\gamma)}_{i(00)}$ are the subcell volume-averaged displacements, and the higher-order terms $W^{(\beta\gamma)}_{i(mn)}$ must be determined by enforcing the equilibrium, interfacial, and periodic conditions (all in an average sense). Thus the total displacement in any subcell is given by,

$$u_i = \bar{u}_i + u^{(\beta\gamma)}_i = \bar{\varepsilon}_{ij}\, x_j + u^{(\beta\gamma)}_i \tag{6.3}$$

The resulting strain components $\varepsilon^{(\beta\gamma)}_{ij}$ in the subcell $(\beta\gamma)$ are given by the sum of the far-field strain plus the perturbation strain,

$$\varepsilon^{(\beta\gamma)}_{ij} = \bar{\varepsilon}_{ij} + \frac{1}{2}\left(\partial_i u^{(\beta\gamma)}_j + \partial_j u^{(\beta\gamma)}_i\right) \tag{6.4}$$

where $\partial_1 = 0$ (since the strains are independent of the fiber direction), $\partial_2 = \partial/\partial x_2 = \partial/\partial y_2 = \partial/\partial \bar{y}^{(\beta)}_2$, and $\partial_3 = \partial/\partial x_3 = \partial/\partial y_3^{(\gamma)} = \partial/\partial \bar{y}^{(\gamma)}_3$. Thus,

$$\varepsilon^{(\beta\gamma)}_{11} = \bar{\varepsilon}_{11} \tag{6.5}$$

$$\varepsilon^{(\beta\gamma)}_{22} = \bar{\varepsilon}_{22} + W^{(\beta\gamma)}_{2(10)} + 3W^{(\beta\gamma)}_{2(20)}\bar{y}^{(\beta)}_2 \tag{6.6}$$

$$\varepsilon^{(\beta\gamma)}_{33} = \bar{\varepsilon}_{33} + W^{(\beta\gamma)}_{3(01)} + 3W^{(\beta\gamma)}_{3(02)}\bar{y}^{(\gamma)}_3 \tag{6.7}$$

$$2\varepsilon^{(\beta\gamma)}_{23} = 2\bar{\varepsilon}_{23} + W^{(\beta\gamma)}_{2(01)} + W^{(\beta\gamma)}_{3(10)} + 3W^{(\beta\gamma)}_{2(02)}\bar{y}^{(\gamma)}_3 + 3W^{(\beta\gamma)}_{3(20)}\bar{y}^{(\beta)}_2 \tag{6.8}$$

$$2\varepsilon^{(\beta\gamma)}_{13} = 2\bar{\varepsilon}_{13} + W^{(\beta\gamma)}_{1(01)} + 3W^{(\beta\gamma)}_{1(02)}\bar{y}^{(\gamma)}_3 \tag{6.9}$$

$$2\varepsilon^{(\beta\gamma)}_{12} = 2\bar{\varepsilon}_{12} + W^{(\beta\gamma)}_{1(10)} + 3W^{(\beta\gamma)}_{1(20)}\bar{y}^{(\beta)}_2 \tag{6.10}$$

6.2   Local governing equations

In the absence of body forces, the equilibrium equations in each subcell are given by:

$$\partial_2 \sigma^{(\beta\gamma)}_{2i} + \partial_3 \sigma^{(\beta\gamma)}_{3i} = 0, \quad i = 1, 2, 3 \tag{6.11}$$

By averaging Eqs. (6.11) over the subcell area, the following expressions are obtained:

$$\frac{1}{h_\beta l_\gamma} \int_{-h_\beta/2}^{h_\beta/2} \int_{-l_\gamma/2}^{l_\gamma/2} \left[\frac{\partial \sigma^{(\beta\gamma)}_{21}}{\partial \bar{y}^{(\beta)}_2} + \frac{\partial \sigma^{(\beta\gamma)}_{31}}{\partial \bar{y}^{(\gamma)}_3}\right] d\bar{y}^{(\gamma)}_3\, d\bar{y}^{(\beta)}_2 = 0 \tag{6.12}$$

$$\frac{1}{h_\beta l_\gamma} \int_{-h_\beta/2}^{h_\beta/2} \int_{-l_\gamma/2}^{l_\gamma/2} \left[\frac{\partial \sigma_{22}^{(\beta\gamma)}}{\partial \bar{y}_2^{(\beta)}} + \frac{\partial \sigma_{32}^{(\beta\gamma)}}{\partial \bar{y}_3^{(\gamma)}} \right] d\bar{y}_3^{(\gamma)} d\bar{y}_2^{(\beta)} = 0 \qquad (6.13)$$

$$\frac{1}{h_\beta l_\gamma} \int_{-h_\beta/2}^{h_\beta/2} \int_{-l_\gamma/2}^{l_\gamma/2} \left[\frac{\partial \sigma_{23}^{(\beta\gamma)}}{\partial \bar{y}_2^{(\beta)}} + \frac{\partial \sigma_{33}^{(\beta\gamma)}}{\partial \bar{y}_3^{(\gamma)}} \right] d\bar{y}_3^{(\gamma)} d\bar{y}_2^{(\beta)} = 0 \qquad (6.14)$$

The average of the stresses over the subcell edges (average tractions) are defined as follows:

$$T_{2i}^{\pm(\beta\gamma)} = \frac{1}{l_\gamma} \int_{-l_\gamma/2}^{l_\gamma/2} \sigma_{2i}^{(\beta\gamma)} \left(\bar{y}_2^{(\beta)} = \pm h_\beta/2 \right) d\bar{y}_3^{(\gamma)} \qquad (6.15)$$

$$T_{3i}^{\pm(\beta\gamma)} = \frac{1}{h_\beta} \int_{-h_\beta/2}^{h_\beta/2} \sigma_{3i}^{(\beta\gamma)} \left(\bar{y}_3^{(\gamma)} = \pm l_\gamma/2 \right) d\bar{y}_2^{(\beta)} \qquad (6.16)$$

where $i = 1,2,3$ and the $\pm$ superscript denotes the positive or negative subcell faces (see Fig. 6.1C). Hence the above three integrated equilibrium equations in the subcell, Eqs. (6.12) through (6.14), can be expressed in terms of $T_{2i}^{\pm(\beta\gamma)}$ and $T_{3i}^{\pm(\beta\gamma)}$ in the following form:

$$\frac{1}{h_\beta} \left[T_{2i}^{+(\beta\gamma)} - T_{2i}^{-(\beta\gamma)} \right] + \frac{1}{l_\gamma} \left[T_{3i}^{+(\beta\gamma)} - T_{3i}^{-(\beta\gamma)} \right] = 0, \qquad i = 1, 2, 3 \qquad (6.17)$$

The constitutive equations of the elastic orthotropic constituent material that fills subcell $(\beta\gamma)$ are given by,

$$\sigma_{ij}^{(\beta\gamma)} = C_{ijkl}^{(\beta\gamma)} \varepsilon_{kl}^{(\beta\gamma)}, \qquad i, j, k, l = 1, 2, 3 \qquad (6.18)$$

where $C_{ijkl}^{(\beta\gamma)}$ are the components of the stiffness tensor for subcell $(\beta\gamma)$. In matrix notation, for orthotropic materials, this constitutive relation can be written as:

$$\begin{bmatrix} \sigma_{11}^{(\beta\gamma)} \\ \sigma_{22}^{(\beta\gamma)} \\ \sigma_{33}^{(\beta\gamma)} \\ \sigma_{23}^{(\beta\gamma)} \\ \sigma_{13}^{(\beta\gamma)} \\ \sigma_{12}^{(\beta\gamma)} \end{bmatrix} = \begin{bmatrix} C_{11}^{(\beta\gamma)} & C_{12}^{(\beta\gamma)} & C_{13}^{(\beta\gamma)} & 0 & 0 & 0 \\ C_{12}^{(\beta\gamma)} & C_{22}^{(\beta\gamma)} & C_{23}^{(\beta\gamma)} & 0 & 0 & 0 \\ C_{13}^{(\beta\gamma)} & C_{23}^{(\beta\gamma)} & C_{33}^{(\beta\gamma)} & 0 & 0 & 0 \\ 0 & 0 & 0 & C_{44}^{(\beta\gamma)} & 0 & 0 \\ 0 & 0 & 0 & 0 & C_{55}^{(\beta\gamma)} & 0 \\ 0 & 0 & 0 & 0 & 0 & C_{66}^{(\beta\gamma)} \end{bmatrix} \begin{bmatrix} \varepsilon_{11}^{(\beta\gamma)} \\ \varepsilon_{22}^{(\beta\gamma)} \\ \varepsilon_{33}^{(\beta\gamma)} \\ 2\varepsilon_{23}^{(\beta\gamma)} \\ 2\varepsilon_{13}^{(\beta\gamma)} \\ 2\varepsilon_{12}^{(\beta\gamma)} \end{bmatrix} \qquad (6.19)$$

By substituting these constitutive relations into Eqs. (6.15) and (6.16), using Eqs. (6.5)–(6.10), and then integrating, the following relations are obtained:

$$T_{21}^{\pm(\beta\gamma)} = C_{66}^{(\beta\gamma)}\left(2\bar{\varepsilon}_{12} + W_{1(10)}^{(\beta\gamma)} \pm \frac{3h_\beta}{2}W_{1(20)}^{(\beta\gamma)}\right) \tag{6.20}$$

$$T_{22}^{\pm(\beta\gamma)} = C_{12}^{(\beta\gamma)}\bar{\varepsilon}_{11} + C_{22}^{(\beta\gamma)}\bar{\varepsilon}_{22} + C_{23}^{(\beta\gamma)}\bar{\varepsilon}_{33} + C_{22}^{(\beta\gamma)}\left(W_{2(10)}^{(\beta\gamma)} \pm \frac{3h_\beta}{2}W_{2(20)}^{(\beta\gamma)}\right)$$
$$+ C_{23}^{(\beta\gamma)}W_{3(01)}^{(\beta\gamma)} \tag{6.21}$$

$$T_{23}^{\pm(\beta\gamma)} = C_{44}^{(\beta\gamma)}\left(2\bar{\varepsilon}_{23} + W_{2(01)}^{(\beta\gamma)} + W_{3(10)}^{(\beta\gamma)} \pm \frac{3h_\beta}{2}W_{3(20)}^{(\beta\gamma)}\right) \tag{6.22}$$

$$T_{31}^{\pm(\beta\gamma)} = C_{55}^{(\beta\gamma)}\left(2\bar{\varepsilon}_{13} + W_{1(01)}^{(\beta\gamma)} \pm \frac{3l_\gamma}{2}W_{1(02)}^{(\beta\gamma)}\right) \tag{6.23}$$

$$T_{32}^{\pm(\beta\gamma)} = C_{44}^{(\beta\gamma)}\left(2\bar{\varepsilon}_{23} + W_{2(01)}^{(\beta\gamma)} + W_{3(10)}^{(\beta\gamma)} \pm \frac{3l_\gamma}{2}W_{2(02)}^{(\beta\gamma)}\right) \tag{6.24}$$

$$T_{33}^{\pm(\beta\gamma)} = C_{13}^{(\beta\gamma)}\bar{\varepsilon}_{11} + C_{23}^{(\beta\gamma)}\bar{\varepsilon}_{22} + C_{33}^{(\beta\gamma)}\bar{\varepsilon}_{33} + C_{23}^{(\beta\gamma)}W_{2(10)}^{(\beta\gamma)}$$
$$+ C_{33}^{(\beta\gamma)}\left(W_{3(01)}^{(\beta\gamma)} \pm \frac{3l_\gamma}{2}W_{3(02)}^{(\beta\gamma)}\right) \tag{6.25}$$

Substituting these expressions into the three equilibrium equations, Eq. (6.17), yields,

$$C_{66}^{(\beta\gamma)}W_{1(20)}^{(\beta\gamma)} + C_{55}^{(\beta\gamma)}W_{1(02)}^{(\beta\gamma)} = 0 \tag{6.26}$$

$$C_{22}^{(\beta\gamma)}W_{2(20)}^{(\beta\gamma)} + C_{44}^{(\beta\gamma)}W_{2(02)}^{(\beta\gamma)} = 0 \tag{6.27}$$

$$C_{44}^{(\beta\gamma)}W_{3(20)}^{(\beta\gamma)} + C_{33}^{(\beta\gamma)}W_{3(02)}^{(\beta\gamma)} = 0 \tag{6.28}$$

Just like edge-average tractions, the edge-average displacements can be defined as,

$$U_{2i}^{\pm(\beta\gamma)} = \frac{1}{l_\gamma}\int_{-l_\gamma/2}^{l_\gamma/2} u_i^{(\beta\gamma)}\left(\bar{y}_2^{(\beta)} = \pm\frac{h_\beta}{2}\right)d\bar{y}_3^{(\gamma)} \tag{6.29}$$

$$U_{3i}^{\pm(\beta\gamma)} = \frac{1}{h_\beta}\int_{-h_\beta/2}^{h_\beta/2} u_i^{(\beta\gamma)}\left(\bar{y}_3^{(\gamma)} = \pm\frac{l_\gamma}{2}\right)d\bar{y}_2^{(\beta)} \tag{6.30}$$

By employing the second-order displacement expansion, Eq. (6.2), the following relations are obtained:

$$
U_{2i}^{\pm(\beta\gamma)} = W_{i(00)}^{(\beta\gamma)} \pm \frac{h_\beta}{2} W_{i(10)}^{(\beta\gamma)} + \frac{h_\beta^2}{4} W_{i(20)}^{(\beta\gamma)} \tag{6.31}
$$

$$
U_{3i}^{\pm(\beta\gamma)} = W_{i(00)}^{(\beta\gamma)} \pm \frac{l_\gamma}{2} W_{i(01)}^{(\beta\gamma)} + \frac{l_\gamma^2}{4} W_{i(02)}^{(\beta\gamma)} \tag{6.32}
$$

Subtracting the first of Eq. (6.31) (+) from the second (-), and subtracting the first (+) of Eq. (6.32) from the second (-), gives:

$$
W_{i(10)}^{(\beta\gamma)} = \frac{1}{h_\beta} \left(U_{2i}^{+(\beta\gamma)} - U_{2i}^{-(\beta\gamma)} \right) \tag{6.33}
$$

$$
W_{i(01)}^{(\beta\gamma)} = \frac{1}{l_\gamma} \left(U_{3i}^{+(\beta\gamma)} - U_{3i}^{-(\beta\gamma)} \right) \tag{6.34}
$$

Adding Eqs. (6.31) and (6.32) gives,

$$
W_{i(20)}^{(\beta\gamma)} = \frac{2}{h_\beta^2} \left(U_{2i}^{+(\beta\gamma)} + U_{2i}^{-(\beta\gamma)} \right) - \frac{4}{h_\beta^2} W_{i(00)}^{(\beta\gamma)} \tag{6.35}
$$

$$
W_{i(02)}^{(\beta\gamma)} = \frac{2}{l_\gamma^2} \left(U_{3i}^{+(\beta\gamma)} + U_{3i}^{-(\beta\gamma)} \right) - \frac{4}{l_\gamma^2} W_{i(00)}^{(\beta\gamma)} \tag{6.36}
$$

To obtain $W_{i(00)}^{(\beta\gamma)}$ in terms of the edge-average displacements, $U_{2i}^{\pm(\beta\gamma)}$ and $U_{3i}^{\pm(\beta\gamma)}$, Eqs. (6.35) and (6.36) are substituted into Eqs. (6.26)–(6.28) obtaining:

$$
C_{66}^{(\beta\gamma)} \left[\frac{2}{h_\beta^2} \left(U_{21}^{+(\beta\gamma)} + U_{21}^{-(\beta\gamma)} \right) - \frac{4}{h_\beta^2} W_{1(00)}^{(\beta\gamma)} \right]
$$
$$
+ C_{55}^{(\beta\gamma)} \left[\frac{2}{l_\gamma^2} \left(U_{31}^{+(\beta\gamma)} + U_{31}^{-(\beta\gamma)} \right) - \frac{4}{l_\gamma^2} W_{1(00)}^{(\beta\gamma)} \right] = 0 \tag{6.37}
$$

$$
C_{22}^{(\beta\gamma)} \left[\frac{2}{h_\beta^2} \left(U_{22}^{+(\beta\gamma)} + U_{22}^{-(\beta\gamma)} \right) - \frac{4}{h_\beta^2} W_{2(00)}^{(\beta\gamma)} \right]
$$
$$
+ C_{44}^{(\beta\gamma)} \left[\frac{2}{l_\gamma^2} \left(U_{32}^{+(\beta\gamma)} + U_{32}^{-(\beta\gamma)} \right) - \frac{4}{l_\gamma^2} W_{2(00)}^{(\beta\gamma)} \right] = 0 \tag{6.38}
$$

$$C_{44}^{(\beta\gamma)}\left[\frac{2}{h_\beta^2}\left(U_{23}^{+(\beta\gamma)}+U_{23}^{-(\beta\gamma)}\right)-\frac{4}{h_\beta^2}W_{3(00)}^{(\beta\gamma)}\right]$$

$$+C_{33}^{(\beta\gamma)}\left[\frac{2}{l_\gamma^2}\left(U_{33}^{+(\beta\gamma)}+U_{33}^{-(\beta\gamma)}\right)-\frac{4}{l_\gamma^2}W_{3(00)}^{(\beta\gamma)}\right]=0 \qquad (6.39)$$

Hence, solving Eqs. (6.37)–(6.39) for $W_{i(00)}^{(\beta\gamma)}$ gives,

$$W_{i(00)}^{(\beta\gamma)}=D_{i2}^{(\beta\gamma)}\left(U_{2i}^{+(\beta\gamma)}+U_{2i}^{-(\beta\gamma)}\right)+D_{i3}^{(\beta\gamma)}\left(U_{3i}^{+(\beta\gamma)}+U_{3i}^{-(\beta\gamma)}\right), \qquad i=1,2,3$$
$$(6.40)$$

where,

$$D_{12}^{(\beta\gamma)}=\frac{C_{66}^{(\beta\gamma)}l_\gamma^2}{2\left(C_{55}^{(\beta\gamma)}h_\beta^2+C_{66}^{(\beta\gamma)}l_\gamma^2\right)} \qquad (6.41)$$

$$D_{13}^{(\beta\gamma)}=\frac{C_{55}^{(\beta\gamma)}h_\beta^2}{2\left(C_{55}^{(\beta\gamma)}h_\beta^2+C_{66}^{(\beta\gamma)}l_\gamma^2\right)} \qquad (6.42)$$

$$D_{22}^{(\beta\gamma)}=\frac{C_{22}^{(\beta\gamma)}l_\gamma^2}{2\left(C_{44}^{(\beta\gamma)}h_\beta^2+C_{22}^{(\beta\gamma)}l_\gamma^2\right)} \qquad (6.43)$$

$$D_{23}^{(\beta\gamma)}=\frac{C_{44}^{(\beta\gamma)}h_\beta^2}{2\left(C_{44}^{(\beta\gamma)}h_\beta^2+C_{22}^{(\beta\gamma)}l_\gamma^2\right)} \qquad (6.44)$$

$$D_{32}^{(\beta\gamma)}=\frac{C_{44}^{(\beta\gamma)}l_\gamma^2}{2\left(C_{33}^{(\beta\gamma)}h_\beta^2+C_{44}^{(\beta\gamma)}l_\gamma^2\right)} \qquad (6.45)$$

$$D_{33}^{(\beta\gamma)}=\frac{C_{33}^{(\beta\gamma)}h_\beta^2}{2\left(C_{33}^{(\beta\gamma)}h_\beta^2+C_{44}^{(\beta\gamma)}l_\gamma^2\right)} \qquad (6.46)$$

Substituting the established expressions of $W_{i(mn)}^{(\beta\gamma)}$ [Eqs. (6.33)–(6.36) along with Eq. (6.40)] into the expressions for $T_{2i}^{\pm(\beta\gamma)}$ and $T_{3i}^{\pm(\beta\gamma)}$ [Eqs. (6.20)–(6.25)] provides the following relations between subcell edge average tractions and subcell edge average displacements, $U_{2i}^{\pm(\beta\gamma)}$ and $U_{3i}^{\pm(\beta\gamma)}$, as follows:

$$
\begin{Bmatrix}
T_{21}^+ \\
T_{21}^- \\
T_{22}^+ \\
T_{22}^- \\
T_{23}^+ \\
T_{23}^- \\
T_{31}^+ \\
T_{31}^- \\
T_{32}^+ \\
T_{32}^- \\
T_{33}^+ \\
T_{33}^-
\end{Bmatrix}^{(\beta\gamma)}
= [\mathbf{K}]^{(\beta\gamma)}
\begin{Bmatrix}
U_{21}^+ \\
U_{21}^- \\
U_{22}^+ \\
U_{22}^- \\
U_{23}^+ \\
U_{23}^- \\
U_{31}^+ \\
U_{31}^- \\
U_{32}^+ \\
U_{32}^- \\
U_{33}^+ \\
U_{33}^-
\end{Bmatrix}^{(\beta\gamma)}
+
\begin{Bmatrix}
2C_{66}^{(\beta\gamma)}\bar{\varepsilon}_{12} \\
2C_{66}^{(\beta\gamma)}\bar{\varepsilon}_{12} \\
C_{12}^{(\beta\gamma)}\bar{\varepsilon}_{11} + C_{22}^{(\beta\gamma)}\bar{\varepsilon}_{22} + C_{23}^{(\beta\gamma)}\bar{\varepsilon}_{33} \\
C_{12}^{(\beta\gamma)}\bar{\varepsilon}_{11} + C_{22}^{(\beta\gamma)}\bar{\varepsilon}_{22} + C_{23}^{(\beta\gamma)}\bar{\varepsilon}_{33} \\
2C_{44}^{(\beta\gamma)}\bar{\varepsilon}_{23} \\
2C_{44}^{(\beta\gamma)}\bar{\varepsilon}_{23} \\
2C_{55}^{(\beta\gamma)}\bar{\varepsilon}_{13} \\
2C_{55}^{(\beta\gamma)}\bar{\varepsilon}_{13} \\
2C_{44}^{(\beta\gamma)}\bar{\varepsilon}_{23} \\
2C_{44}^{(\beta\gamma)}\bar{\varepsilon}_{23} \\
C_{13}^{(\beta\gamma)}\bar{\varepsilon}_{11} + C_{23}^{(\beta\gamma)}\bar{\varepsilon}_{22} + C_{33}^{(\beta\gamma)}\bar{\varepsilon}_{33} \\
C_{13}^{(\beta\gamma)}\bar{\varepsilon}_{11} + C_{23}^{(\beta\gamma)}\bar{\varepsilon}_{22} + C_{33}^{(\beta\gamma)}\bar{\varepsilon}_{33}
\end{Bmatrix}
\tag{6.47}
$$

where $[\mathbf{K}]^{(\beta\gamma)}$ is a 12×12 local stiffness matrix whose elements depend on the dimensions and the material properties of the subcell $(\beta\gamma)$. The elements of $[\mathbf{K}]^{(\beta\gamma)}$ are given in MATLAB Function 6.1 `local`.

MATLAB Function 6.1 – `local` (located within `HFGMC.m`)

This function calculates the local stiffness matrix $[\mathbf{K}]^{(\beta\gamma)}$ and the $D_{ij}^{(\beta\gamma)}$ terms.

```
function [ak, dd] = local(RUC, props)

% +++++++++++++++++++++++++++++++++++++++++++++++++++++++++++++++++++++++++++++++++++
% Copyright 2020 United States Government as represented by the Administrator of the
% National Aeronautics and Space Administration. No copyright is claimed in the
% United States under Title 17, U.S. Code. All Other Rights Reserved. By downloading
% or using this software you agree to the terms of NOSA v1.3. This code was prepared
% by Drs. B.A. Bednarcyk and S.M. Arnold to complement the book "Practical
% Micromechanics of Composite Materials" during the course of their government work.
% +++++++++++++++++++++++++++++++++++++++++++++++++++++++++++++++++++++++++++++++++++

% -- Determine the local stiffness matrix, K, and the terms Dij (Eqs. 6.41 - 6.47)

    NB = RUC.NB;
    NG = RUC.NG;
    nsubs = NB*NG;

    xh = RUC.h;
    xl = RUC.l;

    % -- Preallocate
    ak = zeros(12,12,nsubs);
    dd = zeros(3,3,nsubs);
```

```
for j = 1:NB
    for k = 1:NG

        nsub = j + NB*(k - 1);

        C = props(j, k).C;

        coef1 = 4*(C(6,6)/xh(j)^2 + C(5,5)/xl(k)^2);
        d(1,2) = 2*C(6,6)/(xh(j)^2*coef1);  % -- Eq. 6.41
        d(1,3) = 2*C(5,5)/(xl(k)^2*coef1);  % -- Eq. 6.42

        coef2 = 4*(C(2,2)/xh(j)^2 + C(4,4)/xl(k)^2);
        d(2,2) = 2*C(2,2)/(xh(j)^2*coef2);  % -- Eq. 6.43
        d(2,3) = 2*C(4,4)/(xl(k)^2*coef2);  % -- Eq. 6.44

        coef3 = 4*(C(4,4)/xh(j)^2 + C(3,3)/xl(k)^2);
        d(3,2) = 2*C(4,4)/(xh(j)^2*coef3);  % -- Eq. 6.45
        d(3,3) = 2*C(3,3)/(xl(k)^2*coef3);  % -- Eq. 6.46

        for l = 1:3
            f(1,2) = 2/xh(j)^2 - 4*d(1,2)/xh(j)^2;
            f(1,3) =           - 4*d(1,3)/xh(j)^2;

            g(1,2) =           - 4*d(1,2)/xl(k)^2;
            g(1,3) = 2/xl(k)^2 - 4*d(1,3)/xl(k)^2;
        end

        %-----------------------------------------------------------
        % -- Elements of the K matrix
        %-----------------------------------------------------------

        ak(1,1,nsub) =   C(6,6)/xh(j) + C(6,6)*3*xh(j)*f(1,2)/2;
        ak(1,2,nsub) =  -C(6,6)/xh(j) + C(6,6)*3*xh(j)*f(1,2)/2;
        ak(1,7,nsub) =   C(6,6)*3*xh(j)*f(1,3)/2;
        ak(1,8,nsub) =   C(6,6)*3*xh(j)*f(1,3)/2;

        ak(2,1,nsub) =   C(6,6)/xh(j) - C(6,6)*3*xh(j)*f(1,2)/2;
        ak(2,2,nsub) =  -C(6,6)/xh(j) - C(6,6)*3*xh(j)*f(1,2)/2;
        ak(2,7,nsub) =  -C(6,6)*3*xh(j)*f(1,3)/2;
        ak(2,8,nsub) =  -C(6,6)*3*xh(j)*f(1,3)/2;

        ak(3,3 ,nsub) =   C(2,2)/xh(j) + C(2,2)*3*xh(j)*f(2,2)/2;
        ak(3,4 ,nsub) =  -C(2,2)/xh(j) + C(2,2)*3*xh(j)*f(2,2)/2;
        ak(3,9 ,nsub) =   C(2,2)*3*xh(j)*f(2,3)/2;
        ak(3,10,nsub) =   C(2,2)*3*xh(j)*f(2,3)/2;
        ak(3,11,nsub) =   C(2,3)/xl(k);
        ak(3,12,nsub) =  -C(2,3)/xl(k);

        ak(4,3 ,nsub) =   C(2,2)/xh(j) - C(2,2)*3*xh(j)*f(2,2)/2;
        ak(4,4 ,nsub) =  -C(2,2)/xh(j) - C(2,2)*3*xh(j)*f(2,2)/2;
        ak(4,9 ,nsub) =  -C(2,2)*3*xh(j)*f(2,3)/2;
        ak(4,10,nsub) =  -C(2,2)*3*xh(j)*f(2,3)/2;
        ak(4,11,nsub) =   C(2,3)/xl(k);
        ak(4,12,nsub) =  -C(2,3)/xl(k);

        ak(5,5 ,nsub) =   C(4,4)/xh(j) + C(4,4)*3*xh(j)*f(3,2)/2;
        ak(5,6 ,nsub) =  -C(4,4)/xh(j) + C(4,4)*3*xh(j)*f(3,2)/2;
        ak(5,9 ,nsub) =   C(4,4)/xl(k);
        ak(5,10,nsub) =  -C(4,4)/xl(k);
        ak(5,11,nsub) =   C(4,4)*3*xh(j)*f(3,3)/2;
        ak(5,12,nsub) =   C(4,4)*3*xh(j)*f(3,3)/2;

        ak(6,5 ,nsub) =   C(4,4)/xh(j) - C(4,4)*3*xh(j)*f(3,2)/2;
        ak(6,6 ,nsub) =  -C(4,4)/xh(j) - C(4,4)*3*xh(j)*f(3,2)/2;
        ak(6,9 ,nsub) =   C(4,4)/xl(k);
        ak(6,10,nsub) =  -C(4,4)/xl(k);
        ak(6,11,nsub) =  -C(4,4)*3*xh(j)*f(3,3)/2;
        ak(6,12,nsub) =  -C(4,4)*3*xh(j)*f(3,3)/2;
```

```
ak(7,1 ,nsub) =  C(5,5)*3*xl(k)*g(1,2)/2;
ak(7,2 ,nsub) =  C(5,5)*3*xl(k)*g(1,2)/2;
ak(7,7 ,nsub) =  C(5,5)/xl(k) + C(5,5)*3*xl(k)*g(1,3)/2;
ak(7,8 ,nsub) =  -C(5,5)/xl(k) + C(5,5)*3*xl(k)*g(1,3)/2;

ak(8,1 ,nsub) =  - C(5,5)*3*xl(k)*g(1,2)/2;
ak(8,2 ,nsub) =  - C(5,5)*3*xl(k)*g(1,2)/2;
ak(8,7 ,nsub) =  C(5,5)/xl(k) - C(5,5)*3*xl(k)*g(1,3)/2;
ak(8,8 ,nsub) =  -C(5,5)/xl(k) - C(5,5)*3*xl(k)*g(1,3)/2;

ak(9,3 ,nsub) =  C(4,4)*3*xl(k)*g(2,2)/2;
ak(9,4 ,nsub) =  C(4,4)*3*xl(k)*g(2,2)/2;
ak(9,5 ,nsub) =  C(4,4)/xh(j);
ak(9,6 ,nsub) =  -C(4,4)/xh(j);
ak(9,9 ,nsub) =  C(4,4)/xl(k) + C(4,4)*3*xl(k)*g(2,3)/2;
ak(9,10,nsub) =  -C(4,4)/xl(k) + C(4,4)*3*xl(k)*g(2,3)/2;

ak(10,3 ,nsub) =  -C(4,4)*3*xl(k)*g(2,2)/2;
ak(10,4 ,nsub) =  -C(4,4)*3*xl(k)*g(2,2)/2;
ak(10,5 ,nsub) =  C(4,4)/xh(j);
ak(10,6 ,nsub) =  -C(4,4)/xh(j);
ak(10,9 ,nsub) =  C(4,4)/xl(k) - C(4,4)*3*xl(k)*g(2,3)/2;
ak(10,10,nsub) =  - C(4,4)/xl(k) - C(4,4)*3*xl(k)*g(2,3)/2;

ak(11,3 ,nsub) =  C(2,3)/xh(j);
ak(11,4 ,nsub) =  -C(2,3)/xh(j);
ak(11,5 ,nsub) =  C(3,3)*3*xl(k)*g(3,2)/2;
ak(11,6 ,nsub) =  C(3,3)*3*xl(k)*g(3,2)/2;
ak(11,11,nsub) =  C(3,3)/xl(k) + C(3,3)*3*xl(k)*g(3,3)/2;
ak(11,12,nsub) =  -C(3,3)/xl(k) + C(3,3)*3*xl(k)*g(3,3)/2;

ak(12,3 ,nsub) =  C(2,3)/xh(j);
ak(12,4 ,nsub) =  -C(2,3)/xh(j);
ak(12,5 ,nsub) =  -C(3,3)*3*xl(k)*g(3,2)/2;
ak(12,6 ,nsub) =  -C(3,3)*3*xl(k)*g(3,2)/2;
ak(12,11,nsub) =  C(3,3)/xl(k) - C(3,3)*3*xl(k)*g(3,3)/2;
ak(12,12,nsub) =  -C(3,3)/xl(k) - C(3,3)*3*xl(k)*g(3,3)/2;

% -- Store d per subcell for return
for  l = 1:3
    for  m = 1:3
        dd(l, m, nsub) = d(l, m);
    end
end
            end
        end
    end
end
```

6.3 Interfacial continuity and periodicity conditions

At the interface between the subcells in the RUC, the conditions that the displacements (assuming perfect bonding) and tractions are continuous are imposed as follows:

$$U_{2i}^{+(\beta\gamma)} = U_{2i}^{-(\beta+1,\gamma)}, \quad \beta = 1, ..., N_\beta - 1, \quad \gamma = 1, ..., N_\gamma \tag{6.48}$$

$$T_{2i}^{+(\beta\gamma)} = T_{2i}^{-(\beta+1,\gamma)}, \quad \beta = 1, ..., N_\beta - 1, \quad \gamma = 1, ..., N_\gamma \tag{6.49}$$

and

$$U_{3i}^{+(\beta\gamma)} = U_{3i}^{-(\beta,\gamma+1)}, \quad \beta = 1, ..., N_\beta, \quad \gamma = 1, ..., N_\gamma - 1 \tag{6.50}$$

$$T_{3i}^{+(\beta\gamma)} = T_{3i}^{-(\beta,\gamma+1)}, \quad \beta = 1, ..., N_\beta, \quad \gamma = 1, ..., N_\gamma - 1 \tag{6.51}$$

with $i = 1, 2, 3$. These interfacial conditions provide $6(N_\beta - 1)N_\gamma + 6(N_\gamma - 1)N_\beta$ relations.

The periodicity conditions at the boundaries of the RUC are given by:

$$u_i(y_2 = 0) = u_i(y_2 = h) \tag{6.52}$$

$$\sigma_{2i}(y_2 = 0) = \sigma_{2i}(y_2 = h) \tag{6.53}$$

$$u_i(y_3 = 0) = u_i(y_3 = l) \tag{6.54}$$

$$\sigma_{3i}(y_3 = 0) = \sigma_{3i}(y_3 = l) \tag{6.55}$$

with $i = 1,2,3$, see Fig. 6.1B. Consequently, in terms of subcell edge average displacements and tractions,

$$U_{2i}^{-(1,\gamma)} = U_{2i}^{+(N_\beta,\gamma)}, \quad \gamma = 1, ..., N_\gamma \tag{6.56}$$

$$T_{2i}^{-(1,\gamma)} = T_{2i}^{+(N_\beta,\gamma)}, \quad \gamma = 1, ..., N_\gamma \tag{6.57}$$

and

$$U_{3i}^{-(\beta,1)} = U_{3i}^{+(\beta,N_\gamma)}, \quad \beta = 1, ..., N_\beta \tag{6.58}$$

$$T_{3i}^{-(\beta,1)} = T_{3i}^{+(\beta,N_\gamma)}, \quad \beta = 1, ..., N_\beta \tag{6.59}$$

with $i = 1,2,3$. These periodicity conditions provide $6N_\beta + 6N_\gamma$ additional relations. Consequently, there are $12N_\beta N_\gamma$ relations upon which the HFGMC theory is based [where Eq. (6.47) relates the edge average tractions and displacements]. These relations have been obtained from the fulfillment of the equilibrium conditions and the interfacial continuity and periodicity conditions, all of which have been imposed in the average (integral) sense.

6.4 Overall mechanical system of equations

Equation (6.47) shows that for each subcell $(\beta\gamma)$ there are 12 unknown edge average displacements ($U_{2i}^{\pm(\beta\gamma)}$ and $U_{3i}^{\pm(\beta\gamma)}$) that need to be determined. For an RUC with $\beta = 1$, 2, ..., N_β and $\gamma = 1, 2, ..., N_\gamma$, there are $N_\beta N_\gamma$ subcells, thus providing a total of $12N_\beta N_\gamma$ unknowns. On the other hand, there are a total of $(N_\beta - 1)N_\gamma + N_\beta(N_\gamma - 1)$ internal subcell interfaces on which the continuity of displacements and tractions is imposed, providing $6[(N_\beta - 1)N_\gamma + N_\beta(N_\gamma - 1)]$ equations. In addition, the periodicity conditions that require that the displacements and tractions are identical on the opposite sides of the RUC, provide $6N_\gamma + 6N_\beta$ equations. Thus the total number of equations is exactly equal to the number of unknowns, namely, $12N_\beta N_\gamma$.

However, since in the presence of perfect bonding between the constituents the displacements between neighboring subcells are continuous, it follows that

$$U_{2i}^{+(\beta,\gamma)} = U_{2i}^{-(\beta+1,\gamma)}$$

$$U_{3i}^{+(\beta,\gamma)} = U_{3i}^{-(\beta,\gamma+1)}$$

(6.60)

Therefore the number of unknown edge average displacements can be reduced by one half because $U_{2i}^{+(\beta,\gamma)}$, $U_{3i}^{+(\beta,\gamma)}$ are known from the right-hand-side of Eq. (6.60). Thus, the number of unknowns $U_{2i}^{-(\beta,\gamma)}$, $U_{3i}^{-(\beta,\gamma)}$ in each subcell is reduced to 6, and the total number of unknowns becomes $6N_\beta N_\gamma$. It is noted that this is the identical number of unknowns as the GMC formulation presented in Chapter 5. However, it is possible to reformulate the GMC equations to reduce the number of unknowns considerably, see Aboudi et al. (2013). In the provided MATLAB (2018) implementation of HFGMC, for coding simplicity, in applying the displacement periodicity conditions, a few redundant displacement terms are retained, raising the total number of equations and unknowns to $6N_\beta N_\gamma + 3(N_\beta + N_\gamma)$.

This system of algebraic equations can be symbolically represented by,

$$\mathbf{\Omega}\,\mathbf{U} = \mathbf{f}$$

(6.61)

where the structural-stiffness matrix $\mathbf{\Omega}$ contains information on the geometry and mechanical properties of the materials in the subcells. The vector $\mathbf{U}$ consists of six unknown edge average displacements per subcell, $U_{2i}^{-(\beta\gamma)}$ and $U_{3i}^{-(\beta\gamma)}$, and the force $\mathbf{f}$ terms involve far-field strains, $\bar{\varepsilon}_{ij}$.

The solution of this system of equations provides the unknowns, $U_{2i}^{-(\beta\gamma)}$ and $U_{3i}^{-(\beta\gamma)}$ in all subcells, from which the microvariables $W_{i(mn)}^{(\beta\gamma)}$ can be determined (see Eqs. (6.33)–(6.36) and (6.40)). Hence, with Eqs. (6.5)–(6.10) the strains $\varepsilon_{ij}^{(\beta\gamma)}$ in all subcells are available in terms of the known far-field strains, $\bar{\varepsilon}_{ij}$.

6.5 The mechanical strain concentration tensor

In order to determine the effective elastic constants of the composite, the strain concentration tensors $\mathbf{A}^{(\beta\gamma)}$ in the subcells need to be establish. As with the MOC and GMC, the tensor $\mathbf{A}^{(\beta\gamma)}$ expresses the average strains in the subcell in terms of the externally applied strains $\bar{\varepsilon}_{ij}$, i.e.,

$$\bar{\boldsymbol{\varepsilon}}^{(\beta\gamma)} = \mathbf{A}^{(\beta\gamma)}\bar{\boldsymbol{\varepsilon}} \qquad \text{or} \qquad \bar{\varepsilon}_{ij}^{(\beta\gamma)} = A_{ijkl}^{(\beta\gamma)}\bar{\varepsilon}_{kl}, \qquad i,j,k,l = 1,2,3 \qquad (6.62)$$

where $A_{ijkl}^{(\beta\gamma)}$ are the components of the subcell fourth-order strain concentration tensor and $\bar{\varepsilon}_{ij}^{(\beta\gamma)}$ are the average strain components in the subcell, which are equal to their values at the subcell centroid. These can be simply obtained from Eqs. (6.5)–(6.10) by setting $\bar{y}_2^{(\beta)} = \bar{y}_3^{(\gamma)} = 0$.

In matrix notation, Eq. (6.62) can be represented as below, where the non-zero terms are indicated (for orthotropic constituent materials):

$$
\begin{bmatrix}
\bar{\varepsilon}_{11}^{(\beta\gamma)} \\
\bar{\varepsilon}_{22}^{(\beta\gamma)} \\
\bar{\varepsilon}_{33}^{(\beta\gamma)} \\
\bar{\varepsilon}_{23}^{(\beta\gamma)} \\
\bar{\varepsilon}_{13}^{(\beta\gamma)} \\
\bar{\varepsilon}_{12}^{(\beta\gamma)}
\end{bmatrix}
=
\begin{bmatrix}
1 & 0 & 0 & 0 & 0 & 0 \\
A_{21}^{(\beta\gamma)} & A_{22}^{(\beta\gamma)} & A_{23}^{(\beta\gamma)} & A_{24}^{(\beta\gamma)} & 0 & 0 \\
A_{31}^{(\beta\gamma)} & A_{32}^{(\beta\gamma)} & A_{33}^{(\beta\gamma)} & A_{34}^{(\beta\gamma)} & 0 & 0 \\
A_{41}^{(\beta\gamma)} & A_{42}^{(\beta\gamma)} & A_{43}^{(\beta\gamma)} & A_{44}^{(\beta\gamma)} & 0 & 0 \\
0 & 0 & 0 & 0 & A_{55}^{(\beta\gamma)} & A_{56}^{(\beta\gamma)} \\
0 & 0 & 0 & 0 & A_{65}^{(\beta\gamma)} & A_{66}^{(\beta\gamma)}
\end{bmatrix}
\begin{bmatrix}
\bar{\varepsilon}_{11} \\
\bar{\varepsilon}_{22} \\
\bar{\varepsilon}_{33} \\
\bar{\varepsilon}_{23} \\
\bar{\varepsilon}_{13} \\
\bar{\varepsilon}_{12}
\end{bmatrix}
\tag{6.63}
$$

As with MOC and GMC, the components of the strain concentration tensor can be determined as follows. First apply the global strain component $\bar{\varepsilon}_{11} = 1$, whereas all other global strain components are set equal to zero. The solution of the system of algebraic equations, Eq. (6.61), provides the variables $U_{2i}^{\pm(\beta\gamma)}$ and $U_{3i}^{\pm(\beta\gamma)}$ from which the microvariables $W_{i(mn)}^{(\beta\gamma)}$ can be determined by employing Eqs. (6.33)–(6.36) and (6.40). Therefore the subcell average strains $\bar{\varepsilon}_{ij}^{(\beta\gamma)}$ can be readily determined from Eqs. (6.5)–(6.10) from the microvariables. It immediately follows from Eq. (6.63) that the first-column components of the strain concentration tensor are given by: $A_{21}^{(\beta\gamma)} = \bar{\varepsilon}_{22}^{(\beta\gamma)}$, $A_{31}^{(\beta\gamma)} = \bar{\varepsilon}_{33}^{(\beta\gamma)}$, $A_{41}^{(\beta\gamma)} = \bar{\varepsilon}_{23}^{(\beta\gamma)}$. Next, apply $\bar{\varepsilon}_{22} = 1$ whereas all other global strain components are set equal to zero. Following the same procedure provides the components $A_{i2}^{(\beta\gamma)}$. By continuing the application of this procedure, all components of $\mathbf{A}^{(\beta\gamma)}$ can be established. It is evident from the additional non-zero terms in Eq. (6.63), $A_{24}^{(\beta\gamma)}$, $A_{34}^{(\beta\gamma)}$, $A_{41}^{(\beta\gamma)}$, $A_{42}^{(\beta\gamma)}$, and $A_{43}^{(\beta\gamma)}$, which couple the local and global normal and shear strains, that HFGMC includes shear coupling that is absent in GMC (see Eq. 5.64) and the simpler micromechanics theories. Clearly, the axial shear strains in HFGMC are coupled as well, as indicated by non-zero $A_{56}^{(\beta\gamma)}$ and $A_{65}^{(\beta\gamma)}$ terms in the strain concentration tensor.

6.6 The effective mechanical properties

Once the strain concentration tensors $\mathbf{A}^{(\beta\gamma)}$ in all subcells have been established, it is possible to determine the effective mechanical properties of the composite. The global constitutive equation, given by,

$$
\bar{\sigma} = \mathbf{C}^* \, \bar{\varepsilon}
\tag{6.64}
$$

which relates the composite average stress $\bar{\sigma}$ and strain $\bar{\varepsilon}$, is established by employing the definition of the average stress in the composite:

$$\bar{\sigma} = \frac{1}{hl} \sum_{\beta=1}^{N_\beta} \sum_{\gamma=1}^{N_\gamma} h_\beta l_\gamma \bar{\sigma}^{(\beta\gamma)} \tag{6.65}$$

where $\bar{\sigma}^{(\beta\gamma)}$ are the average stresses in the subcell. They are given by the subcell constitutive equation, expressed in terms of average quantities:

$$\bar{\sigma}_{ij}^{(\beta\gamma)} = C_{ijkl}^{(\beta\gamma)} \bar{\varepsilon}_{kl}^{(\beta\gamma)}, \qquad i, j, k, l = 1, 2, 3 \tag{6.66}$$

In Eq. (6.64), $\mathbf{C}^*$ is the effective elastic stiffness matrix of the composite, which is given by the closed-form expression,

$$\mathbf{C}^* = \frac{1}{hl} \sum_{\beta=1}^{N_\beta} \sum_{\gamma=1}^{N_\gamma} h_\beta l_\gamma \, \mathbf{C}^{(\beta\gamma)} \, \mathbf{A}^{(\beta\gamma)} \tag{6.67}$$

and is identical in form to the GMC effective stiffness expression (Eq. 5.65). Note that, although the HFGMC $\mathbf{A}^{(\beta\gamma)}$ for each subcell includes additional non-zero coupling terms, which influence the effective stiffness matrix, no additional coupling terms (beyond numerical error) are present in the $\mathbf{C}^*$ (for orthotropic constituents).

6.7 Thermomechanical effects

In the presence of a temperature deviation, ΔT, the global constitutive equation takes the form:

$$\bar{\sigma} = \mathbf{C}^* \left(\bar{\varepsilon} - \boldsymbol{\alpha}^* \Delta T \right) \tag{6.68}$$

where $\boldsymbol{\alpha}^*$ is the effective CTE. By employing the Levin Theorem (see Section 3.9.1) (Levin, 1967; Christensen, 1979) the effective coefficient of thermal expansion (CTE) of the composite is given by (see Eq. 3.138),

$$\boldsymbol{\alpha}^* = \sum_{k=1}^{N} v_k [\mathbf{B}_k]^{Tr} \boldsymbol{\alpha}^{(k)} = \frac{1}{hl} \sum_{\beta=1}^{N_\beta} \sum_{\gamma=1}^{N_\gamma} h_\beta l_\gamma \left[\mathbf{B}^{(\beta\gamma)}\right]^{Tr} \boldsymbol{\alpha}^{(\beta\gamma)} \tag{6.69}$$

where, as before, Tr indicates a transpose, $\boldsymbol{\alpha}^{(\beta\gamma)} = \left(\alpha_1^{(\beta\gamma)}, \alpha_2^{(\beta\gamma)}, \alpha_3^{(\beta\gamma)}, 0, 0, 0 \right)^{Tr}$, and the stress concentration tensor is given by:

$$\mathbf{B}^{(\beta\gamma)} = \mathbf{C}^{(\beta\gamma)} \mathbf{A}^{(\beta\gamma)} \mathbf{C}^{*-1} \tag{6.70}$$

Furthermore, in the presence of thermal strains due to temperature changes, ΔT, Eq. (6.19) takes the form

$$
\begin{bmatrix} \sigma_{11}^{(\beta\gamma)} \\ \sigma_{22}^{(\beta\gamma)} \\ \sigma_{33}^{(\beta\gamma)} \\ \sigma_{23}^{(\beta\gamma)} \\ \sigma_{13}^{(\beta\gamma)} \\ \sigma_{12}^{(\beta\gamma)} \end{bmatrix}
=
\begin{bmatrix}
C_{11}^{(\beta\gamma)} & C_{12}^{(\beta\gamma)} & C_{13}^{(\beta\gamma)} & 0 & 0 & 0 \\
C_{12}^{(\beta\gamma)} & C_{22}^{(\beta\gamma)} & C_{23}^{(\beta\gamma)} & 0 & 0 & 0 \\
C_{13}^{(\beta\gamma)} & C_{23}^{(\beta\gamma)} & C_{33}^{(\beta\gamma)} & 0 & 0 & 0 \\
0 & 0 & 0 & C_{44}^{(\beta\gamma)} & 0 & 0 \\
0 & 0 & 0 & 0 & C_{55}^{(\beta\gamma)} & 0 \\
0 & 0 & 0 & 0 & 0 & C_{66}^{(\beta\gamma)}
\end{bmatrix}
\begin{bmatrix}
\varepsilon_{11}^{(\beta\gamma)} - \alpha_1^{(\beta\gamma)}\Delta T \\
\varepsilon_{22}^{(\beta\gamma)} - \alpha_2^{(\beta\gamma)}\Delta T \\
\varepsilon_{33}^{(\beta\gamma)} - \alpha_3^{(\beta\gamma)}\Delta T \\
2\varepsilon_{23}^{(\beta\gamma)} \\
2\varepsilon_{13}^{(\beta\gamma)} \\
2\varepsilon_{12}^{(\beta\gamma)}
\end{bmatrix}
\tag{6.71}
$$

It can be easily shown that the presence of the thermal terms modifies Eq. (6.47) to

$$
\begin{Bmatrix} T_{21}^+ \\ T_{21}^- \\ T_{22}^+ \\ T_{22}^- \\ T_{23}^+ \\ T_{23}^- \\ T_{31}^+ \\ T_{31}^- \\ T_{32}^+ \\ T_{32}^- \\ T_{33}^+ \\ T_{33}^- \end{Bmatrix}^{(\beta\gamma)}
= [\mathbf{K}]^{(\beta\gamma)}
\begin{Bmatrix} U_{21}^+ \\ U_{21}^- \\ U_{22}^+ \\ U_{22}^- \\ U_{23}^+ \\ U_{23}^- \\ U_{31}^+ \\ U_{31}^- \\ U_{32}^+ \\ U_{32}^- \\ U_{33}^+ \\ U_{33}^- \end{Bmatrix}^{(\beta\gamma)}
+
\begin{Bmatrix}
2C_{66}^{(\beta\gamma)}\bar{\varepsilon}_{12} \\
2C_{66}^{(\beta\gamma)}\bar{\varepsilon}_{12} \\
C_{12}^{(\beta\gamma)}\bar{\varepsilon}_{11} + C_{22}^{(\beta\gamma)}\bar{\varepsilon}_{22} + C_{23}^{(\beta\gamma)}\bar{\varepsilon}_{33} \\
C_{12}^{(\beta\gamma)}\bar{\varepsilon}_{11} + C_{22}^{(\beta\gamma)}\bar{\varepsilon}_{22} + C_{23}^{(\beta\gamma)}\bar{\varepsilon}_{33} \\
2C_{44}^{(\beta\gamma)}\bar{\varepsilon}_{23} \\
2C_{44}^{(\beta\gamma)}\bar{\varepsilon}_{23} \\
2C_{55}^{(\beta\gamma)}\bar{\varepsilon}_{13} \\
2C_{55}^{(\beta\gamma)}\bar{\varepsilon}_{13} \\
2C_{44}^{(\beta\gamma)}\bar{\varepsilon}_{23} \\
2C_{44}^{(\beta\gamma)}\bar{\varepsilon}_{23} \\
C_{13}^{(\beta\gamma)}\bar{\varepsilon}_{11} + C_{23}^{(\beta\gamma)}\bar{\varepsilon}_{22} + C_{33}^{(\beta\gamma)}\bar{\varepsilon}_{33} \\
C_{13}^{(\beta\gamma)}\bar{\varepsilon}_{11} + C_{23}^{(\beta\gamma)}\bar{\varepsilon}_{22} + C_{33}^{(\beta\gamma)}\bar{\varepsilon}_{33}
\end{Bmatrix}
+
\begin{Bmatrix}
0 \\ 0 \\ \Gamma_{22}^{(\beta\gamma)} \\ \Gamma_{22}^{(\beta\gamma)} \\ 0 \\ 0 \\ 0 \\ 0 \\ 0 \\ 0 \\ \Gamma_{33}^{(\beta\gamma)} \\ \Gamma_{33}^{(\beta\gamma)}
\end{Bmatrix} \Delta T
\tag{6.72}
$$

where,

$$
\Gamma_{22}^{(\beta\gamma)} = C_{12}^{(\beta\gamma)}\alpha_1^{(\beta\gamma)} + C_{22}^{(\beta\gamma)}\alpha_2^{(\beta\gamma)} + C_{23}^{(\beta\gamma)}\alpha_3^{(\beta\gamma)}
\tag{6.73}
$$

$$
\Gamma_{33}^{(\beta\gamma)} = C_{13}^{(\beta\gamma)}\alpha_1^{(\beta\gamma)} + C_{23}^{(\beta\gamma)}\alpha_2^{(\beta\gamma)} + C_{33}^{(\beta\gamma)}\alpha_3^{(\beta\gamma)}
\tag{6.74}
$$

Consequently, the thermal terms merely modify the force $\mathbf{f}$ in Eq. (6.61), which now includes these additional thermal terms.

As a result of the presence of the thermal terms, Eq. (6.62) takes the form

$$\bar{\boldsymbol{\varepsilon}}^{(\beta\gamma)} = \mathbf{A}^{(\beta\gamma)}\bar{\boldsymbol{\varepsilon}} + \mathbf{A}^{T(\beta\gamma)}\Delta T \tag{6.75}$$

where $\mathbf{A}^{T(\beta\gamma)}$ is the thermal strain concentration tensor of the subcell. In matrix notation, this equation can be written as

$$
\begin{bmatrix} \bar{\varepsilon}_{11}^{(\beta\gamma)} \\ \bar{\varepsilon}_{22}^{(\beta\gamma)} \\ \bar{\varepsilon}_{33}^{(\beta\gamma)} \\ \bar{\varepsilon}_{23}^{(\beta\gamma)} \\ \bar{\varepsilon}_{13}^{(\beta\gamma)} \\ \bar{\varepsilon}_{12}^{(\beta\gamma)} \end{bmatrix}
=
\begin{bmatrix}
1 & 0 & 0 & 0 & 0 & 0 \\
A_{21}^{(\beta\gamma)} & A_{22}^{(\beta\gamma)} & A_{23}^{(\beta\gamma)} & A_{24}^{(\beta\gamma)} & 0 & 0 \\
A_{31}^{(\beta\gamma)} & A_{32}^{(\beta\gamma)} & A_{33}^{(\beta\gamma)} & A_{34}^{(\beta\gamma)} & 0 & 0 \\
A_{41}^{(\beta\gamma)} & A_{42}^{(\beta\gamma)} & A_{43}^{(\beta\gamma)} & A_{44}^{(\beta\gamma)} & 0 & 0 \\
0 & 0 & 0 & 0 & A_{55}^{(\beta\gamma)} & A_{56}^{(\beta\gamma)} \\
0 & 0 & 0 & 0 & A_{65}^{(\beta\gamma)} & A_{66}^{(\beta\gamma)}
\end{bmatrix}
\begin{bmatrix} \bar{\varepsilon}_{11} \\ \bar{\varepsilon}_{22} \\ \bar{\varepsilon}_{33} \\ \bar{\varepsilon}_{23} \\ \bar{\varepsilon}_{13} \\ \bar{\varepsilon}_{12} \end{bmatrix}
+
\begin{bmatrix} A_{11}^{T(\beta\gamma)} \\ A_{22}^{T(\beta\gamma)} \\ A_{33}^{T(\beta\gamma)} \\ 0 \\ 0 \\ 0 \end{bmatrix}
\Delta T
\tag{6.76}
$$

In order to establish the components of $\mathbf{A}^{T(\beta\gamma)}$, as was done in the MOC and GMC, the external loading $\bar{\varepsilon}_{ij} = 0$ together with $\Delta T = 1$ is applied, such that the components of $\mathbf{A}^{T(\beta\gamma)}$ are simply given by $A_{11}^{T(\beta\gamma)} = \bar{\varepsilon}_{11}^{(\beta\gamma)}, A_{22}^{T(\beta\gamma)} = \bar{\varepsilon}_{22}^{(\beta\gamma)}, A_{33}^{T(\beta\gamma)} = \bar{\varepsilon}_{33}^{(\beta\gamma)}$. The values of these average subcell strain components are obtained by solving Eq. (6.61) with $\bar{\varepsilon}_{ij} = 0$ and $\Delta T = 1$, using the $\mathbf{f}$ as modified above to include the thermal terms.

It should be noted that, as was shown for GMC in Section 5.5, HFGMC can calculate the effective CTEs directly, without the need for the Levin theorem. Towards this end, substituting Eq. (6.75) into the average subcell constitutive equation,

$$\bar{\boldsymbol{\sigma}}^{(\beta\gamma)} = \mathbf{C}^{(\beta\gamma)}\big(\bar{\boldsymbol{\varepsilon}}^{(\beta\gamma)} - \boldsymbol{\alpha}^{(\beta\gamma)}\Delta T\big) \tag{6.77}$$

and then volume averaging using Eq. (6.65) yields,

$$\bar{\boldsymbol{\sigma}} = \frac{1}{hl}\sum_{\beta=1}^{N_\beta}\sum_{\gamma=1}^{N_\gamma} h_\beta l_\gamma\, \mathbf{C}^{(\beta\gamma)}\big(\mathbf{A}^{(\beta\gamma)}\bar{\boldsymbol{\varepsilon}} + \mathbf{A}^{T(\beta\gamma)}\Delta T - \boldsymbol{\alpha}^{(\beta\gamma)}\Delta T\big) \tag{6.78}$$

Comparing Eq. (6.78) to Eq. (6.68), it is clear that,

$$\boldsymbol{\alpha}^* = \frac{-\mathbf{C}^{*-1}}{hl}\sum_{\beta=1}^{N_\beta}\sum_{\gamma=1}^{N_\gamma} h_\beta l_\gamma\, \mathbf{C}^{(\beta\gamma)}\big(\mathbf{A}^{T(\beta\gamma)} - \boldsymbol{\alpha}^{(\beta\gamma)}\big) \tag{6.79}$$

Table 6.1 Foundational aspects contrasted between the finite element (FE) method and HFGMC.

Requirements	FE Method	HFGMC
Displacement continuity	Satisfied pointwise at nodes	Satisfied at the interfaces in an average sense
Traction continuity	None	Satisfied at the interfaces in an average sense
Equilibrium	Satisfied through energy/virtual work principle (weak form) expressed using nodal forces	Strong form of equilibrium equations is imposed in a volume average sense within a subcell
Periodicity conditions	Must be applied by user	Imposed as part of the formulation
Discretization	Non-orthogonal elements are standard	Subcells required to be rectangular. See Haj-Ali and Aboudi (2010) for non-rectangular subcell parametric formulation known as PHFGMC.

It can be numerically verified that Eq. (6.79) gives identical predictions for the effective CTE as the Levin Theorem, Eq. (6.69), and thus the HFGMC (like the GMC) is consistent with the Levin Theorem. Therefore either equation can be used.

6.8 Contrast between HFGMC and finite-element analysis

Despite some similarities between the finite-element (FE) method and the HFGMC, the HFGMC contains many unique features not found in the traditional displacement-based FE method. In Table 6.1 these differences are specifically identified. The aim of this section is to draw a clear distinction between the FE and the HFGMC formulation, since one might assume, at first glance, they are essentially the same, as both utilize subvolume discretizations.

6.9 MATLAB implementation

The MATLAB code provided with this text is available from:

   https://github.com/nasa/Practical-Micromechanics

The HFGMC micromechanics theory MATLAB code is not reproduced in this chapter due to its length. The MATLAB function itself, HFGMC.m, is, of course, provided along with the other code. The code assembles the local stiffness matrix $[\mathbf{K}]^{(\beta\gamma)}$ for each subcell (see MATLAB Function 6.1), and then assembles the global structural stiffness matrix $\mathbf{\Omega}$ and the force vector $\mathbf{f}$. Next, Eq. (6.61) is solved for the unknown subcell surface averaged displacements $\mathbf{U}$, from which the strain concentration tensor components can be obtained per subcell (see Section 6.5). The thermal strain concentration tensors (per subcell) are determined in a similar fashion. Note that, as in the GMC MATLAB implementation discussed in Section 5.6, sparse storage for $\mathbf{\Omega}$ is employed:

```
% -- Define Omega in matlab sparse format
Omega = sparse( NR(1:ic), NC(1:ic), A(1:ic) );
```

where A is a vector containing the non-zero components in $\mathbf{\Omega}$ and NR and NC are vectors containing the row and column numbers of each non-zero component in $\mathbf{\Omega}$. The total number of non-zero terms in $\mathbf{\Omega}$ is ic. The solution of Eq. (6.61) is invoked using:

```
% -- Solve the system Omega*U = f (Eq. 6.61)
U = Omega\f;
```

It should be noted that the MATLAB code always treats shear strains as engineering strain (as opposed to tensorial shear strains). However, examining Eq. (6.63) (as well as the analogous equations from all other micromechanics theories, for example see Eq. (5.64)), the shear strain components are tensorial. Therefore the MATLAB implementation of HFGMC utilizes an "engineering" strain concentration tensor, $\mathbf{A}_{Eng}^{(\beta\gamma)}$, where

$$
\begin{bmatrix} \bar{\varepsilon}_{11}^{(\beta\gamma)} \\ \bar{\varepsilon}_{22}^{(\beta\gamma)} \\ \bar{\varepsilon}_{33}^{(\beta\gamma)} \\ \bar{\gamma}_{23}^{(\beta\gamma)} \\ \bar{\gamma}_{13}^{(\beta\gamma)} \\ \bar{\gamma}_{12}^{(\beta\gamma)} \end{bmatrix} = \mathbf{A}_{Eng}^{(\beta\gamma)} \begin{bmatrix} \bar{\varepsilon}_{11} \\ \bar{\varepsilon}_{22} \\ \bar{\varepsilon}_{33} \\ \bar{\gamma}_{23} \\ \bar{\gamma}_{13} \\ \bar{\gamma}_{12} \end{bmatrix}
\tag{6.80}
$$

This equation is used to obtain the average subcell strains from the global strains in MATLAB Function 3.10 – MicroFields.m. $\mathbf{A}_{Eng}^{(\beta\gamma)}$ is easily determined from $\mathbf{A}^{(\beta\gamma)}$ (which is based on tensorial shear strain) by transforming the tensorial shear stains in Eq. (6.63) to engineering shear strains ($\gamma_{ij} = 2\varepsilon_{ij}$, $i \neq j$),

$$
\begin{bmatrix} \bar{\varepsilon}_{11}^{(\beta\gamma)} \\ \bar{\varepsilon}_{22}^{(\beta\gamma)} \\ \bar{\varepsilon}_{33}^{(\beta\gamma)} \\ 2\bar{\varepsilon}_{23}^{(\beta\gamma)} \\ 2\bar{\varepsilon}_{13}^{(\beta\gamma)} \\ 2\bar{\varepsilon}_{12}^{(\beta\gamma)} \end{bmatrix} = \underbrace{\begin{bmatrix} 1 & 0 & 0 & 0 & 0 & 0 \\ A_{21}^{(\beta\gamma)} & A_{22}^{(\beta\gamma)} & A_{23}^{(\beta\gamma)} & \frac{1}{2}A_{24}^{(\beta\gamma)} & 0 & 0 \\ A_{31}^{(\beta\gamma)} & A_{32}^{(\beta\gamma)} & A_{33}^{(\beta\gamma)} & \frac{1}{2}A_{34}^{(\beta\gamma)} & 0 & 0 \\ 2A_{41}^{(\beta\gamma)} & 2A_{42}^{(\beta\gamma)} & 2A_{43}^{(\beta\gamma)} & A_{44}^{(\beta\gamma)} & 0 & 0 \\ 0 & 0 & 0 & 0 & A_{55}^{(\beta\gamma)} & A_{56}^{(\beta\gamma)} \\ 0 & 0 & 0 & 0 & A_{65}^{(\beta\gamma)} & A_{66}^{(\beta\gamma)} \end{bmatrix}}_{\mathbf{A}_{Eng}^{(\beta\gamma)}} \begin{bmatrix} \bar{\varepsilon}_{11} \\ \bar{\varepsilon}_{22} \\ \bar{\varepsilon}_{33} \\ 2\bar{\varepsilon}_{23} \\ 2\bar{\varepsilon}_{13} \\ 2\bar{\varepsilon}_{12} \end{bmatrix}
\tag{6.81}
$$

For all other theories presented (other than HFGMC), which all lack shear coupling, $\mathbf{A}_{Eng}^{(\beta\gamma)} = \mathbf{A}^{(\beta\gamma)}$ since $A_{24}^{(\beta\gamma)} = A_{34}^{(\beta\gamma)} = A_{41}^{(\beta\gamma)} = A_{42}^{(\beta\gamma)} = A_{43}^{(\beta\gamma)} = 0$, see Eq. (6.63). Consequently, $\mathbf{A}_{Eng}^{(\beta\gamma)}$ and $\mathbf{A}^{(\beta\gamma)}$ do not need to be considered separately in the MATLAB code for these other theories.

The `HFGMC.m` function can be used in stand-alone micromechanics analyses, as well as micromechanics-based CLT problems. No additional input is required beyond that of the GMC theory (as described in Section 5.6). The `ory='HFGMC'` is used in `GetEffProps.m` to invoke the HFGMC theory, as shown below:

```
Mat = 141;
Theory = 'HFGMC';
name = "26x26 HFGMC - Sun & Vaidya Carbon Epoxy";
Vf = 0.6;
Constits.Fiber = constitprops{13};
Constits.Matrix = constitprops{14};
RUCid = 26;
[props{Mat}] = RunMicro(Theory, name, Vf, Constits, props{Mat}, RUCid);
```

6.10 Example problems

6.10.1 Effective elastic properties and CTE

To illustrate the predictive capability of the HFGMC theory, the predictions for the effective moduli of a unidirectional continuous fiber-reinforced carbon/epoxy composite are compared with the FE results reported by Sun and Vaidya (1996) and GMC results. The fiber volume fraction of the carbon/epoxy composite is 0.60. Fig. 6.2 shows the 26×26 subcell discretization of the RUC used in the analysis of the composite by both HFGMC and GMC. Table 6.2 provides the constituent material properties.

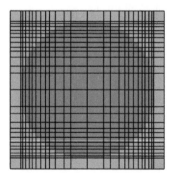

Figure 6.2 Volume discretization of the RUC employed in the analysis of a carbon/epoxy unidirectional composite with a fiber volume fraction of 0.60.

Table 6.2 Elastic moduli of constituent fiber and matrix materials considered by Sun and Vaidya (1996).

Material	Mat	E_{11} (GPa)	E_{22} (GPa)	G_{12} (GPa)	ν_{12}	ν_{23}	α_{11}^* (10^{-6}/C)	α_{22}^* (10^{-6}/C)
Carbon fiber	13	235.0	14.0	28.0	0.20	0.25	−0.9	9.0
Epoxy matrix	14	4.8	4.8	1.8	0.34	0.34	54	54

*CTEs assumed, not provided by Sun and Vaidya (1996).

To execute the GMC and HFGMC using `MicroAnalysis.m`, the problems must be defined in `MicroProblemDef.m` (see MATLAB Function 3.6) as follows:

```
NP = NP + 1;
OutInfo.Name(NP) = "Example Section 6.10.1 - 26x26 GMC - Sun & Vaidya Carbon Epoxy";
MicroMat{NP} = 140;
Loads{NP}.DT = 0;
Loads{NP}.Type  = [  S,  S,  S,  S,  S,  S];
Loads{NP}.Value = [  0,  0,  0,  0,  0,  0];

NP = NP + 1;
OutInfo.Name(NP) = "Example Section 6.10.1 - 26x26 HFGMC - Sun & Vaidya Carbon Epoxy";
MicroMat{NP} = 141;
Loads{NP}.DT = 0;
Loads{NP}.Type  = [  S,  S,  S,  S,  S,  S];
Loads{NP}.Value = [  0,  0,  0,  0,  0,  0];
```

where `MicroMat = 140` and `141` are defined in `GetEffProps.m` (see MATLAB Function 3.7) as follows:

```
Mat = 140;
Theory = 'GMC';
name = "26x26 GMC - Sun & Vaidya Carbon Epoxy";
Vf  = 0.6;
Constits.Fiber = constitprops{13};
Constits.Matrix = constitprops{14};
RUCid = 26;
[props{Mat}] = RunMicro(Theory, name, Vf, Constits, props{Mat}, RUCid);

Mat = 141;
Theory = 'HFGMC';
name = "26x26 HFGMC - Sun & Vaidya Carbon Epoxy";
Vf  = 0.6;
Constits.Fiber = constitprops{13};
Constits.Matrix = constitprops{14};
RUCid = 26;
[props{Mat}] = RunMicro(Theory, name, Vf, Constits, props{Mat}, RUCid);
```

Clearly, the only change needed to invoke the HFGMC theory rather than GMC is specifying `Theory = 'HFGMC'` in `GetEffProps.m`.

Comparison of the HFGMC and GMC predictions of the effective properties and the FE results of Sun and Vaidya (1996) for the carbon/epoxy composite are shown in Table 6.3. The agreement between HFGMC and FE is excellent for all properties, with a difference of less than 1%. The GMC results are generally very good, but the G_{12} prediction differs from the FE result by 6.3%. Note that Sun and Vaidya (1996) did not provide effective CTE predictions, but the GMC and HFGMC CTE prediction are in good agreement.

Table 6.3 Comparison of predicted effective elastic moduli of a carbon/epoxy unidirectional composite ($v_f = 0.60$).

Effective elastic moduli	E_{11} (GPa)	E_{22} (GPa)	G_{12} (GPa)	G_{23} (GPa)	v_{12}	v_{23}	α_{11} (10^{-6}/C)	α_{22} (10^{-6}/C)
HFGMC	142.9	9.61	6.04	3.10	0.252	0.350	−0.072	32.6
FEA – Sun and Vaidya (1996)	142.6	9.60	6.00	3.10	0.250	0.350	—	—
GMC	142.9	9.47	5.62	3.03	0.253	0.358	−0.091	32.9

Consider a unidirectional glass/epoxy composite, which, recall, has greater transverse property mismatch compared to carbon/epoxy, whose constituent properties are given in Table 3.4. Using the 26×26 RUC shown in Fig. 6.2, but now varying the fiber volume fraction, the effective properties predicted by GMC and HFGMC can be compared. Towards this end, the problem can be defined in `MicroProblemDef.m` as follows

```
NP = NP + 1;
OutInfo.Name(NP) = "Example Section 6.10.1 - 26x26 GMC vs. HFGMC - glass-epoxy";
MicroMat{NP} = 142;
Loads{NP}.DT = 0;
Loads{NP}.Type  = [ S,  S,  S,  S,  S,  S];
Loads{NP}.Value = [ 0,  0,  0,  0,  0,  0];
```

where `MicroMat = 142` is defined in `GetEffProps.m` as follows:

```
Mat = 142;
name = "All theories glass-epoxy Vf = x";
Constits.Fiber = constitprops{2};
Constits.Matrix = constitprops{4};
% -- Call function (contained in this file) to get eff props across all Vf
RunOverVf(name, Constits, props, Mat);
props{Mat}.Quit = true; % -- Tells MicroAnalysis to quit after getting eff props
```

and the function `RunOverVf` (contained within `GetEffProps.m`) has been expanded to include HFGMC property predictions using the 26×26 RUC. This will create a text-based output file: `Props vs Vf - All theories glass-epoxy Vf=x ... .txt`. Note that the flag `IncludeHFGMC` must be set to `true` in the function `RunOverVf` to obtain HFGMC results.

The effective properties predicted by GMC and HFGMC are shown in Fig. 6.3 as a function of fiber volume fraction, where the 26×26 RUC shown in Fig. 6.2 has again been used. Note that the fiber volume fraction is controlled by altering the thickness of the matrix subcell around the perimeter of the RUC. Fig. 6.3 generally indicates that, as the fiber volume fraction increases, the discrepancy between GMC and HFGMC increases for all properties shown, except for α_1, which is aligned with the fiber direction.

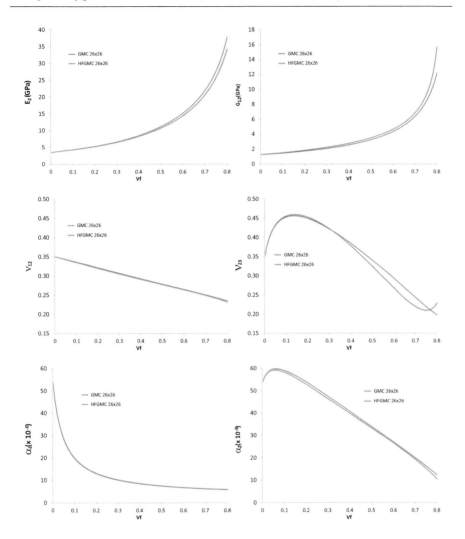

Figure 6.3 Effective properties of a glass/epoxy unidirectional composite as a function of fiber volume fraction predicted by GMC and HFGMC for the 26x26 subcell RUC shown in Fig. 6.2.

6.10.2   Local fields: HFGMC versus FEA

Similar to the results presented in Chapter 5, the stress fields predicted by HFGMC have been compared to those of GMC and FE for a 0.5 fiber volume fraction glass/epoxy composite subjected to 0.02 strain in the transverse (x_3-) direction. The constituent material properties are the same as those given in Section 5.7.2. The FE mesh and GMC and HFGMC RUCs considered are shown in Fig. 6.4. As shown, two HFGMC RUCs are considered. The first (Fig. 6.4C) is the same 26×26 RUC used for GMC (and for HFGMC in the previous example problem, see Fig. 6.4B), except that the matrix region between the fiber and the RUC border has been subdivided into four subcells.

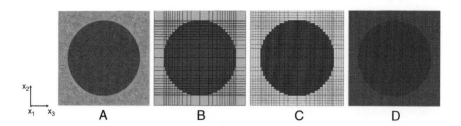

Figure 6.4 (A) Finite element (FE) mesh of a fiber/matrix RUC. (B) 26×26 GMC RUC. (C) 32×32 HFGMC RUC. (D) 100×100 HFGMC RUC.

This was done in order to provide more grid refinement in this region and thus more accurately capture the stress gradients between adjacent fibers. This is accomplished using the Extra option for RUCid = 26 in GetRUC.m (see MATLAB Function 5.2). A portion of the applicable section of code in GetRUC.m is as follows:

```
% -- 26x26 RUC with option to add Extra subcells between fibers,
%     modifies edge subcell dimensions to obtain desired Vf
elseif RUCid == 26
    % -- Extra = # of subcells between fiber and edge of RUC (for HFGMC)
    %Extra = 1;
    Extra = 4;

    RUC.NB = 26 + 2*(Extra - 1);
    RUC.NG = 26 + 2*(Extra - 1);
    .
    .
    .
```

This enables any number of extra subcells to be added to the matrix region between fibers in this RUC.

The 100×100 subcell RUC (Fig. 6.4D) has been introduced in GetRUC.m as the default user-defined RUC (RUCid = 99). This RUC is different from all other RUCs considered previously in that it is composed of all square subcells. A portion of the applicable section of code in GetRUC.m is as follows:

```
% -- Spot for user to put custom RUC
elseif RUCid == 99

    RUC.NB = 100;
    RUC.NG = 100;
    RUC.Vf_max = 0.99;
    RUC.h = ones(RUC.NB,1);
    RUC.l = ones(RUC.NG,1);

    % -- 100x100
    RUC.mats = [M,M,M,M,M,M,M,M,M,M,M,M,M,M,M,M,M,M,M,M,M,M,M,M,M,M,M,M,M,M,M,M,M,M,
    .
    .
    .
```

As shown, any RUC can be defined by specifying the number of subcells in each direction, along with the subcell dimensions and the placement of the fiber (F), matrix (M), and interface (I) constituents in the RUC (stored in the array RUC.mats).

The two HFGMC problems can be defined in `MicroProblemDef.m` as follows (while ensuring `Extra = 4` is specified in `GetRUC.m`, as shown above):

```
NP = NP + 1;
OutInfo.Name(NP) = "Example Section 6.10.2 - 32x32 HFGMC - contour glass-epoxy";
MicroMat{NP} = 143;
Loads{NP}.DT = 0;
Loads{NP}.Type  = [ S,  S,   E,  S,  S,  S];
Loads{NP}.Value = [ 0,  0, 0.02,  0,  0,  0];

NP = NP + 1;
OutInfo.Name(NP) = "Example Section 6.10.2 - 100x100 HFGMC - contour glass-epoxy";
MicroMat{NP} = 144;
Loads{NP}.DT = 0;
Loads{NP}.Type  = [ S,  S,   E,  S,  S,  S];
Loads{NP}.Value = [ 0,  0, 0.02,  0,  0,  0];
```

where `Mat = 143` and `Mat = 144` are defined in `GetEffProps.m` as follows:

```
Mat = 143;
Theory = 'HFGMC';
name = "32x32 Glass-Epoxy Contour HFGMC";
Vf  = 0.50;
Constits.Fiber = constitprops{8};
Constits.Matrix = constitprops{9};
RUCid = 26;
[props{Mat}] = RunMicro(Theory, name, Vf, Constits, props{Mat}, RUCid);

Mat = 144;
Theory = 'HFGMC';
name = "100x100 Glass-Epoxy Contour HFGMC";
Vf  = 0.50;
Constits.Fiber = constitprops{8};
Constits.Matrix = constitprops{9};
RUCid = 99;
[props{Mat}] = RunMicro(Theory, name, Vf, Constits, props{Mat}, RUCid);
```

The resulting local fields are compared in Fig. 6.5. The GMC and FE results are identical to those presented in Figure 5.10. As before, the limits of the colorbar scales for each component are identical. However, although care has been taken to make the colorbars similar, they are not identical as the plots of the FE results have been produced using different software. The most important observation is the ability of HFGMC to predict a nonzero transverse shear stress (σ_{23}) distribution in the composite RUC. This "shear coupling" is absent in GMC as zero transverse shear stresses are predicted throughout the RUC in this case. The HFGMC prediction of the transverse shear stresses matches well with the FE prediction in both magnitude and distribution, particularly for the more refined subcell grid. Examining the individual stress component fields in Fig. 6.5 it is clear that HFGMC captures the stress concentrations evident in the FE results to a much greater extent than does GMC. This is particularly evident in the two transverse stress components, where, as discussed in Chapter 5, the GMC results are constant in rows and columns of subcells. Similarly, in the von Mises stress and pressure distributions, HFGMC provides a better match with the refined FE results, although the predictions of GMC are quite accurate as well. The ability of GMC to provide a reasonably accurate prediction of the von Mises stress state enables the theory to make accurate predictions in the presence of inelastic constituent materials

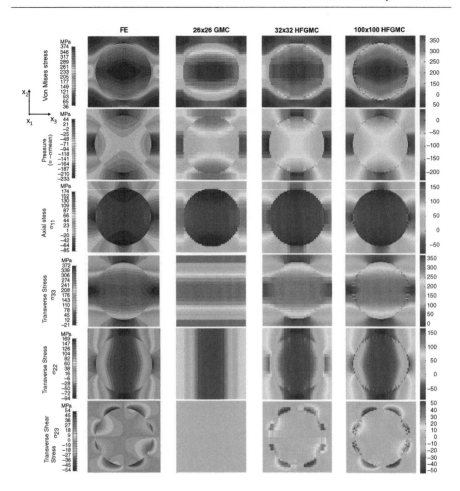

Figure 6.5 Comparison of local stress field distributions using a finite element RUC (first column), a 26×26 GMC RUC (second column), a 32×32 HFGMC RUC (third column) and 100×100 HFGMC RUC (fourth column) for a glass/epoxy composite subjected to 0.02 global transverse (x_3-direction) strain. Note that the same scales are applicable for GMC and HFGMC.

(see Aboudi et al., 2013). Additional detailed comparisons of HFGMC with FE results are available in the literature (e.g., Aboudi et al., 2013; Haj-Ali and Aboudi, 2009) that confirm that the HFGMC local fields compare well with FE under various conditions and loadings.

Comparing the local fields predicted by HFGMC using the two different subcell grid refinements in Fig. 6.5, as expected, the local fields match the FE predictions even better as the subcell grid is refined. It should be noted that the HFGMC fields are plotted based on subcell average values; the linear variations of the stresses across each subcell are not included and no smoothing operations are performed. Furthermore, the 32×32 RUC has a single subcell per step around the diameter of the fiber, whereas

Figure 6.6 Three RUCs employed to examine the effect of fiber packing arrangement.
(A) Square fiber packing (32 × 32 subcells). (B) Hexagonal fiber packing (68 × 120 subcells),
(C) Random fiber packing (79 × 79 subcells).

in the 100×100 RUC (which is comprised of all square subcells), most steps consist of multiple subcells. As a result, the corners of the steps are more resolved in the 100×100 RUC and thus stress concentrations from the step corners are visible in the local fields at the fiber/matrix interface. This is an artifact of the rectangular subcells employed in standard HFGMC and was a motivating factor in the development of Parametric HFGMC (PHFGMC) (Haj-Ali and Aboudi, 2009). PHFGMC, like the FE model shown in Fig. 6.5A (which employed isoparametric elements), is not limited to rectangular subvolumes.

6.10.3 Influence of fiber-packing arrangement – GMC vs. HFGMC

While GMC is capable of discretizing an RUC that appears to represent different geometric arrangements of the fibers, the average (first-order) nature of the GMC theory renders the method quite ineffective for modeling such geometric effects. To illustrate this, consider the glass/epoxy composite considered in the previous example (with fiber volume fraction 0.5) with three different fiber-packing arrangements: square fiber packing, hexagonal fiber packing, and random fiber packing (see Fig. 6.6). The deterministic fiber-packing arrangements (square and hexagonal) can be classified as ordered, whereas random fiber packing is classified as disordered. As mentioned in Chapter 3, square fiber packing results in cubic effective material properties, with 6 independent stiffness terms, as $G_{23} \neq E_{22}/2\left(1 + \nu_{23}\right)$. In contrast, perfect hexagonal fiber packing has six planes of symmetry (Cowin and Doty, 2007), and thus should provide transversely isotropic effective properties. The 32×32 RUC representing square fiber packing, shown in Fig. 6.6A, is identical to that used in the previous example. The hexagonal (Fig. 6.6B) and random (Fig. 6.6C) packing RUCs were generated using two additional MATLAB functions, RUCHex.m and RUCRand.m, both of which are called from the function GetRUC.m. These functions automatically generate hexagonal and randomly packed GMC and HFGMC RUCs with square subcells given the fiber volume fraction, and are associated with RUCid = 200 and RUCid = 300, respectively. They are based on MATLAB functions developed by Murthy et al. (2017). A portion of the applicable code within GetRUC.m is as follows:

```
% -- Hex pack RUC with square subcells
elseif RUCid == 200
    RUC.Vf_max = 0.85;
    Target_Ni = 2;
    Target_NG = 120;
    Plot = true;

    [RUC.Vf,RUC.Vi,RawRUC]= RUCHex(Vf,Vi,Target_Ni,Target_NG,Plot);
    .
    .
    .


% -- Random pack RUC with square subcells
elseif RUCid == 300
    RUC.Vf_max = 0.7;
    Nfibers = 10; % -- Number of fibers, default is 10
    Radf = 10; % -- Subcell in radius, default is 10
    if (Vi > 0)
        Tint = 2;
    else
        Tint = 0;
    end
    touching = false; % -- Allow or disallow touching fibers/interfaces
    MaxTries = 4000;
    Plot = true;

    [RUC.Vf,RUC.Vi,RawRUC]= RUCRand(Vf,Radf,Nfibers,Tint,touching,MaxTries,Plot);
    .
    .
    .
```

As shown, RUCHex.m requires a target value of N_γ (set in GetRUC.m as Target_NG) and will attempt to obtain as close as possible to the specified fiber volume fraction while maintaining hexagonal fiber packing and square subcells. A target value for the number of subcells across the thickness of the interface material (if one is present) is also specified as Target_Ni, as is a logical variable, Plot, to tell RUCHex.m whether or not to plot the generated hexagonally packed RUC. This plot will display immediately after the RUC is generated, giving the opportunity to inspect the RUC and abort the simulation if necessary.

RUCRand.m is nondeterministic and will generate a different RUC each time it is called (using a random number seed based on ten-thousandths of a second of the current time). RUCRand.m requires specification of the number of fibers in the RUC (Nfibers), the fiber radius (in terms of number of subcells, Radf), a logical variable (touching) that, when true, enables fibers to touch each other (with no matrix subcells in between), and a maximum number of tries (MaxTries) the function is permitted to attempt to place all fibers within the RUC. Obviously, as the fiber volume fraction increases, it becomes more difficult to place all specified fibers in the RUC without creating overlaps, and for high volume fractions, it is possible that the algorithm will fail to place all fibers. In such a case, execution of the MATLAB code will stop. RUCRand.m also correctly mirrors fibers across the RUC boundaries to maintain periodic RUC geometry and attempts to obtain the specified fiber volume fraction while maintaining square subcells. Finally, if an interfacial material is present, Tint specifies the interfacial thickness in terms of number of subcells, while this variable is set to zero in cases without an interface.

The three RUCs shown in Fig. 6.6 have been analyzed using both GMC and HFGMC, wherein a global transverse load of $\bar{\sigma}_{22} = 100$ MPa has been applied. For the case with hexagonal packing, `Target_NG = 120` has been specified, resulting in the 68×120 subcell RUC and a fiber volume fraction of exactly 0.5. For the random microstructure, an RUC containing 10 fibers, each with a radius of 10 subcells, has been chosen. This results in an RUC with 79×79 subcells and a fiber volume fraction of 0.5063.

As mentioned, each time the random RUC generator is executed, a different RUC geometry will result. To save and retrieve a particular RUC, functionality has been added to `MicroAnalysis.m` (and `LamAnalysis.m`) to save the RUC information to a JavaScript Object Notation (JSON) file (see ECMA International, 2013) as a final step as follows:

```
%----------------------------------------------------------------------
% 9) Write RUC
%----------------------------------------------------------------------
WriteRUC(OutInfo, OutInfo.Name(NP), effprops)
```

The function `WriteRUC.m` generates a JSON file containing all information about the RUC. For example, for a problem using `Mat = 152` in `GetEffProps.m`, which represents an IM7/8552 composite with a four fiber random 25×25 RUC, the following text-based JSON file "`RUC - … .json`" is generated (where the filename contains the problem name, date, and time):

```
{
    "RUC": [
        {
            "id": 300,
            "Mat": 152,
            "Vf": 0.512,
            "Vf_max": 0.7,
            "Vi": 0,
            "NB": 25,
            "NG": 25,
            "h": [1,1,1,1,1,1,1,1,1,1,1,1,1,1,1,1,1,1,1,1,1,1,1,1,1],
            "l": [1,1,1,1,1,1,1,1,1,1,1,1,1,1,1,1,1,1,1,1,1,1,1,1,1],
            "mats": [
                [1,3,3,3,3,1,1,1,1,1,1,1,3,3,3,1,1,1,1,1,1,1,1,1,1],
                [3,3,3,3,1,1,1,1,1,1,1,1,1,1,3,3,3,1,1,1,1,1,1,1,1],
                [3,3,3,3,1,1,1,1,1,1,1,1,1,1,3,3,3,1,1,1,1,1,1,1,1],
                [3,3,3,3,1,1,1,1,1,1,1,1,1,1,3,3,3,3,3,1,1,1,1,3,3],
                [3,3,3,3,1,1,1,1,1,1,1,1,1,1,3,3,3,3,3,3,3,3,3,3,3],
                [3,3,3,3,3,1,1,1,1,1,1,1,1,3,3,3,3,3,3,3,3,3,3,3,3],
                [3,3,3,3,3,1,1,1,1,1,1,1,1,3,3,3,3,3,3,3,3,3,3,3,3],
                [3,3,3,3,3,3,3,1,1,1,1,3,3,3,3,1,1,1,3,3,3,3,3,3,3],
                [3,3,3,3,3,3,3,3,3,3,3,3,1,1,1,1,1,1,1,1,3,3,3,3,3],
                [3,3,3,3,3,3,3,3,3,3,3,3,1,1,1,1,1,1,1,1,3,3,3,3,3],
                [3,3,1,1,1,1,3,3,3,3,3,1,1,1,1,1,1,1,1,1,1,3,3,3,3],
                [1,1,1,1,1,1,1,1,3,3,3,1,1,1,1,1,1,1,1,1,1,1,3,3,3],
                [1,1,1,1,1,1,1,1,1,3,3,3,1,1,1,1,1,1,1,1,1,1,3,3,3],
                [1,1,1,1,1,1,1,1,1,1,3,3,1,1,1,1,1,1,1,1,1,1,3,3,3,1],
                [1,1,1,1,1,1,1,1,1,1,3,3,3,1,1,1,1,1,1,1,1,1,3,3,3,1],
                [1,1,1,1,1,1,1,1,1,1,3,3,3,1,1,1,1,1,1,1,1,1,3,3,3,1],
                [1,1,1,1,1,1,1,1,1,1,3,3,3,3,3,1,1,1,1,1,3,3,3,3,3,1],
                [1,1,1,1,1,1,1,1,1,3,3,3,3,3,3,3,3,3,3,3,3,3,3,3,3],
                [1,1,1,1,1,1,1,1,1,3,3,3,3,3,3,3,3,3,3,3,3,3,3,3,3],
                [3,3,1,1,1,1,1,3,3,3,3,3,3,3,3,3,3,3,3,3,1,1,1,3,3],
                [3,3,3,3,3,3,3,3,3,3,3,3,3,3,3,3,3,3,3,1,1,1,1,1,1],
                [3,3,3,3,3,3,3,3,3,3,3,3,3,3,3,3,3,3,3,3,1,1,1,1,1,1],
                [1,3,3,3,3,3,3,3,3,3,3,3,3,3,3,3,3,1,1,1,1,1,1,1,1],
                [1,3,3,3,3,3,3,1,1,1,1,3,3,3,3,3,1,1,1,1,1,1,1,1,1],
                [1,3,3,3,3,1,1,1,1,1,1,1,1,3,3,3,1,1,1,1,1,1,1,1,1]
            ],
```

```
"matsCh": [
    "FMMMMFFFFFFFFFMMMFFFFFFFFFF",
    "MMMMFFFFFFFFFFFMMMFFFFFFFFF",
    "MMMMFFFFFFFFFFFFMMMFFFFFFFFF",
    "MMMMFFFFFFFFFFFFMMMMMFFFFMM",
    "MMMMFFFFFFFFFFFFMMMMMMMMMMM",
    "MMMMMFFFFFFFFFMMMMMMMMMMMM",
    "MMMMMFFFFFFFFMMMMMMMMMMMM",
    "MMMMMMMFFFFMMMFFFFMMMMMMM",
    "MMMMMMMMMMMMFFFFFFFFFMMMMM",
    "MMMMMMMMMMMMFFFFFFFFFMMMMM",
    "MMFFFFMMMMMFFFFFFFFFFFMMMM",
    "FFFFFFFFFMMFFFFFFFFFFFMMMM",
    "FFFFFFFFMMMFFFFFFFFFFFMMMM",
    "FFFFFFFFFMMFFFFFFFFFFFFMMMF",
    "FFFFFFFFFFMMFFFFFFFFFFFMMMF",
    "FFFFFFFFFFMMMFFFFFFFFFFMMMF",
    "FFFFFFFFFFMMMMMFFFFMMMMMMF",
    "FFFFFFFFFMMMMMMMMMMMMMMMM",
    "FFFFFFFFFMMMMMMMMMMMMMMMM",
    "MMFFFFMMMMMMMMMMMMMFFFFMM",
    "MMMMMMMMMMMMMMMMMFFFFFFFFF",
    "MMMMMMMMMMMMMMMMMFFFFFFFFF",
    "FMMMMMMMMMMMMMMMFFFFFFFFFF",
    "FMMMMMMFFFFMMMMMFFFFFFFFFF",
    "FMMMMFFFFFFFFFMMMFFFFFFFFFF"
  ]
 }
]
}
```

Note that the constituent materials in the JSON file (see "mats" in the above example) are not used. Rather, the constituent materials specified in the composite material definition in GetEffProps.m (i.e., Constits.Fiber, Constits.Matrix, Constits.Interface) are assigned to the RUC according to the arrangement in "matsCh" in the JSON file (F, M, I).

To enable use of an RUC stored in a JSON file, RUCid = 1000 has been added to GetRUC.m, with the applicable portion of code as follows:

```
% -- Read RUC from json file
elseif RUCid == 1000

    % -- Add jsonlab folder to path
    addpath('Functions/Utilities/jsonlab');

    % -- prompt for RUC json file
    [file,path] = uigetfile('*.json','Select json file containing RUC(s)');
    if isequal(file,0)
        error('        *** RUC json file selection canceled ***')
    end
    RUCFileName = [path, file];
    jsondata = loadjson(RUCFileName);
    .
    .
    .
```

Note that only RUCid = 300 and RUCid = 1000 are written to the JSON files (although this can be easily changed in WriteRUC.m).

When executing a problem using an RUC from a JSON file with RUCid = 1000, the user is prompted by the MATLAB code to choose the desired JSON file, and then a dialog box enables the user to choose the desired RUC contained within the chosen JSON file, as shown in Fig. 6.7.

One can construct any RUC and save it in a JSON file using a text editor. Multiple RUCs can also be stored in the same JSON file. For the case shown in Fig. 6.7, where

Figure 6.7 Dialog box that appears to enable selection of the RUC contained within an existing JSON file.

the JSON contains four RUCs, the JSON file is as follows (where some lines have been omitted and replaced with three dots to save space):

```
{
    "RUC": [
        {
            "id": 300,
            "Mat": 152,
            "Vf": 0.512,
            "Vf_max": 0.7,
            "Vi": 0,
            "NB": 25,
            "NG": 25,
            "h": [1,1,1,1,1,1,1,1,1,1,1,1,1,1,1,1,1,1,1,1,1,1,1,1,1],
            "l": [1,1,1,1,1,1,1,1,1,1,1,1,1,1,1,1,1,1,1,1,1,1,1,1,1],
            "mats": [
                [3,3,1,1,1,1,1,1,1,3,3,3,3,3,3,3,3,3,3,3,3,3,3,3,3],
                    .
                    .
                    .
                [3,3,3,3,1,1,1,1,3,3,3,3,3,3,3,3,3,1,1,1,1,3,3,3,3]
            ],
            "matsCh": [
                "MMFFFFFFFFMMMMMMMMMMMMMMMM",
                    .
                    .
                    .
                "MMMMFFFMMMMMMMMMFFFMMMM"
            ]
        },
        {
            "id": 300,
            "Mat": 164,
            "Vf": 0.512,
            "Vf_max": 0.7,
            "Vi": 0,
            "NB": 25,
            "NG": 25,
            "h": [1,1,1,1,1,1,1,1,1,1,1,1,1,1,1,1,1,1,1,1,1,1,1,1,1],
```

```
        "l":  [1,1,1,1,1,1,1,1,1,1,1,1,1,1,1,1,1,1,1,1,1,1,1,1,1],
        "mats":  [
            [4,4,2,2,2,2,2,2,2,2,4,4,4,4,2,2,2,2,2,2,2,2,2,2,4],
            .
            .

            .
            [4,4,2,2,2,2,2,2,2,2,4,4,4,4,4,2,2,2,2,2,2,2,2,4,4]
        ],
            "matsCh":  [
                "MMFFFFFFFFFMMMMFFFFFFFFFFFM",
                .
                .

                .
                "MMFFFFFFFFFMMMMMFFFFFFFFFMM"
            ]
        },
        {
            "id":  300,
            "Mat":  165,
            "Vf":  0.512,
            "Vf_max":  0.7,
            "Vi":  0,
            "NB":  25,
            "NG":  25,
            "h":  [1,1,1,1,1,1,1,1,1,1,1,1,1,1,1,1,1,1,1,1,1,1,1,1,1],
            "l":  [1,1,1,1,1,1,1,1,1,1,1,1,1,1,1,1,1,1,1,1,1,1,1,1,1],
            "mats":  [
                [4,4,4,4,4,4,4,4,4,4,4,2,2,2,2,2,2,2,4,4,4,4,4,4,4],
                .
                .

                .
                [4,4,4,4,4,4,4,4,4,4,4,2,2,2,2,2,2,2,4,4,4,4,4,4,4]
            ],
            "matsCh":  [
                "MMMMMMMMMMMFFFFFFFFMMMMMM",
                .
                .

                .
                "MMMMMMMMMMMFFFFFFFFMMMMMM"
            ]
        },
        {
            "id":  1000,
            "Mat":  184,
            "Vf":  0.355556,
            "Vf_max":  0.7,
            "Vi":  0,
            "NB":  30,
            "NG":  30,
            "h":  [1,1,1,1,1,1,1,1,1,1,1,1,1,1,1,1,1,1,1,1,1,1,1,1,1,1,1,1,1,1],
            "l":  [1,1,1,1,1,1,1,1,1,1,1,1,1,1,1,1,1,1,1,1,1,1,1,1,1,1,1,1,1,1],
            "mats":  [
                [14,14,14,14,14,14,14,14,14,2,2,2,2,2,2,2,2,2,2,2,2,2,2,2,2,2,2,2,2,14],
                .
                .

                .
                [14,14,14,14,14,14,14,14,14,2,2,2,2,2,2,2,2,2,2,2,2,2,2,2,2,2,2,2,2,14]
            ],
            "matsCh":  [
                "FFFFFFFFFMMMMMMMMMMMMMMMMMMMMMF",
                .
                .

                .
                "FFFFFFFFFMMMMMMMMMMMMMMMMMMMMMF"
            ]
        }
    }
  ]
}
```

The specific random RUC employed in this example, shown in Fig. 6.6C, has been provided in the JSON file "RUCs - Example Section 6.10.3 - 79x79 Random.json" in the "Output" folder.

To analyze the three RUCs shown Fig. 6.6 using both GMC and HFGMC, the following six problems are defined in MicroProblemDef.m:

```
NP = NP + 1;
OutInfo.Name(NP) = "Example Section 6.10.3 - 26x26 Sq GMC - contour glass-epoxy";
MicroMat{NP} = 145;
Loads{NP}.DT = 0;
Loads{NP}.Type  = [  S,   S,  S,  S,  S,  S];
Loads{NP}.Value = [  0, 100,  0,  0,  0,  0];

NP = NP + 1;
OutInfo.Name(NP) = "Example Section 6.10.3 - 32x32 Sq HFGMC - contour glass-epoxy";
MicroMat{NP} = 143;
Loads{NP}.DT = 0;
Loads{NP}.Type  = [  S,   S,  S,  S,  S,  S];
Loads{NP}.Value = [  0, 100,  0,  0,  0,  0];

NP = NP + 1;
OutInfo.Name(NP) = "Example Section 6.10.3 - 68x120 Hex GMC - contour glass-epoxy";
MicroMat{NP} = 146;
Loads{NP}.DT = 0;
Loads{NP}.Type  = [  S,   S,  S,  S,  S,  S];
Loads{NP}.Value = [  0, 100,  0,  0,  0,  0];

NP = NP + 1;
OutInfo.Name(NP) = "Example Section 6.10.3 - 68x120 Hex HFGMC - contour glass-epoxy";
MicroMat{NP} = 147;
Loads{NP}.DT = 0;
Loads{NP}.Type  = [  S,   S,  S,  S,  S,  S];
Loads{NP}.Value = [  0, 100,  0,  0,  0,  0];

NP = NP + 1;
OutInfo.Name(NP) = "Example Section 6.10.3 - 79x79 Random (json) GMC - contour glass-
epoxy";
MicroMat{NP} = 148;
Loads{NP}.DT = 0;
Loads{NP}.Type  = [  S,   S,  S,  S,  S,  S];
Loads{NP}.Value = [  0, 100,  0,  0,  0,  0];

NP = NP + 1;
OutInfo.Name(NP) = "Example Section 6.10.3 - 79x79 Random (json) HFGMC - contour glass-
epoxy";
MicroMat{NP} = 149;
Loads{NP}.DT = 0;
Loads{NP}.Type  = [  S,   S,  S,  S,  S,  S];
Loads{NP}.Value = [  0, 100,  0,  0,  0,  0];
{NP}.Value = [  0,   0, 100,  0,  0,  0];
```

where Mat = 143 was defined in the previous example, and the following five additional materials have been defined in GetEffProps.m:

```
Mat = 145;
Theory = 'GMC';
name = "26x26 Glass-Epoxy Contour GMC";
Vf  = 0.50;
Constits.Fiber = constitprops{8};
Constits.Matrix = constitprops{9};
RUCid = 26;
[props{Mat}] = RunMicro(Theory, name, Vf, Constits, props{Mat}, RUCid);
```

```
Mat = 146;
Theory = 'GMC';
name = "68x120 Hex Glass-Epoxy Contour GMC";
Vf   = 0.50;
Constits.Fiber = constitprops{8};
Constits.Matrix = constitprops{9};
RUCid = 200;
[props{Mat}] = RunMicro(Theory, name, Vf, Constits, props{Mat}, RUCid);

Mat = 147;
Theory = 'HFGMC';
name = "68x120 Hex Glass-Epoxy Contour HFGMC";
Vf   = 0.50;
Constits.Fiber = constitprops{8};
Constits.Matrix = constitprops{9};
RUCid = 200;
[props{Mat}] = RunMicro(Theory, name, Vf, Constits, props{Mat}, RUCid);

Mat = 148;
Theory = 'GMC';
name = "79x79 Random (json) Glass-Epoxy Contour GMC";
Vf   = 0.50;
Constits.Fiber = constitprops{8};
Constits.Matrix = constitprops{9};
RUCid = 1000;
[props{Mat}] = RunMicro(Theory, name, Vf, Constits, props{Mat}, RUCid);

Mat = 149;
Theory = 'HFGMC';
name = "79x79 Random (json) Glass-Epoxy Contour HFGMC";
Vf   = 0.50;
Constits.Fiber = constitprops{8};
Constits.Matrix = constitprops{9};
RUCid = 1000;
[props{Mat}] = RunMicro(Theory, name, Vf, Constits, props{Mat}, RUCid);
```

For the GMC and HFGMC random RUC problems, when prompted, the following JSON file (provided in the Output folder) should be selected: RUCs - Example Section 6.10.3 - 79x79 Random.json. Also, to plot the GMC and HFGMC local fields using the same colorbars, as done in Fig. 6.8 and Fig. 6.9 for the hexagonal and random packing, the color bar limits must be specified in PlotMicroFields.m.

The predicted effective mechanical properties and CTEs, using the three fiber-packing arrangements, are compared in Table 6.4 and Table 6.5, respectively. In all cases GMC is in better agreement with HFGMC for Young's moduli compared to shear moduli. For the square fiber-packing arrangement, GMC matches HFGMC reasonably well, with a maximum difference between the two methods in predicted properties of 8.5% in the case of G_{12} and G_{13}. For the hexagonal fiber packing, the difference is much greater, with a maximum of 26.8% for G_{13}. Furthermore, it is clear that GMC does not predict transversely isotropic properties for hexagonal packing as E_{22} and E_{33} differ by 11%, G_{12} and G_{13} differ by 15%, ν_{12} and ν_{13} differ by 7%, and α_{22} and α_{33} differ by 25%. Calculating a transversely isotropic G_{23} for GMC using the average of E_{22} and E_{33}, that is, $G_{23}^{TI} = (E_{22} + E_{33})/4\,(1 + \nu_{23}) = 2887$ MPa, which differs from the predicted G_{23} by 15%. In contrast, for HFGMC, all the properties that should be equal for transverse isotropy differ by less than 2.1%; furthermore

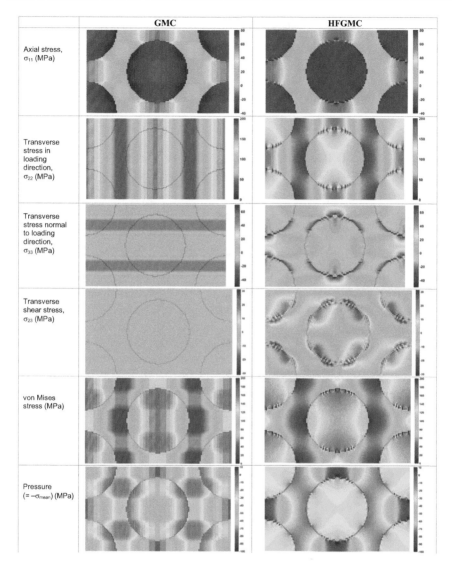

Figure 6.8 Comparison of local stress fields predicted by GMC and HFGMC for a 0.5 fiber volume fraction glass/epoxy composite with hexagonal fiber packing subjected to $\bar{\sigma}_{22} = 100$ MPa.

$G_{23}^{TI} = 3253$ MPa (calculated as shown above), differs from the predicted G_{23} by only 1.6%. Consequently, HFGMC is able to provide a very good approximation of the transverse isotropy imparted by the hexagonal fiber packing arrangement. In contrast, the GMC results indicate the theory's inability to correctly capture the effects of fiber packing arrangement on the predicted composite properties (due to its lack of shear

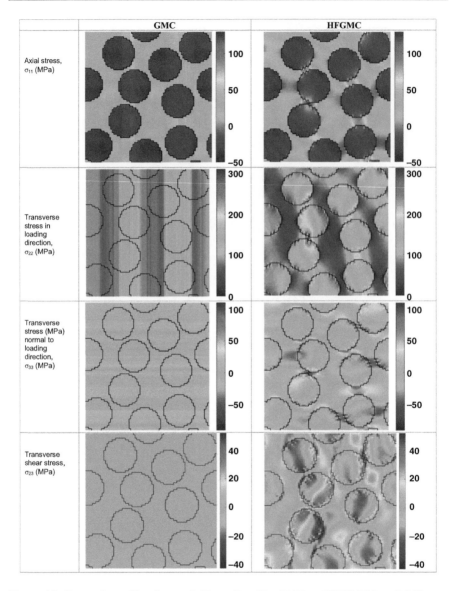

Figure 6.9 Comparison of local stress fields predicted by GMC and HFGMC for a 0.5 fiber volume fraction glass/epoxy composite with random fiber packing (RUC with 10 fibers and a fiber radius of 10 subcells) subjected to $\bar{\sigma}_{22} = 100$ MPa.

coupling). In fact, comparing the predictions of GMC with both square and hexagonal fiber packing to the HFGMC hexagonal packing results, it is clear that GMC with square packing is in better agreement. *Therefore it is best practice when using GMC to always employ RUCs with square fiber packing.*

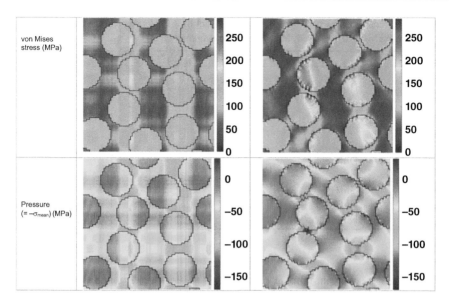

Figure 6.9 Continued

Table 6.4 Effective mechanical properties of a 0.5 fiber volume fraction glass/epoxy composite predicted by GMC and HFGMC with different fiber packing arrangements.

		E_{11} (MPa)	E_{22} (MPa)	E_{33} (MPa)	G_{12} (MPa)	G_{13} (MPa)	G_{23} (MPa)	ν_{12}	ν_{13}	ν_{23}
Square	GMC	36229	10381	10381	3189	3189	2444	0.262	0.262	0.323
	HFGMC	36233	10790	10790	3487	3487	2636	0.261	0.261	0.309
	Difference	0.01%	3.8%	3.8%	8.5%	8.5%	7.3%	0.5%	0.5%	4.4%
Hexagonal	GMC	36218	8875	7874	2959	2513	2444	0.258	0.275	0.450
	HFGMC	36232	9232	9131	3476	3432	3199	0.260	0.262	0.412
	Difference	0.04%	3.9%	13.8%	14.9%	26.8%	23.6%	0.9%	5.1%	9.4%
Random	GMC	36630	7885	7898	2629	2635	2472	0.267	0.267	0.453
	HFGMC	36652	9781	9295	3856	3627	3536	0.255	0.261	0.422
	Difference	0.1%	19.4%	15.0%	31.8%	27.3%	30.1%	4.7%	2.1%	7.4%

This point can be further illustrated by considering the random fiber packing geometry shown in Fig. 6.6C. This packing arrangement is one random instance of many possible fiber packing arrangements (given the specified 10 fibers and 10 subcell fiber radius with square subcells within the RUC). As such, a different instance (resulting from generating a new random RUC (RUCid = 300) with the MATLAB code) would be expected to give different predicted properties. However, if the RUC were large enough

Table 6.5 Effective CTEs of a 0.5 fiber volume fraction glass/epoxy composite predicted by GMC and HFGMC with different fiber packing arrangements.

		α_{11} $(10^{-6}/°C)$	α_{22} $(10^{-6}/°C)$	α_{33} $(10^{-6}/°C)$
Square	GMC	7.561	33.46	33.46
	HFGMC	7.606	32.79	32.79
	Difference	0.6%	2.1%	2.1%
Hexagonal	GMC	7.418	31.63	39.62
	HFGMC	7.594	32.63	33.30
	Difference	2.3%	3.1%	19.0%
Random	GMC	7.328	35.81	35.69
	HFGMC	7.601	30.09	33.07
	Difference	3.6%	19.0%	7.9%

and contained a sufficient number of fibers (e.g. > 100, see Liu and Goshal, 2014a), each instance should provide nearly identical effective properties. Such a geometry is referred to as a statistical representative volume element (SRVE) (Gitman et al., 2007 and Ostoja-Starzewski, 2006).

Comparing GMC to HFGMC for the random fiber packing arrangement (Table 6.4 and Table 6.5), the difference is even greater than in the hexagonally packed case, with a maximum difference of 31.8% (G_{12}). This again underscores GMC's inability to capture the effects of fiber packing arrangement, which of course includes any disordered microstructure. It is interesting to note that, for this instance of random packing with 10 fibers, the predicted HFGMC properties are not that close to transversely isotropic (for example E_{22} and E_{33} differ by 5%). However, as more fibers are included within the RUC (or if many instances are averaged), the HFGMC predictions approach transverse isotropy.

Figs. 6.8 and 6.9 compare the GMC and HFGMC local stress fields for the hexagonal and random fiber packing, respectively. It is clear that GMC's lack of shear coupling ($\sigma_{23} = 0$), resulting in constant σ_{22} and σ_{33} in rows and columns of subcells, makes it impossible for GMC to capture correctly the effects of the fiber packing arrangement. In the hexagonal packing, the concentrations that occur between fibers in HFGMC (for σ_{22} and σ_{33}) are only captured in a very average, or smeared, sense by GMC, so the concentrations are considerably lower in magnitude. In the random packing, the HFGMC stress fields are dominated by peaks occurring between closely spaced fibers. For example, the peak local stress in the loading direction, σ_{22}, is under predicted by approximately 35% by GMC (compared to HFGMC). Furthermore, the peak local transverse shear stress, σ_{23}, predicted by HFGMC reaches 50 MPa, which is one-half of the global applied stress magnitude, while GMC (without shear coupling) is not capable of predicting this local stress component at all. Clearly, this significant difference in local fields will also be important if failure (damage) or other material nonlinearities (e.g., inelasticity) are included.

In summary, this example has illustrated that the approximate nature of GMC renders it unsuitable for predicting the effect of different fiber packing arrangements. In fact, Liu and Goshal (2014b) showed that, in GMC, the columns and rows of subcells can be swapped arbitrarily with no impact on the predicted results. Consequently, *when using GMC, best practice is to use an RUC with square fiber packing.* It is also advisable to use more subcells for refining the circular fiber, which more accurately represents each row and column of subcells and enables better approximation of the peak stresses in the matrix (see Fig. 6.5).

6.10.4 SiC/SiC composite with hexagonal fiber packing

Consider a SiC/SiC CMC whose constituent material properties were given in Table 5.2, with a fiber volume fraction of 0.28 and a BN interface volume fraction of 0.13. This composite was modeled using GMC in Sections 5.7.3 and 5.7.4. Now, modeling this composite with HFGMC with hexagonal fiber packing, the RUC shown in Fig. 6.10 was employed. The subcell boundaries have been removed and the RUC has been tiled nine times in a three by three pattern in Fig. 6.11, with the boundary of the center RUC indicated. This is done to illustrate that, in the x_2-direction, the composite microstructure has continuous straight paths through the matrix while, in the x_3-direction, any straight path intersects the compliant BN interface material.

To analyze the hexagonally packed RUC shown in Fig. 6.10 using HFGMC with tensile loading of 100 MPa applied uniaxially in both the x_2- and x_3-directions, the following problems are defined in `MicroProblemDef.m`:

```
NP = NP + 1;
OutInfo.Name(NP) = "Example Section 6.10.4 - 136x236 Hex SiC-SiC HFGMC - S22";
MicroMat{NP} = 150;
Loads{NP}.DT = 0;
Loads{NP}.Type  = [ S,   S,   S,  S,  S,  S];
Loads{NP}.Value = [ 0, 100,   0,  0,  0,  0];

NP = NP + 1;
OutInfo.Name(NP) = "Example Section 6.10.4 - 136x236 Hex SiC-SiC HFGMC - S33";
MicroMat{NP} = 150;
Loads{NP}.DT = 0;
Loads{NP}.Type  = [ S,   S,   S,  S,  S,  S];
Loads{NP}.Value = [ 0,   0, 100,  0,  0,  0];
```

`Mat = 150` has been defined in `GetEffProps.m` as follows:

```
Mat = 150;
Theory = 'HFGMC';
name = "136x236 Hex SiC-SiC HFGMC";
Vf = 0.28;
Vi = 0.13;
Constits.Fiber = constitprops{10};
Constits.Matrix = constitprops{11};
Constits.Interface = constitprops{12};
RUCid = 200;
[props{Mat}] = RunMicro(Theory,name,struct('Vf',Vf,'Vi',Vi),Constits,props{Mat},RUCid);
```

and a target N_γ of 240, along with a target interface thickness of eight subcells, has been specified in `GetRUC.m`. Note that the actual N_γ chosen by `RUCHex.m` for this example is 236.

Figure 6.10 136 × 236 subcell RUC with hexagonal fiber packing used to represent a 0.28 fiber volume fraction SiC/SiC CMC with a 0.13 volume fraction BN coating.

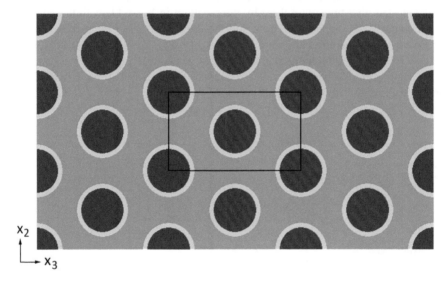

Figure 6.11 136 × 236 subcell RUC shown in Fig. 6.10 has been tiled nine time to show the hexagonal packing arrangement.

The effective properties of the SiC/SiC composite predicted by HFGMC are given in Table 6.6. With this level of RUC refinement, HFGMC does an excellent job of predicting transversely isotropic effective properties. The properties that should be equal for transverse isotropy are all within 0.06%, and the G_{23} calculated from E_{22} and v_{23} using the transversely isotropic relationship is within 0.3% of the predicted G_{23}. Hence, the linear thermoelastic behavior of the material is predicted to be nearly identical in the x_2- and x_3-directions. However, if loading is applied in the x_2-direction

Table 6.6 Effective mechanical properties of a 0.28 fiber volume fraction SiC/SiC composite with a 0.13 volume fraction BN interface predicted by HFGMC using the 136×236 subcell RUC shown in Fig. 6.10.

E_{11} (MPa)	E_{22} (MPa)	E_{33} (MPa)	G_{12} (MPa)	G_{13} (MPa)	G_{23} (MPa)	ν_{12}	ν_{13}	ν_{23}	α_{11} $(10^{-6}/°C)$	α_{22} $(10^{-6}/°C)$	α_{33} $(10^{-6}/°C)$
302116	150168	150103	70884	70846	58927	0.214	0.214	0.270	3.138	3.149	3.149

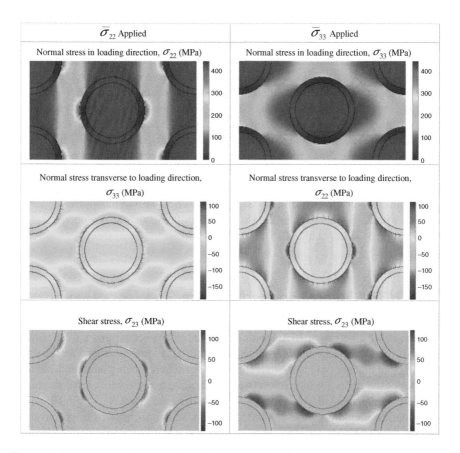

Figure 6.12 Local stress fields predicted by HFGMC for a 0.28 fiber volume fraction SiC/SiC composite with a 0.13 volume fraction BN interface with hexagonal fiber packing subjected to $\bar{\sigma}_{22} = 100$ MPa (left side) and $\bar{\sigma}_{33} = 100$ MPa (right side).

versus the x_3-direction, the resulting local stress fields are vastly different. Figure 6.12 compares the local stresses in the x_2-x_3 plane when the composite is subjected to separate uniaxial load cases, for example $\bar{\sigma}_{22} = 100$ MPa and $\bar{\sigma}_{33} = 100$ MPa. When applying $\bar{\sigma}_{22}$, the aforementioned continuous paths of stiff SiC matrix material results in columns of high local σ_{22}, with higher concentrations at the boundaries between the matrix and the compliant interface. In contrast, when applying $\bar{\sigma}_{33}$, the local σ_{33} in the matrix is quite diffuse and the magnitudes of the interfacial stress concentrations are much lower. There are also major differences in the normal stress component transverse to the loading direction, with the continuous straight matrix paths in the x_2-direction leading to higher local σ_{22} stress concentrations when $\bar{\sigma}_{33}$ is applied. However, an even starker difference is present when comparing the local σ_{23} fields. For instance in the case of $\bar{\sigma}_{33}$ loading, with no continuous straight matrix paths, the stress transfer through the RUC must involve significant shear stresses over a large area of the matrix. Alternatively, when applying $\bar{\sigma}_{22}$, significant local σ_{23} only arises at the boundary between the matrix and the compliant BN interface.

Despite the vast differences in the local stress fields shown in Figure 6.12, the fact that the effective properties in Table 6.6 are so nearly transversely isotropic requires that the averages of the appropriate component-wise strain fields be nearly identical. In addition, regardless of the direction of the applied loading, the average of each stress component must equal the applied stress component. For example, for the applied loading of $\bar{\sigma}_{22} = 100$ MPa, the average of the local σ_{22} must be 100 MPa, while the average of the local σ_{33} must be zero, and the average of the local σ_{23} must be zero. However, because the point-wise local stresses are so different, strength and allowable predictions, which are, of course, based on local, not average stresses, will be quite different in the two directions. Only the effective thermoelastic properties associated with hexagonal packing are isotropic. Similar results can be generated using FE modeling (e.g., Stier et al., 2015).

6.10.5 HFGMC subcell grid convergence

Like the FE method, HFGMC is dependent on the geometric discretization employed. In this example, an array of RUCs with hexagonal fiber packing and a fiber volume fraction of 0.6 are considered with increasing discretization refinement, as shown in Fig. 6.13. Note that not only is the overall number of subcells being increased, but also the shapes of the fibers are more closely approximating a circle with each refinement. The RUCs were generated using RUCHex.m. by altering the target N_y.

Considering a unidirectional IM7/8552 composite, whose properties are given in Table 3.4, the MATLAB script MicroAnalysis.m was used to determine the effective properties for each of the eight RUCs shown in Fig. 6.13. The problem is defined in MicroProblemDef.m as follows:

```
NP = NP + 1;
OutInfo.Name(NP) = "Example Section 6.10.5 - HFGMC Subcell Grid Convergence";
% -- Target_NG must be altered in GetRUC.m to change subcell grid
MicroMat{NP} = 151;
Loads{NP}.DT = 0;
Loads{NP}.Type  = [  S,   S,  S,  S,  S,  S];
Loads{NP}.Value = [  0, 100,  0,  0,  0,  0];
Loads{NP}.CriteriaOn = [1,0,0,0];
```

with MAT = 151 defined in GetEffProps.m as follows:

```
Mat = 151;
Theory = 'HFGMC';
name = "Grid convergence Hex Vf = 0.6 IM7-8552 HFGMC";
Vf = 0.6;
Constits.Fiber = constitprops{1};
Constits.Matrix = constitprops{3};
RUCid = 200;
[props{Mat}] = RunMicro(Theory, name, Vf, Constits, props{Mat}, RUCid);
```

Before executing each problem, the RUC grid refinement must be adjusted by specifying `Target_NG` in `GetRUC.m` (i.e., `Target_NG` =20, 40, 60, 80, 120, 160, 200, 240).

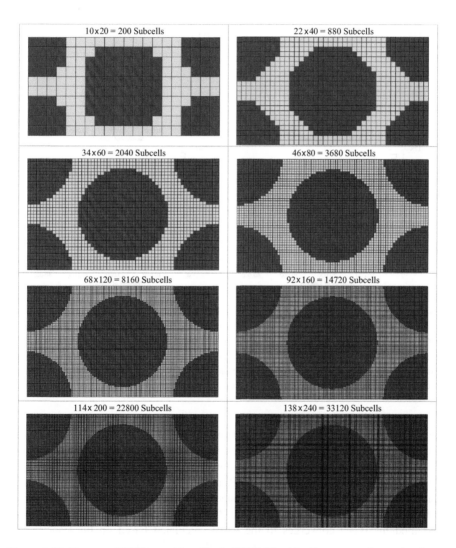

Figure 6.13 Hexagonal RUCs considered in the HFGMC grid refinement study ranging from 200 to 33,120 subcells.

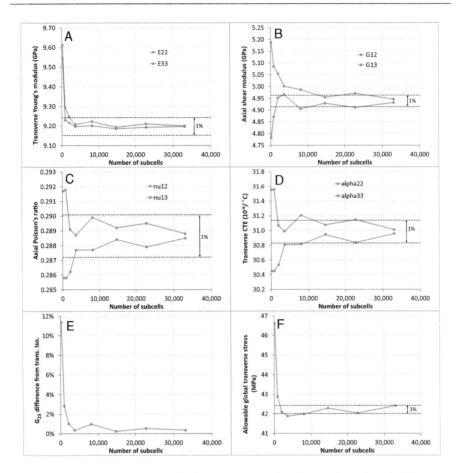

Figure 6.14 Comparisons of the effective properties predicted by HFGMC for a 0.6 fiber volume fraction unidirectional IM7/8552 composite represented by the eight RUCs shown in Fig. 6.13. (A) Transverse Young's modulus. (B) Axial shear modulus. (C) Axial Poisson's ratio. (D) Transverse CTE. (E) Difference between predicted G_{23} and the transversely isotropic relation $(E_{22} + E_{33})/(4(1 + v_{23}))$. (F) Allowable global transverse (x_2-direction) stress based on exceeding the allowable of the first subcell based on the max stress criterion.

Fig. 6.14 shows plots of the predicted effective properties as a function of the number of subcells within the RUCs and compares the properties that should be equal in the case of transverse isotropy. For instance, in Fig. 6.14A, E_{22} and E_{33} are shown to converge rapidly to approximately 9.20 GPa as the RUC is refined. Likewise, the axial shear moduli, axial Poisson's ratios, and transverse CTEs all converge to within 1% for the 46×80 subcell RUC (see Fig. 6.13). Fig. 6.14E shows that the predicted transverse shear modulus (G_{23}) converges to the value calculated based on the transversely isotropic relation (using the average of E_{22} and E_{33}). The predicted global transverse allowable stress (based on the first subcell exceeding the max stress criterion) also

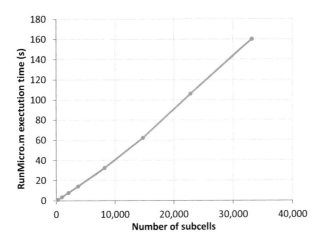

Figure 6.15 Execution time for the core HFGMC micromechanics calculations (within RunMicro.m) as a function of number of subcells for the IM7/8552 composite modeled using the eight RUCs shown in Fig. 6.13.

converges to within 1% of approximately 42.2 MPa, as shown in Fig. 6.14F. One should not expect perfect convergence with HFGMC because of the inability of an orthogonal discretization (i.e., rectangular subcells) to perfectly represent a circular fiber. As previously mentioned, this restriction of HFGMC to orthogonal discretizations has been eliminated by Haj-Ali and Aboudi (2010, 2013) through the development of an isoparametric formulation of HFGMC, known as parametric HFGMC (PHFGMC).

Fig. 6.15 shows a plot of the time to execute RunMicro.m vs. the number of subcells in the RUC. Note that the execution times vary based on the computer used, and only a single instance is plotted (as opposed to an average execution time value) per subcell refinement. The curve is nonlinear, although not strongly nonlinear as the MATLAB code takes advantage of efficient algorithms and the sparse nature of the HFGMC structural stiffness matrix. Clearly, from Figs. 6.14 and 6.15, a trade off exists between achieving convergence (and more generally "accuracy") and the computational cost to achieve a given level of accuracy. Furthermore, there is a diminishing return on the accuracy improvement associated with further refinement of the HFGMC subcell discretization. This is true not only for HFGMC, but is also typical of methods that exhibit geometric discretization dependence.

6.10.6 HFGMC effective property statistics based on random RUC packing

In this example, 1000 realizations of random RUCs (generated using RUCRand.m) containing 10 fibers (with a radius of 10 subcells) were analyzed with HFGMC to obtain effective properties of an IM7/8552 composite with a fiber volume fraction of 0.5. The effective properties from each case are written to a file and statistically processed (histograms, means, standard deviations). This can be accomplished utilizing MATLAB Script 6.2 – RandomRUCHist.m, which is quite similar to the first half of MicroAnalysis.m. The setup of each of the 1000 problems is identical and is done in MicroProblemDef.m using a for loop as follows,

```
NUMBER = 1000; % -- Note, it will take a long time to run 1000 random cases
for NP1 = 1: NUMBER
    NP = NP + 1;
    OutInfo.Name(NP) = "Example Section 6.10.6 - HFGMC Random 1000 Instances";
    MicroMat{NP} = 152;
    Loads{NP}.DT = 0;
    Loads{NP}.Type  = [  S,  S,  S,  S,  S,  S];
    Loads{NP}.Value = [  0,  0,  0,  0,  0,  0];
end
```

The material referenced for every case is the same, and it is defined in GetEffProps.m as follows,

```
Mat = 152;
Theory = 'HFGMC';
name = "Random Vf = 0.5 IM7-8552 HFGMC";
Vf = 0.5;
Constits.Fiber = constitprops{1};
Constits.Matrix = constitprops{3};
RUCid = 300;
[props{Mat}] = RunMicro(Theory, name, Vf, Constits, props{Mat}, RUCid);
```

The parameters Nfibers and Radf (in GetRUC.m) were both set to 10, while touching fibers were disallowed (touching = false) and plotting of the RUC for every case was turned off (Plot = false). The resulting random RUCs consist of 79×79 subcells with an actual fiber volume fraction of 0.5063 (similar to Fig. 6.6C).

When RandomRUCHist.m runs the micromechanics analysis for each of the 1000 problems, a new random RUC is generated for every case, and the properties are written to a text file "… - Random Results.txt". The code at the end of RandomRUCHist.m reads in the results from the text file and makes histogram plots for each of the properties, while also calculating the mean and standard deviation. This section of code can also be used to generate histograms from existing text files from previous runs of RandomRUCHist.m. Note that running all 1000 HFGMC problems will take several hours.

The resulting histograms are shown in Fig. 6.16. It is clear that the mean values of the predicted properties are close to isotropic, with E_{22} and E_{33}, ν_{12} and ν_{13}, and G_{12} and G_{13} all differing by less than 0.1%. The predicted G_{23} value is within 0.22% of the transversely isotropic value, $G_{23}^{TI} = (E_{22} + E_{33})/4\,(1 + \nu_{23}) = 2720$ MPa. The standard deviations are quite small, with a maximum of approximately 2% of the mean for G_{13}. This indicates that the effective properties are relatively insensitive to the random fiber packing for this composite at this fiber volume fraction. As shown by Arnold et al. (2016), were the fiber volume fraction lower, there would be more room

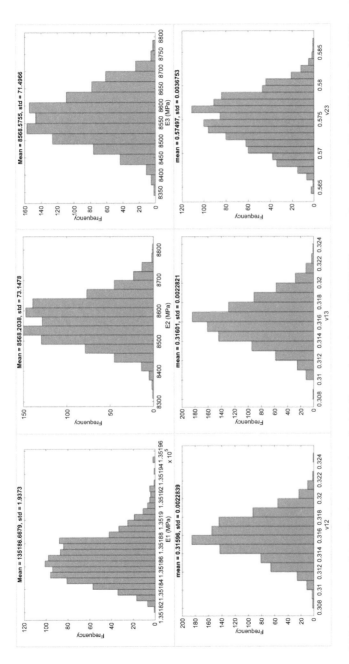

Figure 6.16 Histograms of the effective properties of a 0.5 fiber volume fraction IM7/8552 composite as predicted by HFGMC using 1000 random RUCs with 10 fibers. The mean and standard deviation are indicated on each plot.

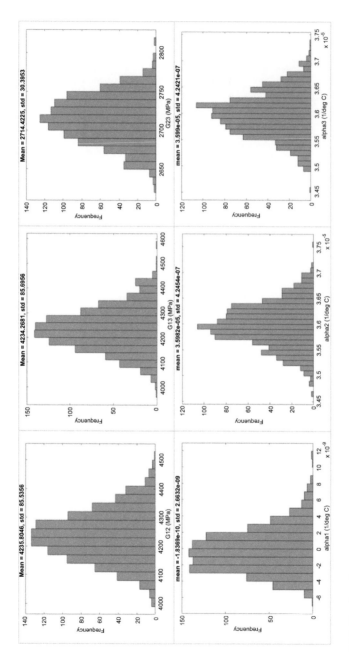

Figure 6.16 Continued.

in the RUC for variation of the fiber locations, and one would expect higher sensitivity of the effective properties to the random fiber packing. Finally, it is interesting to note that the mean of α_{11} is very close to zero, with some RUCs predicting positive α_{11} and some predicting negative α_{11}. MATLAB Script 6.2.

MATLAB Script 6.2 – `RandomRUCHist.m`

This function runs multiple stand-alone micromechanics problems to determine and output the effective properties to a text file. Then it produces histograms for each effective property.

```
% ++++++++++++++++++++++++++++++++++++++++++++++++++++++++++++++++++++++++++++++
% Copyright 2020 United States Government as represented by the Administrator of the
% National Aeronautics and Space Administration. No copyright is claimed in the
% United States under Title 17, U.S. Code. All Other Rights Reserved. By downloading
% or using this software you agree to the terms of NOSA v1.3. This code was prepared
% by Drs. B.A. Bednarcyk and S.M. Arnold to complement the book "Practical
% Micromechanics of Composite Materials" during the course of their government work.
% ++++++++++++++++++++++++++++++++++++++++++++++++++++++++++++++++++++++++++++++
%
%
% Purpose: This script runs multiple stand-alone micromechanics problems, stores the
%          calculated effective properties, and plots a histogram for each property.
%          The composite materials are defined in GetEffProps.m and the loading and
%          problem setup is specified in MicroProblemDef.m
%
% ++++++++++++++++++++++++++++++++++++++++++++++++++++++++++++++++++++++++++++++


% -- Clear memory and close files
clear;
close all;
clc;

% -- Add needed function locations to the path
addpath('Functions/Utilities');
addpath('Functions/WriteResults');
addpath('Functions/Utilities/WordFunctions');
addpath('Functions/Micromechanics');
addpath('Functions/Margins');

%-----------------------------------------------------------------
% I) Define Problems
%-----------------------------------------------------------------
[NProblems, OutInfo, MicroMat, Loads] = MicroProblemDef();

%-----------------------------------------------------------------
% II) Get constituent properties
%-----------------------------------------------------------------
[constitprops] = GetConstitProps;

%-----------------------------------------------------------------
% III) Get effective properties from micromechanics
%-----------------------------------------------------------------
effprops = cell(1,300);
```

```
% -- Setup output
[OutInfo] = SetupOutput(OutInfo, 1);


Filename = [OutInfo.OutFile,' - Random Results.txt'];


fid = fopen(Filename,'wt');
Fmf3e2 = '%d \t %E \t %E \t %E \t %E \t %E \t %E \t %E \t %E \t %E \t %E \t %E \t ...
          %E \n';
Fmst = '%s \t %s \t %s \t %s \t %s \t %s \t %s \t %s \t %s \t %s \t %s \t %s \t %s ...
          \t %s \n';
Fmn = '\n';
TSHead = ["NP ,", "Vf ,", "E1 (MPa) ,", "E2 (MPa) ,", "E3 (MPa) ,", "v12 ,", "v13 ,", ...
          "v23 ,", "G12 (MPa) ,", "G13 (MPa) ,", "G23 (MPa) ,", ...
          "alpha1 (1/deg C) ,", "alpha2 (1/deg C) ,", "alpha3 (1/deg C) ,"];

fprintf(fid,Fmst, TSHead);
fprintf(fid,Fmn, ' ');


Vf = zeros(NProblems, 1);
Results = zeros(NProblems, 12);

% -- Loop through problems
for NP = 1: NProblems
    disp(' ');
    disp(['***** Problem Number ', char(num2str(NP)), ' *****']);
    disp(' ');
    mat = MicroMat{NP};
    effprops{mat}.used = true;
    [effprops] = GetEffProps(constitprops, effprops);
    Vf(NP) = effprops{mat}.RUC.Vf;
    Results(NP, 1) = effprops{mat}.E1;
    Results(NP, 2) = effprops{mat}.E2;
    Results(NP, 3) = effprops{mat}.E3;
    Results(NP, 4) = effprops{mat}.v12;
    Results(NP, 5) = effprops{mat}.v13;
    Results(NP, 6) = effprops{mat}.v23;
    Results(NP, 7) = effprops{mat}.G12;
    Results(NP, 8) = effprops{mat}.G13;
    Results(NP, 9) = effprops{mat}.G23;
    Results(NP, 10) = effprops{mat}.a1;
    Results(NP, 11) = effprops{mat}.a2;
    Results(NP, 12) = effprops{mat}.a3;

    fprintf(fid, Fmf3e2, NP, Vf(NP), Results(NP, 1), Results(NP, 2), Results(NP, 3), ...
                         Results(NP, 4), Results(NP, 5), Results(NP, 6), ...
                         Results(NP, 7), Results(NP, 8), Results(NP, 9), ...
                         Results(NP, 10), Results(NP, 11), Results(NP, 12));
end

% -- Code to produce histograms from text output file
% -- NOTE: This can be run separately in command window or saved and run as separate
%          script, where 'filename.txt' below is the file containing the data
% Filename = 'filename.txt'; % -- Replace with actual filename
A = importdata(Filename);
Results = A.data(:, 3:14);
TSHead = split(A.textdata, ',');
for i = 1: 12
    figure
    set(gcf,'color','white')
    histogram(Results(:,i));
    txt = char(TSHead(i + 2));
    xlabel(txt(1:end - 1));
    ylabel('frequency');
    Mean = mean(Results(:,i));
    Std = std(Results(:,i));
    Title = ['mean = ', num2str(Mean), ', std = ', num2str(Std)];
    title(Title);
end
```

Table 6.7 Glass and epoxy constituent allowables (taken from Exercises 4.16 and 5.18).

Mat	Material Name	XT (MPa)	XC (MPa)	YT (MPa)	YC (MPa)	Q (MPa)	R (MPa)	S (MPa)
2	Glass	2358	−1653	2358	−1653	1000	1000	1000
4	Epoxy	1×10^6	-1×10^6	42	−181	81	81	81

6.10.7 Transverse shear stress allowable predictions

In this example, the transverse shear stress allowable is predicted using GMC and HFGMC based on only the max stress criterion. Consider a 0.6 fiber volume fraction glass/epoxy composite represented by a 26×26 RUC (RUCid = 26, with Extra = 1) whose constituent allowables are given in Table 6.7, subjected to an applied global transverse shear stress of $\bar{\sigma}_{23} = 100$ MPa (with all other global stress components zero). Using GMC and HFGMC, the following two problems can be defined in MicroProblemDef.m:

```
NP = NP + 1;
OutInfo.Name(NP) = "Example Section 6.10.7 - GMC";
MicroMat{NP} = 153;
Loads{NP}.DT = 0;
Loads{NP}.Type  = [  S,  S,  S,   S,  S,  S];
Loads{NP}.Value = [  0,  0,  0, 100,  0,  0];
Loads{NP}.CriteriaOn = [1,0,0,0];

NP = NP + 1;
OutInfo.Name(NP) = "Example Section 6.10.7 - HFGMC";
MicroMat{NP} = 154;
Loads{NP}.DT = 0;
Loads{NP}.Type  = [  S,  S,  S,   S,  S,  S];
Loads{NP}.Value = [  0,  0,  0, 100,  0,  0];
Loads{NP}.CriteriaOn = [1,0,0,0];
```

with the following two materials defined in GetEffProps.m:

```
Mat = 153;
Theory = 'GMC';
name = "26x26 Glass-Epoxy GMC";
Vf = 0.6;
Constits.Fiber = constitprops{2};
Constits.Matrix = constitprops{4};
RUCid = 26;
[props{Mat}] = RunMicro(Theory, name, Vf, Constits, props{Mat}, RUCid);

Mat = 154;
Theory = 'HFGMC';
name = "26x26 Glass-Epoxy HFGMC";
Vf = 0.6;
Constits.Fiber = constitprops{2};
Constits.Matrix = constitprops{4};
RUCid = 26;
[props{Mat}] = RunMicro(Theory, name, Vf, Constits, props{Mat}, RUCid);
```

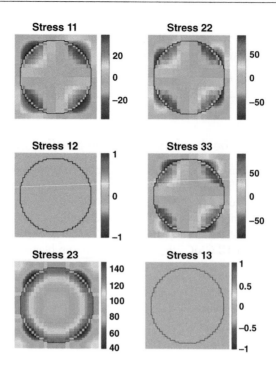

Figure 6.17 Local stress fields (MPa) predicted by HFGMC for a 0.6 fiber volume fraction glass/epoxy composite represented by a 26 × 26 RUC subjected to $\bar{\sigma}_{23} = 100$ MPa.

The GMC theory predicts all zero local subcell stresses, aside from a constant local σ_{23} in all subcells that is equal to the applied $\bar{\sigma}_{23}$ (100 MPa). This is GMC's lack of shear coupling; applying $\bar{\sigma}_{23}$ results in only local σ_{23} in the subcells. In contrast, using HFGMC, the stress fields shown in Fig. 6.17 result from the applied global transverse shear loading. HFGMC captures the normal stresses induced by the applied $\bar{\sigma}_{23}$ along with the varations of the local σ_{23} field in the fiber and matrix subcells. Note that the other local shear stress components (σ_{12} and σ_{13}) are zero as there is no coupling with these components due to the continuous fiber microstructure.

It is also interesting to examine the MoS and effective allowable predictions of HFGMC. The minimum MoS predicted by HFGMC is -0.54096, resulting in a composite allowable shear stress of $\bar{\sigma}_{23} = 45.9$ MPa (compared to 81 MPa for GMC, which is equal to the matrix shear allowable stress). However, this minimum MoS predicted by HFGMC is associated with reaching the normal (σ_{22} and σ_{33}) stress allowables in the matrix subcells $(\beta, \gamma) = (5,9), (9,5), (18,22),$ and $(22,18)$, rather than (as one might expect) reaching the transverse shear (σ_{23}) stress allowable locally. The following data were extracted from the results file:

Subcell or Mat	MoS11	MoS22	MoS33	MoS23	MoS13	MoS12
5,9 - M	26505.2855	1.5774	−0.54096	0.84181	99999	99999
9,5 - M	26505.2855	−0.54096	1.5774	0.84181	99999	99999
18,22 - M	26505.2855	−0.54096	1.5774	0.84181	99999	99999
22,18 - M	26505.2855	1.5774	−0.54096	0.84181	99999	99999

```
Controlling Subcell is (18,22) - M
Controlling Criterion is MaxStress Component 2
Minimum MoS: -0.54096
```

HFGMC is predicting that the controlling MoS is due to transverse *normal* stress rather than transverse *shear* stress, even though only global transverse shear stress has been applied. The reason for this is that, as shown in Table 6.7, the matrix shear stress allowable (81 MPa) is nearly double the matrix transverse tensile stress allowable (42 MPa) and, as shown in Fig. 6.17, significant local tensile σ_{22} and σ_{33} are induced in the matrix. Obviously, GMC is not capable of predicting this (nor are standard composite level failure theories, which are usually based on global load components only). It is also interesting to note that in a case such as this, where the minimum MoS is due to normal tensile matrix stresses rather than transverse shear stresses, improvements of the matrix shear allowable stress would make no difference on the composite performance (which is contrary to what one might expect). A clear benefit of micromechanics is its ability to identify these types of underlying mechanisms, thus providing the understanding necessary to improve the composite effective properties.

6.11   Summary

This chapter introduced the HFGMC micromechanics theory. By employing a second-order displacement field within the subcells (as compared to the first-order displacement field used in the MOC and the GMC), the HFGMC exhibits coupling between its predicted normal and shear local stress and strain fields. This shear coupling is lacking in models that are more analytical in nature, like the MOC, the GMC, and the MT Method, and, in certain circumstances, can make a major difference in the fidelity of the model's predictions. For typical continuously reinforced polymer matrix composites, example problems illustrated that the local fields predicted by the HFGMC compare much more favorably to FE generated local fields than do those predicted by the GMC. It should be noted, however, that these improved fields come at the cost of computational efficiency, while HFGMC (like the FE method) also adds the complication of subcell grid dependence that is absent in the GMC. The HFGMC is thus closer to the FE method, in terms of fidelity, but has better computational efficiency and is much better suited for inclusion in multiscale analyses. Additional example problems illustrated the ability of HFGMC (and inability of GMC) to capture correctly the effects of fiber packing arrangement for both ordered and disordered microstructures. Finally, the convergence of HFGMC results, in terms of subcell grid refinement, and the construction of predicted property histograms based on a number of random fiber arrangement instances, were demonstrated.

6.12   Advanced topics

The HFGMC micromechanics theory has been extensively generalized, extended, and applied to analyze a wide variety of technologically relevant composite problems. The following is a list of some of these extensions.

- Three-dimensional (triply-periodic) microstructures (e.g., for short fiber and woven composites) (Aboudi et al., 2013)
- Inelastic composites – plastic and viscoplastic constituents (Aboudi et al., 2013)
- Fiber/matrix interfacial debonding (Aboudi et al., 2013)
- Parametric HFGMC – non-rectangular subcells (Haj-Ali and Aboudi, 2013)
- Finite strain composites – constituents undergoing large deformation including plasticity and viscoplasticity (Aboudi et al., 2013)
- Smart composites – piezo-electro-magnetic behavior of the constituents (Aboudi et al., 2013)
- Shape memory alloy composites (Aboudi et al., 2013)
- Effective thermal conductivity calculations (Bednarcyk et al., 2017)
- Multiscale HFGMC – recursive micromechanics wherein each subcell is represented by a composite RUC (Bednarcyk et al., 2015)
- Progressive damage analysis (Aboudi et al., 2013, Pineda, et al., 2013)
- Dynamic formulation for simulation of impact (Aboudi et al., 2013; Bednarcyk et al., 2019)
- Clustering of fibers (Bednarcyk et al., 2015)
- Two-way thermomechanical coupling (Aboudi et al., 2013)
- Statistical misalignment of fibers (Bednarcyk et al., 2014)
- Effective damping properties (Bednarcyk et al., 2016)

6.13   Exercises

Exercise 6.1. Explain the difference between sparse and non-sparse (full) storage of a matrix. Consider a unidirectional 0.6 volume fraction IM7/8552 composite (constituent properties in Table 3.4) represented by a 3×3 RUC (RUCid = 3) and a 26×26 RUC (RUCid = 26 with Extra = 1 in GetRUC.m). Determine the number of non-zeros and sparsity, which equals (1 − number of non-zeros/number of elements in full matrix), for the GMC M matrix and the HFGMC Omega matrix for these two RUCs. Hint: There are MATLAB functions that return the number of non-zeros in, and size of, a matrix.

Exercise 6.2. For the four problems defined in Exercise 6.1, calculate the number of equations in the overall mechanical system of equations (as implemented in the MATLAB code) using the expressions in Section 6.4. Note that, as presented herein, both GMC and HFGMC have the same number of equations. However, as described in Section 6.4, the MATLAB implementation of HFGMC has retained additional unknowns for coding simplicity. What is the percentage increase in number of equations for HFGMC (compared to GMC), for each RUC, associated with this coding simplification?

Exercise 6.3. Consider a unidirectional 0.6 volume fraction IM7/8552 composite (constituent properties in Table 3.4) represented by a 26×26 RUC with square fiber arrangement (RUCid = 26 with Extra = 1 in GetRUC.m) and a 32×58 RUC with hexagonal fiber packing (RUCid = 200) analyzed using the HFGMC theory. Temporarily alter the HFGMC.m MATLAB function to solve the system of equations using a *nonsparse* Omega matrix. Show this altered code and compare the time it takes to solve the system of equations one time with the equations stored as sparse vs. non-sparse using the tic and toc MATLAB functions around the solution for u. Calculate the number of times speed up provided by the sparse storage for the two RUCs. Hint: use the MATLAB function full() to transform from sparse storage to non-sparse storage.

Exercise 6.4. Consider a unidirectional 0.6 volume fraction IM7/8552 composite (constituent properties in Table 3.4) represented by a 26×26 RUC (RUCid = 26) with an applied global transverse stress of $\bar{\sigma}_{22} = 100\,\text{MPa}$, and all other stress components zero, using HFGMC. Plot the local σ_{23} stress component field for the 26×26 RUC (RUCid = 26 with Extra = 1 in GetRUC.m). Compare the results to the same problem using the RUCid = 26 with 4 subcells between the fiber and the edge of the RUC generated using the parameter Extra = 4 in the MATLAB function GetRUC.m (resulting in a 32×32 RUC). Repeat this for a fiber volume fraction of 0.4.

Exercise 6.5. Consider a glass/epoxy composite (constituent properties in Table 3.4) with fiber volume fraction 0.4167 represented by the three RUCs shown below, with discontinuous, overlap, and continuous reinforcement in the x_2-direction, respectively. Note that all three reinforcements are continuous in the x_1-direction, thus these RUCs do not represent fiber-reinforced composites. Analyze these three RUCs using both GMC and HFGMC with an applied $\bar{\sigma}_{22} = 100$ MPa and all other stress components zero. Provide the effective mechanical properties and plot the local σ_{22} (for GMC and HFGMC) and σ_{23} (for HFGMC) stress component fields for each RUC. Hint: see RUCid = 8 in the MATLAB function GetRUC.m. Which RUC results in the biggest difference between GMC and HFGMC for the effective E_{22}? Which RUC results in the largest local σ_{22} and σ_{23} stress concentrations for HFGMC?

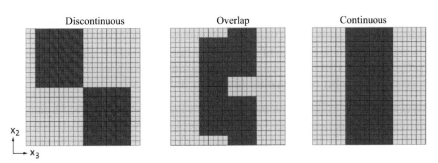

Exercise 6.6. Consider a unidirectional 0.6 volume fraction glass/epoxy composite (constituent properties in Table 3.4) represented by a 26×26 RUC with an applied temperature change of $\Delta T = 100$ °C (with all global stresses zero) using GMC and HFGMC. Use the extra 4 subcells between the fiber and the edge of the RUC generated using RUCid = 26 with the parameter Extra = 4 in the MATLAB function GetRUC.m. Plot the local σ_{23} stress component field for the HFGMC case. Plot and compare the local von Mises stress and pressure fields between GMC and HFGMC (using the same scale for GMC and HFGMC in PlotMicroFields.m; also ensure that PlotSvm = true). Is the difference in the local von Mises stress and pressure fields between GMC and HFGMC greater in the fiber or in the matrix?

Exercise 6.7. Consider a unidirectional 0.6 volume fraction glass/epoxy composite (constituent properties in Table 3.4) represented by (a) a 68×120 RUC with hexagonal packing (RUCid = 200) and (b) a 26×26 RUC (RUCid = 26 with Extra = 4 in GetRUC.m). Apply a temperature change of $\Delta T = 100$ °C using HFGMC. Plot and compare the local σ_{23} stress component fields and the local von Mises stress and pressure fields between the two fiber packing arrangements (using the same scale for each packing in PlotMicroFields.m; also ensure that PlotSvm = true). For each of the plotted fields, which packing arrangement exhibits higher stress concentrations in the matrix?

Exercise 6.8. Based on Eq. 6.63, when using HFGMC, which local strain components will arise if only a global $\bar{\varepsilon}_{23}$ is applied with all other global strain components = 0. Which local strain components will arise in the RUC if only a global $\bar{\varepsilon}_{13}$ is applied with all other global strain components = 0. Which local strain components will arise in the RUC if only a global $\bar{\varepsilon}_{12}$ is applied with all other global strain components = 0. How does this differ from GMC?

Exercise 6.9. Consider a unidirectional 0.6 volume fraction IM7/8552 composite (constituent properties in Table 3.4) represented by a 26×26 RUC (RUCid = 26 with Extra = 4 in GetRUC.m). Analyze this composite using HFGMC and plot all six components of the local <u>strain</u> field when a global $\bar{\gamma}_{23} = 0.01$ is applied with all other global stress components = 0. Also plot all six components of the local strain field when global $\bar{\gamma}_{12} = 0.01$ is applied with all other global stress components = 0.

Exercise 6.10. Consider a unidirectional 0.5 volume fraction glass/epoxy composite (constituent properties in Table 3.4) with a 10 fiber random RUC. Use the MATLAB script RandomRUCHist.m to simulate 100 realizations of the random RUC using HFGMC. Set the parameters Nfibers and Radf (in GetRUC.m) to 10 and disallow touching fibers (touching = false) – this should result in a 79×79 subcell RUC with an actual fiber volume fraction of 0.5063. Plot histograms of the effective properties E_{11}, E_{22}, and E_{33}. Plot the running mean (cumulative moving average) of E_{22} and of E_{33} versus the number of data points (from 1 to 100).

Exercise 6.11. Consider a unidirectional 0.6 volume fraction IM7/8552 composite (constituent properties in Tables 3.4 and 4.3) represented by a 26×26 RUC (RUCid = 26 with the Extra = 4 in GetRUC.m). Compare the allowable $\bar{\sigma}_{22}$, $\bar{\sigma}_{23}$, and $\bar{\sigma}_{12}$ (assuming uniaxial applied stress loading is used to determine each) predicted by GMC and

HFGMC based on the first subcell exceeding its allowable using only the Tsai-Wu criterion. Calculate the percent difference of the GMC result from the HFGMC result for each allowable.

Exercise 6.12. Consider a unidirectional 0.6 volume fraction glass/epoxy composite (constituent properties in Tables 3.4 and 6.7). Generate and plot $\bar{\sigma}_{22} - \bar{\sigma}_{33}$ failure envelopes using only the max stress criterion for this composite using an angle increment of $5°$ and HFGMC with (a) a 68×120 RUC with hexagonal packing (RUCid = 200) and (b) a 26×26 RUC (RUCid = 26 with Extra = 4 in GetRUC.m). Are the failure envelopes symmetric about the line $\bar{\sigma}_{22} = \bar{\sigma}_{33}$? Why or why not?

Exercise 6.13. Consider a unidirectional 0.5 volume fraction glass/epoxy composite (constituent properties in Tables 3.4 and 6.7) with a 10 fiber random RUC (RUCid = 300). In GetRUC.m, set the parameters Nfibers and Radf to 10 and disallow touching fibers (touching = false) – this should result in a 79×79 subcell RUC with an actual fiber volume fraction of 0.5063. Generate and plot three $\bar{\sigma}_{22} - \bar{\sigma}_{33}$ failure envelopes using only the max stress criterion for this composite using HFGMC, wherein three different random RUCs are used. Use an angle increment of $5°$.

Exercise 6.14. Consider a regular 0.5 fiber volume fraction, four ply glass/epoxy laminate $[+\theta, -\theta]_s$ with ply thicknesses of 1 mm subjected to an N_x load (with all other force and moment resultants zero). Determine the percent difference in the effective laminate properties and the allowable N_x load (based on only the Tsai-Wu criterion) using the GMC and HFGMC theories as a function of the ply angle θ. Use a 26×26 RUC (RUCid = 26 with Extra = 4 in GetRUC.m). Vary the ply angle θ from $0°$ to $90°$ in increments of $15°$. Use the constituent properties given in Table 3.4 and the stress allowables as given in Table 6.7. Provide a table showing the percent differences, and highlight the maximum difference for each property in the table.

Exercise 6.15. Consider a regular 0.6 fiber volume fraction, four ply glass/epoxy laminate $[+45°, -45°]_s$ with ply thicknesses of 0.25 mm using the GMC and HFGMC theories to represent the plies. Use a 26×26 RUC (RUCid = 26 with Extra = 4 in GetRUC.m). Using the constituent properties given in Table 3.4 and stress allowables given in Table 6.7, plot N_x-N_y failure envelopes for the two theories (using only the Tsai-Wu criterion and an angle increment of $5°$). Explain the cause of the difference in quadrant III.

References

Aboudi, J., Arnold, S.M., Bednarcyk, B.A., 2013. Micromechanics of Composite Materials A Generalized Multiscale Analysis Approach. Elsevier, New York, United States.

Arnold, S.M., Mital, S., Murthy, P.L., Bednarcyk, B.A., 2016. Multiscale Modeling of Random Microstructures in SiC/SiC Ceramic Matrix Composites within MAC/GMC Framework" American Society for Composites 31st Annual Technical Conference, Williamsburg, VA, United States, Sept 19-22, 2016.

Bednarcyk, B.A., Aboudi, J., Arnold, S.M, 2014. The effect of general statistical fiber misalignment on damage initiation in composites. Composites: Part B, 66, 97–108.

Bednarcyk, B.A., Aboudi, J., Arnold, S.M, 2015. Analysis of Fiber Clustering in Composite Materials Using High-Fidelity Multiscale Micromechanics. Int. J. Solids Struct. 69-70, 311–327.

Bednarcyk, B.A., Aboudi, J., Arnold, S.M, 2016. Enhanced composite damping through engineered interfaces. Int. J. Solids Struct. 92-93, 91–104.

Bednarcyk, B.A, Aboudi, J., Arnold, S.M, 2017. Micromechanics of composite materials governed by vector constitutive laws. Int. J. Solids Struct. 100-111, 137–151.

Bednarcyk, B.A, Aboudi, J., Arnold, S.M., Pineda, E.J., 2019. A multiscale two-way thermomechanically coupled micromechanics analysis of the impact response of thermo- elastic-viscoplastic composites. Int. J. Solids Struct. 161, 228–242.

Christensen, R.M., 1979. Mechanics of Composite Materials. John Wiley and Sons, New York, United States.

Cowin, S.C., Doty, S, 2007. Modeling Material Symmetry, Chapter 5" in Tissue Mechanics. Springer, New York, United States.

ECMA International, 2013. Standard ECMA-404, The JSON Data Interchange Format, European Computer Manufacturers Association. Geneva, Switzerland. https://www.ecma-international.org/publications/files/ECMA-ST-ARCH/ECMA-404%201st%20edition%20 October%202013.pdf.

Gitman, IM, Askes, H, Sluys, LJ., 2007. Representative volume: Existence and size determination. Eng. Fract. Mech. 74 (16), 2518–2534.

Haj-Ali, R., Aboudi, J., 2009. Nonlinear micromechanical formulation of the high fidelity generalized method of cells. Int. J. Solids Struct. 46, 2577–2592.

Haj-Ali, R., Aboudi, J., 2010. Reformulation of the high-fidelity generalized method of cells with arbitrary cell geometry for refined micromechanics and damage in composites. Int. J. Solids Struct. 47, 3447–3461.

Haj-Ali, R., Aboudi, J., 2013. A new and general formulation of the parametric HFGMC micromechanical method for two and three-dimensional multi-phase composites. Int. J. Solids Struct. 50, 907–919.

Liu, K.C., Ghoshal, A., 2014a. Validity of Random Microstructures Simulation in Fiber-reinforced Composite Materials. Compos.s Part B: Eng. 57, 56–70.

Liu, K.C., Goshal, A., 2014b. Inherent symmetry and microstructure ambiguity in micromechanics. Compos. Struct. 108, 311–318.

Levin, V.M., 1967. On the coefficients of thermal expansion of heterogeneous materials. Mech. Solids 2, 58–61.

MATLAB, 2018. version R2018. The MathWorks Inc, Natick, Massachusetts.

Murthy, P.L.N., Bednarcyk, B.A., and Mital, S.K. (2017) "A Compilation of MATLAB Scripts and Functions for MAC/GMC Analysis" NASA/TM-2017-219500.

Ostoja-Starzewski, M., 2006. Material spatial randomness: From statistical to representative volume element. Probab. Eng. Mech. 21 (2), 112–132.

Pineda, E.J., Bednarcyk, B.A., Waas, A.M., Arnold, S.M, 2013. Progressive failure of a unidirectional fiber-reinforced composite using the method of cells: discretization objective computational results. Int. J. Solids Struct. 50, 1203–1216.

Stier, B., Bednarcyk, B.A., Simon, J.-W., and Reese, S. (2015) "Investigation of Micro-Scale Architectural Effects on Damage of Composites" NASA/TM-2015-218740.

Sun, C.T., Vaidya, R.S., 1996. Prediction of Composite Properties from a Representative Volume Element. Compos. Sci. Technol. 56, 171–179.

Progressive damage and failure

Chapter outline

In Chapter 4 the concept of stress allowables was introduced in conjunction with the calculation of margins of safety. If any failure criterion was met at any point within the composite (resulting in a negative margin of safety at that point), "failure" of the entire unidirectional composite or laminate was indicated. In reality, this first-point definition of failure should be associated with damage initiation rather than final failure. Many composites and laminates can carry additional load beyond the point of damage initiation before actual failure. Between damage initiation and actual failure is damage progression (or propagation). Damage progression is often gradual and distributed, while failure is catastrophic and results in the inability of the composite to carry additional load. This is illustrated for an example cross-ply E-glass/epoxy laminate from the World-Wide Failure Exercise (Soden et al., 2004) in Fig. 7.1. The laminate stress-strain curve exhibits clear nonlinearity, and the ultimate strength is approximately four times the damage initiation stress (indicated by deviation from linearity). The subject of this chapter is the modeling of this type of damage progression after damage initiation has occurred.

The term "damage" can be used very broadly to denote some degradation of performance that can ultimately lead to failure of a given material, component, or structure. However, one must realize that damage has an underlying defect/mechanistic origin within the material, such as nucleation and growth of voids, cavities, microcracks, fiber/matrix debonding, and other microscopic or mesoscopic defects. These defects can later coalesce into distinct fracture modes at the culmination of the failure process (e.g., propagating meso/macro cracks or softening/localization zones). Therefore

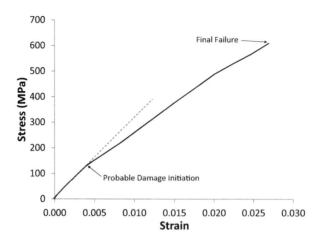

Figure 7.1 x-direction stress-strain curve for a cross-ply $[0/90]_s$ E-glass/epoxy laminate (see Table 14 in Soden et al., 2004).

damage will manifest itself phenomenologically in some form of degraded behavior (e.g., loss of stiffness or strength).

Continuum damage mechanics (CDM), which seeks to mathematically represent damage in materials, was introduced by Kachanov (1958) and later modified by Rabotnov (1969). CDM has now reached a stage at which it can be used to solve practical engineering problems on a regular basis (e.g., Krajcinovic, 1996; Voyiadjis et al., 1998; Lemaitre and Desmorat, 2005; Aboudi, 2011). Fracture mechanics considers the process of initiation and growth of cracks and microcracks as discrete phenomena. In contrast, CDM uses a continuous variable, which is related to the density of the previously mentioned defects (e.g., microcracks), to describe the deterioration of the material before, and possibly after, the initiation of macrocracks. Based on this damage variable, constitutive equations of damage evolution are developed to predict propagation of damage (e.g., see Lemaitre and Chaboche, 1990).

The most typical CDM approach appearing in the literature is of the stiffness reduction type, where, most often, the concept of an effective stress is utilized. Kachanov (1958) introduced the idea of a scalar damage variable, D, that effectively accounted for the change in the material's effective resisting cross-sectional area, $\bar{A}$, corresponding to the damaged area. This effective area $\bar{A}$ is obtained from the original, undamaged area, A, by removing the surface intersections of the defects (e.g., microcracks and cavities). To accomplish this, Kachanov (1958) considered a volume element at the macroscale that was of a size large enough to contain many defects, yet small enough to be considered as a material point within a continuum (see Section 3.2.1 for a discussion on the related concept of a representative volume element). The effective area is defined as,

$$\bar{A} = (1 - D)A \tag{7.1}$$

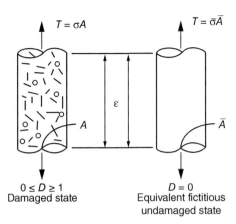

Figure 7.2 Schematic illustrating isotropic damage under uniaxial tension (concept of effective stress).

where $D = 0$ corresponds to an undamaged state and $D = 1$ corresponds to complete damage (failure) of the material point. The cross-sectional areas A and $\bar{A}$ are shown in Fig. 7.2 for a cylindrical material element in the damaged and effective states, respectively. Next the strain equivalence hypothesis states that the strains in the damaged state and the equivalent state (see Fig. 7.2) are the same. Therefore, the usual stress, σ, is replaced by an effective stress, $\bar{\sigma}$ (Lemaitre and Chaboche, 1990), which is defined as,

$$\bar{\sigma} = \frac{\sigma}{(1 - D)} \tag{7.2}$$

Thus, in the presence of damage ($0 < D < 1$), the effective area is reduced by a factor of $(1 - D)$, while the effective stress is increased by this same factor, such that the force (denoted as T in Fig. 7.2), $T = \sigma A = \bar{\sigma} \bar{A}$, is preserved. The effective uniaxial Hooke's Law can then be written as $\bar{\sigma} = E\varepsilon$ or $\sigma = (1 - D)E\varepsilon$, and thus the material's Young's modulus is, in effect, reduced by the factor $(1 - D)$, but the strain remains unchanged.

The simplest approach to continuum damage involves eliminating the stiffness contribution of a material's subvolume when a chosen failure criterion is satisfied. Note that, in the case of GMC and HFGMC, the applicable subvolume is the subcell, while in the case of MT it is the applicable constituent material. If the damage is characterized by the scalar variable D that alters the material's elastic stiffness tensor, the material's multiaxial thermoelastic constitutive equation can be written as,

$$\sigma = (1 - D)\mathbf{C}(\varepsilon - \alpha \Delta T) \tag{7.3}$$

Thus, in Eq. (7.3), every component of the stiffness tensor is impacted identically by the damage variable D. The subvolume elimination method then treats the damage evolution as a step function ($D = 0$ or $D = 1$). As long as the adopted failure criterion is not satisfied, $D = 0$. Once the adopted failure criterion is satisfied in the subvolume in question, D instantaneously changes to a value close to 1, and thus this subvolume's stiffness contribution is effectively eliminated. If the structure or composite contains many such subvolumes, this will not necessarily cause global failure. Rather, the damage will progress as additional subvolumes damage and stresses redistribute until the structure or composite reaches global failure. Standard stress- and strain-based multiaxial failure criteria can be readily employed in this method. This chapter considers one or more of the four failure criteria introduced in Chapter 4: (1) maximum stress, (2) maximum strain, (3) Tsai-Hill, and (4) Tsai-Wu. Note that, in practice, to avoid numerical issues, some minimum damage factor ($\mathtt{DamageFactor} = 1 - D$) close to zero, but never exactly zero, is introduced (see MATLAB functions $\mathtt{MicroSimDamage.m}$ and $\mathtt{LamSimDamage.m}$).

The discussion above, and the implementation discussed in this chapter, are limited to isotropic scalar damage. That is, there is one scalar damage variable that reduces *all stiffness components* of a material subvolume [see Eq. (7.3)]. In general, the damage formulation can include multiple scalar damage variables (e.g., Bednarcyk et al., 2010) or tensorial damage variables (e.g., Lemaitre and Desmorat, 2005) to account for the directionality/orientation of the damage at a material point. For instance, for loading as shown in Fig. 7.2, microcracks would tend to orient normal to the applied loading direction. Therefore the reduction of the stiffness in the loading direction would be greater than that in the other directions. This effect, which is neglected in the isotropic scalar damage approach, can be captured using multiple scalar damage variables or a tensorial damage variable, which enable the damage to impact the different stiffness components differently. Of course, such formulations tend to be more complex than the isotropic scalar damage used herein. Note, however, that directional effective stiffness reduction at the composite (or ply) scale can be approximated by implementing isotropic scalar damage at the microscale within GMC and HFGMC. As such, the damage can, and often does, initiate and propagate directionally within the RUC, resulting in directional effective stiffness reduction at the composite (or ply) scale.

Modeling continuum damage is notoriously complex in terms of numerical and discretization issues (see Lemaitre and Desmorat, 2005) and, as of 2020, is still an area of active research in micromechanics. This is particularly true when local material softening, which involves a decrease in local stress while local strain still increases, is present. Bažant and co-workers (Bažant and Cedolin, 1979, 1991; Bažant and Oh, 1983; Bažant, 1994) have examined the continuum damage problem extensively and given strategies to alleviate some of the common issues, including ways to improve the consistency of the energy dissipated as the geometric discretization used in the simulation is altered. These concepts have been applied to HFGMC by Pineda and co-workers (Pineda and Waas, 2012; Pineda et al., 2012, 2013), but they are beyond the scope of this text. Note that nonlocal models (e.g., Eringen, 1966; Bažant, 1994; Jirásek, 1998) as well as discrete damage approaches (e.g., Sukumar et al., 2000; Belytschko et al., 2001; Garikipati and Hughes, 1998; Ortiz and Pandolfi, 1999; Garikipati, 2002; Xie and Waas, 2006; Rudraraju et al., 2010) have also been used to address inconsistencies in the presence of local material softening damage.

Herein, the progressive damage simulations are intended to account for the effective stiffness loss of the composite in an average sense. The details of the damage distribution within the RUC, particularly as the damage in the RUC becomes extensive, should not be viewed as predictive. This is to say that the damage zones that arise in the simulations do not truly represent individual cracks. The tip of even a "crack-like" damage zone in HFGMC that is one subcell wide is still not a crack as there is no stress singularity at the crack tip, and the linear elastic fracture mechanics equations are not in effect. Further, because the RUC is doubly-periodic, each damage zone that arises is repeated infinitely in two directions and is continuous in the third. Therefore, the progressive damage considered is more representative of distributed damage that impacts the effective stiffness of the composite. The numerical and discretization issues that affect progressive damage simulations will, however, be illustrated. A fundamental example, involving progressive damage of a monolithic material with a center notch using HFGMC, is discussed in Section 7.2.

The implementation of subvolume elimination within the MATLAB scripts discussed earlier involves four main additions. First, the loading must be applied in an incremental fashion so that, second, the subvolumes can be checked for damage initiation at each load level. Third, if damage does initiate in one or more subvolumes, the scripts must determine new effective composite properties that reflect the composite stiffness reduction caused by local failure in the subvolumes. These reduced stiffness properties are then used for the next increment of the applied loading. Fourth, once damage has initiated, iterations are needed at a given load level to enable damage to progress within an increment. This procedure has been implemented in a script for simulating progressive damage in unidirectional composites (`MicroSimDamage.m`) and a script for simulating progressive damage in laminates (`LamSimDamage.m`). These are described in detail in the next section. Subsequent sections illustrate the factors that influence progressive damage of unidirectional and laminated composites, including constituent material strengths, loading direction, micromechanics theory, fiber packing arrangement, failure initiation theory, loading increment size, subcell grid density, and number of fibers in a random RUC. These examples focus on glass/epoxy and SiC/SiC composites.

7.1 MATLAB implementation

7.1.1 `MicroSimDamage.m` and `LamSimDamage.m` MATLAB scripts

The MATLAB code provided with this text is available from:

https://github.com/nasa/Practical-Micromechanics

Two additional MATLAB scripts have been provided to perform progressive damage simulations: `MicroSimDamage.m`, which simulates progressive damage in a unidirectional composite, and `LamSimDamage.m`, which simulates micromechanics-based progressive damage in a laminate. Both scripts are based on subvolume elimination, which, as described previously, reduces all stiffness components of a constituent/subcell once a specified failure criterion is fulfilled. Note, although possible, no damage evolution law is employed in the current version of these scripts, that is, the stiffness components are instantaneously reduced by a large factor (`DamageFactor = 0.0001`) once the failure criterion is met. The capability has been implemented for the GMC, HFGMC, and MT micromechanics theories. In the case of MT, because only average fiber and matrix stresses and strains (as opposed to multiple subcell stresses and strains) are calculated, failure of the entire matrix or entire fiber will happen all at once. If MT is used within lamination theory with LamSimDamage.m, however, the fiber and matrix in each ply can still damage independently.

It should also be noted that the LamSimDamage.m implementation is only truly correct for loading that does not induce bending (symmetric laminates subjected to only force resultants or midplane strains). This is because if the laminate is subjected to bending, the stress and strain states vary through the thickness within each ply, and thus damage could occur differently at different through-thickness points in each ply. Since damage reduces the stiffness of the composite, this would mean that the material properties would need to vary within each ply. While material property variations within each ply can be handled using multiple integration points per ply, *this has not been implemented in the current version of the MATLAB code.*

To simulate progressive damage that occurs as loading is applied to a composite or laminate, it is necessary to apply the specified loading in an incremental fashion so failure criteria can be checked at each load level, and stress redistribution resulting from failure can take place. Towards this end, in progressive damage simulations, the loads specified in MicroProblemDef.m and LamProblemDef.m are considered to be final values, with the increment size of the loading calculated by dividing these final values by a number of increments (Loads{NP}.NINC). The number of loading increments can be specified in either MicroProblemDef.m or LamProblemDef.m and defaults to 100 if not specified. The progressive simulation begins at a load level and temperature change value of zero. Note that *thermal loading has not been implemented in the MATLAB code for progressive damage*, but it can be added. MicroSimDamage.m and LamSimDamage.m both include a loop over which all the loading increments are applied. The applied loading is gradually increased incrementally (resulting in new local stresses and strains), and margins of safety are calculated at the microscale (for each subcell for GMC and HFGMC, and for each material for MT) using the same MATLAB functions that were introduced in Chapter 4. If any margin of safety of a subcell (or material for MT) is negative, it means that the failure criterion has been exceeded, and the subcell (or material) "fails" and its stiffness is immediately reduced. If any subcell (or material) fails at a given load level/increment level, it means the properties of the composite (or ply within a laminate) have changed, and the micromechanics theory must calculate new effective properties of the composite (or ply) and new strain concentration tensors for use in the next increment. If no new micro-scale failures have occurred at a given load level/increment, then the composite stiffness has not changed, and the same effective properties and strain concentration tensors for each subcell can be used again for the next increment. As such, for the sake of execution speed, the progressive damage MATLAB scripts only calculate new effective properties and strain concentration tensors when there are new micro scale failures.

In addition, thanks to the relatively simple subcell elimination approach, the damaged properties of each constituent material can be calculated in advance rather than during the incremental loading loop. A MATLAB function DamageConstit.m is used to calculate a reduced, damaged stiffness matrix for each constituent material based on a specified damage factor (DamageFactor). Thus, the code stores the two sets of possible material properties for each constituent material: undamaged and completely damaged. Then, during the incremental loading loop, if a subcell (or material in the case of MT) fails, the damaged material is substituted for the original, undamaged material. Note that, in the case of laminates, unique materials must be created and stored per ply so that the each ply may damage independently (see step 3 in Fig. 7.3B).

It is also important to include an iteration loop within the incremental loading loop to enable a converged damage and stress state to be reached at each loading increment. This allows the stresses to redistribute when a subcell fails such that additional damage may occur, at the same loading increment, due to the new local stress state. In this way, damage can grow across many subcells without any additional loading being applied due to the changes in local stresses caused by the progression of damage. Furthermore, at any given iteration, only subcells with the minimum margin of safety are permitted to fail (actually, all those within a specified tolerance of the minimum margin of safety, see variable tol = 0.001 in DamageMicro.m). This is because, when incrementing to the next load level, many subcells can attain negative margins, regardless of whether stress redistribution would change the damage path such that they would never be negative. By limiting subcell failure to only those subcells that are very close to the minimum margin of safety in each iteration, the damage naturally progresses only to the subcells that should fail first. Then stress redistribution occurs to see which additional subcells should fail in the next iteration. Once no additional subcells fail at a given loading increment, the iterations stop, and the simulation progresses to the next loading increment. The ramifications of omitting the iteration loop and limiting subcell failure to the subcells with the minimum margin within each iteration is shown in a notched epoxy HFGMC example in Section 7.2.

While progressive damage has been implemented in the MATLAB code with an eye toward computational efficiency, each time additional damage occurs, the code must recalculate the strain concentration tensors. In previous chapters, this calculation only needed to be done once per problem (or per ply in a laminate). Now this may have to be done tens, hundreds, or even thousands of times per progressive damage simulation during the incremental loading and the iterations. Therefore, for HFGMC with large RUCs, especially in the case of laminate simulations (where the strain concentration tensors are calculated per ply), simulation times can reach hours.

Flow charts for MicroSimDamage.m and LamSimDamage.m are given in Fig. 7.3A and B. The additions, compared to MicroAnalysis.m and LamAnalysis.m (described earlier in Chapter 4), are highlighted in red. Two additional features have been included in both MicroSimDamage.m and LamSimDamage.m to speed their execution. First, early in the simulation, after calculating the margins of safety in step 7, the minimum margin throughout the entire composite or laminate is determined. As described in Chapter 4, if the minimum margin (of all subcells) is positive, the applied loads can be scaled to determine the load at which the first subcell (or material for MT) failure will occur. The scripts then skip the loading increments until just before this first subcell (or material) failure will occur. This same procedure is followed throughout the simulation such that, if for any increment all margins are positive, the scripts attempt to skip increments until additional failures will occur (i.e., when negative margins will next occur). If no additional failure will occur up to the final applied load level, the code will jump to

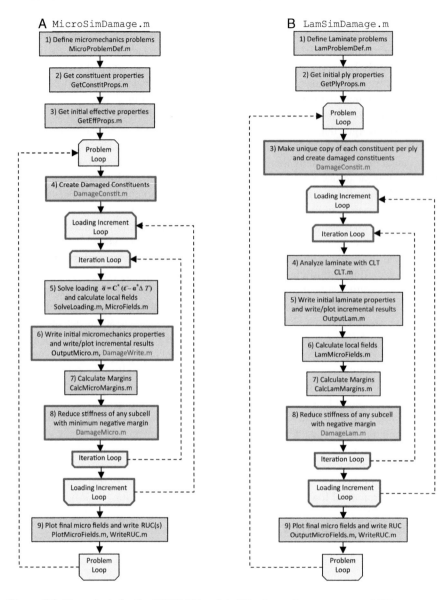

Figure 7.3 Flow charts for the MATLAB scripts (A) MicroSimDamage.m and (B) LamSimDamage.m, which enable progressive damage simulations based on the micro scale subvolume elimination approach, for unidirectional composites and laminates, respectively. The additions, compared to MicroAnalysis.m and LamAnalysis.m (see Chapter 4), are highlighted in red.

the final load level and terminate. Second, after writing and plotting results (step 6 in `MicroSimDamage.m`, step 5 in `LamSimDamage.m`), the scripts check if the global stiffness has been reduced to a low value, and if so, the applied loading loop is terminated so the script can end. This is essentially a check for global failure, where the composite or laminate can no longer support much load, and it is intended to correspond to the point at which a similar experiment would be terminated. To monitor the global stiffness reduction, the sum of global compliance components $\left(S_{11}^* + S_{22}^* + S_{33}^*\right)$ is checked at each increment for its change from its initial value, in the case of `MicroSimDamage.m`. Alternatively, for `LamSimDamage.m`, the sum of the 11 and 22 components of the inverse ABD matrix is monitored. Summing these components provides a measure of the overall compliance of the composite or laminate. When this compliance measure has increased by a specified factor (`Loads{NP}.TerminationFactor`), the MATLAB scripts exit the applied loading loop and the simulation ends. To disable this feature, a high termination factor can be specified, or it can simply be omitted from the problem definition, see `MicroProblemDef.m` and `LamProblemDef.m`.

In step 6 of `MicroSimDamage.m` (see Fig. 7.3A) a stress-strain plot, as well as the local von Mises stress field in the RUC and the subcell failure progression, are displayed and updated during the simulation. This enables one to see the simulation's progress while the code is executing. The von Mises stress and subcell failure plots are only updated after an additional subcell failure has occurred, and no such plots are made for the MT theory (which will execute significantly quicker than GMC and HFGMC). When these plots are updated, the combined figure is also saved in a "`DamSnapShots`" folder within the standard `Output` folder so the progressive damage results can be viewed after the simulation. In the case of `LamSimDamage.m`, a progressive load vs. midplane strain plot is displayed, along with damage plots for each ply, in step 5. The plies to be plotted can be specified in `Loads{NP}.PlyDamWatchOn`, by specifying a 1 (on) or 0 (off) per ply. For example, `Loads{NP}.PlyDamWatchOn = [1,1,0,0]` would indicate that the damage progression in plies 1 and 2 would be plotted, while that in plies 3 and 4 would not. A typical use case for this is only plotting one half of the plies in a symmetric laminate. A history of which subcells fail at which increment and iteration is also written to the MATLAB command window and to the problem output file. Finally, stress vs. strain and load vs. midplane strain histories are written to files, both with data points included for every iteration (e.g., `ITER Load vs eps0-`'`name`'`.txt`) and with only the final converged data points written per increment (e.g., `Load vs eps0-`'`name`'`.txt`). The converged stress-strain data are the true predictions of a given simulation, whereas the data per iteration is for informational purposes only.

Note that, of the three output related options specified in `MicroProblemDef.m` and `LamProblemDef.m` (`OutInfo.Format`, `OutInfo.MakePlots`, and `OutInfo.WriteMargins`) only `OutInfo.MakePlots` is active. If it is set to true, the appropriate plots will be made and written at the final applied load level.

7.1.2 Simple progressive damage problems

As a first example of using the MATLAB progressive damage capability, consider a unidirectional 0.55 fiber volume fraction glass/epoxy composite, whose constituent properties and allowables are given in Tables 3.4 and 6.7. To apply a transverse strain of 0.02 and simulate the progressive damage response using GMC (with a 2×2 RUC) and only the max stress criterion, the following problem can be defined in MicroProblemDef.m:

```
NP = NP + 1;
OutInfo.Name(NP) = "Section 7.1.2 - GMC 2x2 progressive damage";
MicroMat{NP} = 171;
Loads{NP}.Type  = [ S,     E,   S,   S,   S,   S];
Loads{NP}.Value = [ 0,  0.02,   0,   0,   0,   0];
Loads{NP}.CriteriaOn = [1, 0, 0, 0];
Loads{NP}.NINC = 400;
```

Here Loads{NP}.NINC = 400 has been specified to divide the applied $\bar{\varepsilon}_{22}$ loading into 400 increments such that each increment, $\Delta\bar{\varepsilon}_{22} = 0.02/400 = 0.00005$. The composite material can be defined as usual in GetEffProps.m as follows:

```
Mat = 171;
Theory = 'GMC';
name = "2x2 Glass-Epoxy GMC Vf = 0.55";
Vf = 0.55;
Constits.Fiber = constitprops{2};
Constits.Matrix = constitprops{4};
RUCid = 2;
[props{Mat}] = RunMicro(Theory, name, Vf, Constits, props{Mat}, RUCid);
```

The progressive damage problem is then executed using MicroSimDamage.m.

When running progressive damage problems such as this, a figure will appear on the screen that shows the problem progress in the form of a stress-strain plot, along with plots of the local von Mises stress field and the damage within the RUC, as shown in Fig. 7.4. Damaged (or failed) subcells are shown as dark red in the far right damage plot. Note that, for MT progressive damage simulations, only a plot of the stress-strain response is shown. These plots are updated during the simulation, and they are also written as .bmp files to a new "DamSnapShots" folder that is created within the Output folder. This is done because, for more complex progressive damage simulations, many files are written and it is helpful to have the output organized by problem. The damage progress, or "snapshot" figure is written to this folder for the final (converged) iteration of each loading increment in which damage has progressed, in addition to the very first loading increment and the first iteration that shows any damage (Initiation Location Damage Snapshot ... .bmp). This initiation location figure (as the name implies) enables the easy identification of where damage initiated in the progressive damage simulation.

In addition, four text-based output files are written to the problem's output folder, including the standard-problem-output file. Two text files are written containing the stress-strain response, one including only the final converged points for each loading increment, and one (for information purposes only) containing points for every iteration at each loading increment. Finally, a file is written containing the composite effective properties for every iteration at each loading increment (because the effective properties change as the composite is damaged). Note that information on the loading increments, iterations, and subcell (or material) failure events is also written to the MATLAB command window and to the standard problem output file.

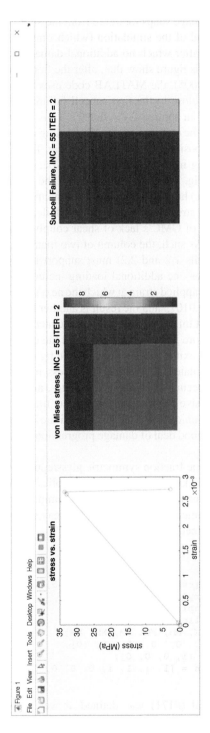

Figure 7.4 Example figure that appears on the screen showing the 55th increment and 2nd iteration during a micromechanics progressive damage simulation.

Fig. 7.5 shows the final damage snapshot figure from the simulation defined above. Note that this is not at the end of the simulation (which runs to an applied strain of 0.02), but rather at the point after which no additional damage occurs. The symbols in the stress-strain curve in this figure show that, after the first loading increment (the second symbol at $\bar{\varepsilon}_{22} = 0.00005$), the MATLAB code uses the minimum margin of safety to project the loading close to the first subcell failure. The next point that is plotted is for increment 53, at an applied $\bar{\varepsilon}_{22} = 0.0027$. Thus, by using this projection (increment jumping) method, the code has skipped 52 loading increments during which no failure occurs and the composite response remains linear. This obviously speeds the problem execution. At loading increment 54, no failure occurs, but at increment 55 ($\bar{\varepsilon}_{22} = 0.0028$), as shown in Fig. 7.4, the subcell between the fibers in the x_2-direction (vertical direction in the figure) has failed (according to the max stress criterion). This causes an abrupt drop in the stress-strain curve as the failed subcell can no longer carry any stress and, because of GMC's lack of shear coupling, neither can the fiber subcell (in the x_2-direction). As such, the column of two matrix subcells (on the right hand side of the RUC, subcells 1,2 and 2,2) must support nearly all of the applied loading. The code then applies one additional loading increment and is then able to once again project close to the applied strain at which the next subcell failure will occur, see Fig. 7.5. This is at $\bar{\varepsilon}_{22} = 0.01075$ and the result is another large drop in stress as all matrix subcells have failed. At this point, the composite cannot support any additional load in the x_2-direction, no more failures can occur, and the code projects to the final applied strain level (0.02) and exits (without updating the figure).

Clearly, the predicted ultimate strength (highest stress reached) during the progressive damage simulation occurs when damage initiates, and thus could have been predicted without the progressive damage simulation. But this is only when, as in the case of transverse tension on a unidirectional PMC, the failure is brittle. Next a laminate will be considered wherein a good deal of damage progression occurs prior to reaching the ultimate strength.

Consider a 0.55 fiber volume fraction symmetric glass/epoxy quasi-isotropic laminate [-45,0,45,90]$_s$ with ply thicknesses of 0.15 mm and plies modeled using the same 2×2 RUC as in the previous example. To subject this laminate to a midplane strain, $\varepsilon_x^0 = 0.1$ using a strain increment of 0.00005 and the max stress criterion only, the following problem can be defined in LamProblemDef.m:

```
NP = NP + 1;
OutInfo.Name(NP) = "Section 7.1.2 - glass-epoxy quasi-iso";
Geometry{NP}.Orient = [-45,0,45,90,90,45,0,-45];
nplies = length(Geometry{NP}.Orient);
Geometry{NP}.plymat(1:nplies) = 171;
Geometry{NP}.tply(1:nplies)   = 0.15;
Loads{NP}.Type  = [ EK, NM, NM, NM, NM, NM];
Loads{NP}.Value = [0.1,  0,  0,  0,  0,  0];
Loads{NP}.CriteriaOn = [1, 0, 0, 0];
Loads{NP}.PlyDamWatchOn = [1, 1, 1, 1, 0, 0, 0, 0];
Loads{NP}.NINC = 2000;
```

where the composite material (#171) was defined above in GetEffProps.m. The progressive damage simulation can then be executed using LamSimDamage.m, with

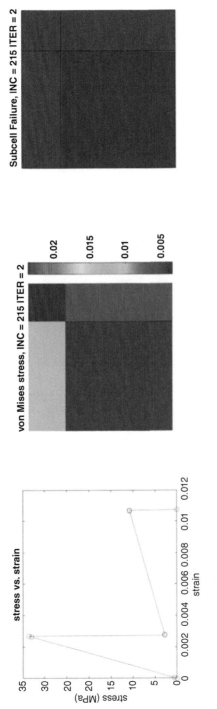

Figure 7.5 $\bar{\sigma}_{22} - \bar{\varepsilon}_{22}$ stress-strain curve, along with the local von Mises stress field (MPa) and the damage distribution at the final output snapshot (INC = 215, ITER = 2) for the progressive damage simulation of a unidirectional 0.55 fiber volume fraction glass/epoxy composite modeled using GMC, the max stress criterion only, and a strain increment of 0.00005.

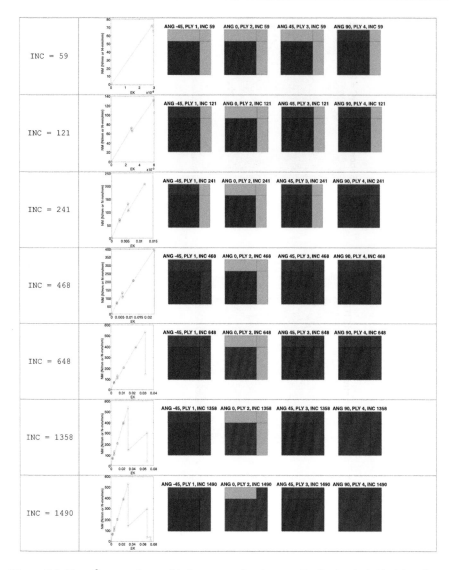

Figure 7.6 $N_x - \varepsilon_x^0$ curve, along with the progressive damage distribution in half of the plies, for the progressive damage simulation of a [-45,0,45,90]$_s$ 0.55 fiber volume fraction glass/epoxy laminate whose plies are modeled using GMC, the max stress criterion only, and a strain increment of 0.00005.

results shown in Figs. 7.6 and 7.7. Because the laminate is symmetric, only one half (four plies) of the laminate is chosen for plotting.

As Fig. 7.6 shows, damage initiates in the 90 plies in the matrix subcells between fibers in the loading (vertical, x_2-) direction at a loading increment of 59. This causes only a small drop in the load as little stress is carried in the matrix of the 90 plies because there is a fiber in 0 plies aligned with the loading direction. As the applied midplane

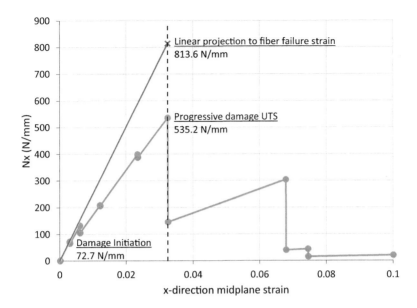

Figure 7.7 $N_x - \varepsilon_x^0$ curve for the progressive damage simulation of a [-45,0,45,90]$_s$ 0.55 fiber volume fraction glass/epoxy laminate whose plies are modeled using GMC, the max stress criterion only, and a strain increment of 0.00005 compared to a linear projection.

strain loading continues to increase, so too does the load. The next subcells to fail, at loading increment 121, are matrix subcells in the ±45 plies. The load has nearly doubled compared to the failure initiation load (132.6 N/mm vs. 72.7 N/mm). Next the load again increases significantly, when, at loading increment 241, the remaining matrix subcells in the 90 plies fail. Another large increase in loading causes the remaining matrix subcells in the ±45 plies to fail at loading increment 468, and then the 0 ply fibers fail at increment 648 and a load of 535.2 N/mm, signaling that the laminate has reached its simulated ultimate strength. The load, however has not yet dropped close to zero because some subcells remain unfailed. The applied midplane strain continues to increase until the fibers in the 90 plies fail at increment 1358, followed closely by failure of most of the 0 ply matrix at increment 1490. Now the laminate has virtually no remaining stiffness and no additional failures occur. Because the laminate never exceeded the load just before failure of the fibers in the 0 plies (535.2 N/mm, INC = 468), the appropriate predicted ultimate strength is associated with the failure of the 0 ply fibers.

Unlike the unidirectional glass/epoxy composite, this quasi-isotropic laminate exhibits a great deal of damage progression prior to the predicted ultimate tensile strength (UTS), as shown in the load vs. midplane strain plot in Fig. 7.7. Matrix damage initiated in the 90 plies at a load level of $N_x = 72.7$ N/mm, which would have been predicted simply by using the minimum MoS as in Chapter 4. However, using progressive damage, it was not until $N_x = 535.2$ N/mm that the laminate's predicted UTS was reached. Furthermore, by the time the laminate reaches its UTS, its effective Young's modulus has decreased from an initial value of 20.9 GPa to 13.8 GPa. If

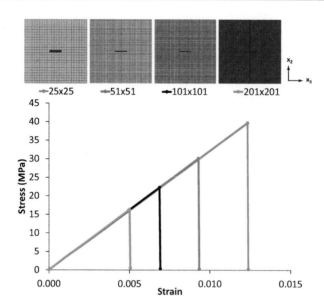

Figure 7.8 Simulated tensile (x_2-direction) stress-strain response of a center notched epoxy RUC using HFGMC wherein four subcell grid refinements have been considered, illustrating pathological mesh dependence in the presence of material softening.

one were to assume that no damage occurred in the laminate, and the initial stiffness were maintained until failure at the midplane strain at which the 0 ply fibers failed ($\varepsilon_x^0 = 0.03245$), the projected laminate UTS would be 813.6 N/mm (52% higher). This is shown in Fig. 7.7.

7.2   Pathological mesh dependence in HFGMC

A fundamental progressive damage example is depicted in Fig. 7.8 involving a monolithic epoxy RUC with a center notch modeled using HFGMC with successively more refined subcell grids. Note that this is a doubly-periodic RUC, thus the notch repeats infinitely in the x_2- and x_3-directions, and it is infinitely long in the x_1-direction. The notch length is 20% of the total RUC width and one subcell in height, and it is assigned the identical material properties as the epoxy material, but with all stiffness terms decreased by a factor of 10,000. This stiffness decrease essentially eliminates the notch material's ability to develop any significant stress, and the factor of 10,000 is identical to the stiffness decrease applied to failed subcells in the progressive damage simulations. Hence these subcells are representative of a preexisting notch in the RUC. As Fig. 7.8 shows, four different subcell discretizations have been considered, with the starting notch length being maintained by altering the aspect ratios of the three middle columns of subcells. Progressive damage simulations of the notched epoxy, using these four RUCs, were conducted by applying tensile loading in the x_2-direction under strain control with a strain increment of 0.00005. The maximum stress criterion (only) was used to determine the point at which subcell failure will occur.

These problems can be defined in `MicroSimDamage.m` as follows:

```
% -- 7.2 - Epoxy with notch
MicroMat{NP + 1} = 172; % -- HF 25x25
MicroMat{NP + 2} = 173; % -- HF 51x51
MicroMat{NP + 3} = 174; % -- HF 101x101
MicroMat{NP + 4} = 175; % -- HF 201x201

for I = 1:4
    NP = NP + 1;
    OutInfo.Name(NP) = "Section 7.2 - Epoxy with notch";
    Loads{NP}.Type  = [S,    E,    S,    S,  S,    S];
    Loads{NP}.Value = [0,    0.02, 0,    0,  0,    0];
    Loads{NP}.NINC = 400;
    Loads{NP}.CriteriaOn = [1,0,0,0];
end
```

where the notched RUCs are defined in `GetEffProps.m` as follows:

```
Mat = 172;
Theory = 'HFGMC';
name = "notch in monolithic epoxy 25x25";
Vf = 0.2;
Constits.Fiber = constitprops{16};
Constits.Matrix = constitprops{15};
RUCid = 30;
[props{Mat}] = RunMicro(Theory, name, Vf, Constits, props{Mat}, RUCid);

Mat = 173;
Theory = 'HFGMC';
name = "notch in monolithic epoxy 51x51";
Vf = 0.2;
Constits.Fiber = constitprops{16};
Constits.Matrix = constitprops{15};
RUCid = 31;
[props{Mat}] = RunMicro(Theory, name, Vf, Constits, props{Mat}, RUCid);

Mat = 174;
Theory = 'HFGMC';
name = "notch in monolithic epoxy 101x101";
Vf = 0.2;
Constits.Fiber = constitprops{16};
Constits.Matrix = constitprops{15};
RUCid = 32;
[props{Mat}] = RunMicro(Theory, name, Vf, Constits, props{Mat}, RUCid);

Mat = 175;
Theory = 'HFGMC';
name = "notch in monolithic epoxy 201x201";
Vf = 0.2;
Constits.Fiber = constitprops{16};
Constits.Matrix = constitprops{15};
RUCid = 33;
[props{Mat}] = RunMicro(Theory, name, Vf, Constits, props{Mat}, RUCid);
```

Table 7.1 Glass/epoxy ($V_f = 0.55$) constituent strengths backed out as stress allowables using the MT method in Chapter 4 with only the Tsai-Wu criterion.

Mat	Material name	XT (MPa)	XC (MPa)	YT (MPa)	YC (MPa)	Q (MPa)	R (MPa)	S (MPa)
2	Glass - MT	2358	−1653	2358	−1653	1000	1000	1000
4	Epoxy - MT	1×10^6	-1×10^6	42	−181	81	81	81

Note that the matrix (constituent material #15) is the epoxy material, but with allowables backed out from HFGMC in the next Section 7.3 (see Table 7.1). Constituent material #16 is this same material with stiffnesses decreased by a factor of 10,000 to represent the notch. Both of these materials are in GetConstitProps.m. In addition, specialized RUCs for this problem, designated as RUCid = 30 through 33, are defined in GetRUC.m. In this RUC, the specified fiber volume fraction is taken as the width of the notch divided by the RUC width, and the specified fiber material is assigned to subcells such that the notch is one subcell tall and the correct width. The widths of the center three subcells in the notch are also adjusted to ensure the total width of the notch is as intended, while the subcell grid through which the damage progresses is composed of all square subcells. The grid refinement is chosen by specifying the RUCid (30 – 33), which in turn selects the appropriate number of subcells in the two directions (RUC.NB and RUC.NG).

The stress-strain curves resulting from this notched epoxy progressive damage simulation are plotted in Fig. 7.8. Clearly, these results display significant subcell grid refinement dependence, with the damage initiation stress (which corresponds to the UTS) decreasing as the subcell grid is refined. This type of discretization dependence in CDM progressive damage simulations involving softening materials is referred to as *pathological mesh dependence* because it is an artifact of the numerical procedure. In fact, with continued subcell grid refinement, the local stresses (Fig. 7.9) would never converge, and the predicted UTS would go to zero, which is clearly incorrect and non-physical. Of course, a model must exhibit mesh dependence in order for there to be pathological mesh dependence. Therefore simpler models like MT, MOC, and GMC do not exhibit pathological mesh dependence, whereas models like finite element analysis (FEA) and HFGMC do.

The pathological mesh dependence can be understood by examining the damage progression in the problem, which is depicted for the 25×25 RUC case in Fig. 7.9. Snapshots from the simulation of the stress-strain curve, the von Mises stress, and the growth of the notch are shown. The notch begins to grow when the stress in the two subcells at the notch tips reaches the tensile strength of the epoxy. This occurs at increment (INC) number 248 of the applied loading, when the global stress is approximately 40 MPa. As the subcells adjacent to the notch tips fail and their stiffness is reduced by a factor of 10,000, the code must iterate (while remaining at loading increment 248) to allow the stress in the newly failed subcells to redistribute. This essentially allows the newly failed subcells to become part of the notch, and subsequently, the next adjacent subcells experience the peak stress and they fail. The model continues to iterate and, as shown in Fig. 7.9, the notch continues to grow all the

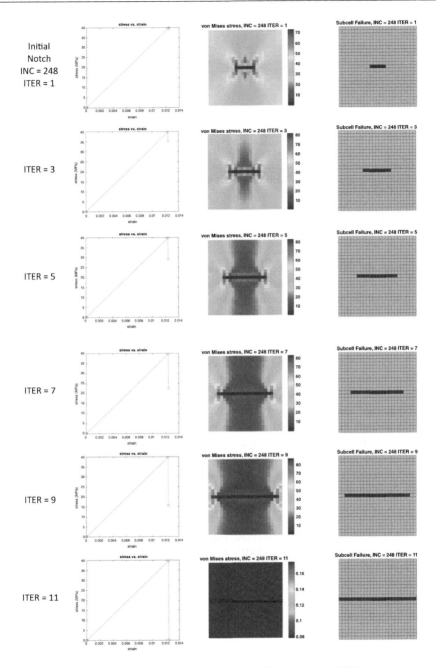

Figure 7.9 Stress-strain curve, local von Mises stress (MPa), and subcell failure pattern snapshots during an HFGMC simulation of the transverse tensile (x_2-direction) response of a center notched epoxy RUC. Results are shown for every other iteration as the notch grows through the RUC in a brittle fashion, all at the same loading increment (INC).

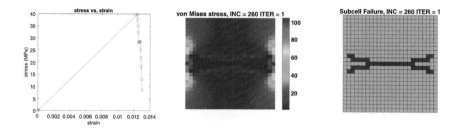

Figure 7.10 A snapshot of the stress-strain curve, local von Mises stress (MPa), and subcell failure pattern during an HFGMC simulation of the transverse tensile (x_2-direction) response of a center notched epoxy RUC. In this example, iterations are not performed and all subcells with negative margins are failed at each loading increment. This does not ensure converged results at each loading increment and can lead to nonphysical damage paths as shown.

way through the RUC (in a self-similar fashion), and the failure is completely brittle. That is, as soon as the damage began to progress, it immediately grew across the RUC without any additional loading being applied.

The damage progression for the more refined RUCs is the same; once the damage initiates in the subcells adjacent to the notch, it grows in a brittle manner. However, now, with the more refined subcell grid relative to the notch, for a given applied strain, the stress in the subcells at the notch tips becomes higher and higher. In fact, as the subcell grid becomes more and more dense, the notch becomes more and more like a true crack, and the stress at the notch tip continues to grow without bound. Thus, in the limit of an infinitely refined subcell grid, any applied loading would lead to infinite stress at the notch tip, and the RUC would fail completely with even an infinitesimal applied load. This is the essence of pathological mesh dependence. If the softening material damage localizes to the scale of the discretization (as in FEA and HFGMC), the damage will grow while dissipating no energy in the limit as the discretization becomes smaller and smaller. As mentioned previously, Bažant et al. (1979, 1983, 1991, 1994) developed the crack band approach to continuum damage modeling, which builds in a length scale to the damage, to regularize the dissipated energy and minimize pathological mesh dependence. Pineda and Waas (2012) and Pineda et al. (2012, 2013) have incorporated the crack band model within HFGMC, but this work is beyond the scope of this text. Therefore, when using the subcell elimination method presented in this chapter in conjunction with HFGMC, one must remain cognizant of the presence of pathological mesh dependence. As mentioned, because MT has no localization and GMC has no subcell grid dependence, these theories are not subject to pathological mesh dependence.

An illustration of the need for iterations within a given loading increment, coupled with allowing only the subcells that are very close to the minimum margin to fail per iteration, is shown in Fig. 7.10. Here, for the 25×25 notched epoxy case modeled with HFGMC, only one iteration per increment is permitted and, at each increment, any subcell with a negative margin is failed. A snapshot of the stress-strain curve, von Mises stress, and damage pattern is shown near final failure in Fig. 7.10. Clearly the damage path is unrealistic as the damage zone has branched due to multiple subcells having negative margins during a particular increment. By limiting the failures to only the subcells that have the minimum margin and iterating, as done in the Fig. 7.9 results, stresses are permitted to redistribute as the damage advances, and a converged damage state is attained at the completion of each increment (with ITER=11 being the final iteration). In the context of progressive damage modeling, it is important to ensure a converged damage state throughout the simulation to have a valid solution.

7.3   Constituent strength consistency

Recall that in Chapter 4 it was demonstrated that constituent material allowables can be adjusted to match known composite allowables and that these "backed out" allowables are micromechanics theory dependent. In this section constituent material allowables are backed out for HFGMC. These allowables, as well as those backed out previously using MT, are then employed within progressive damage simulations using both theories.

The glass/epoxy constituent stress allowables given in Table 7.1 (replicated from Table 6.7) were obtained using MT and the Tsai-Wu failure criterion such that the resulting composite stress allowables matched those of a glass/epoxy ply (Table 4.2). These allowables have now been used as the damage initiation strengths in a progressive damage simulation of a 0.55 fiber volume fraction glass/epoxy composite loaded in the x_2-direction with a strain increment of 0.00005 (see Fig. 7.11). The MT method

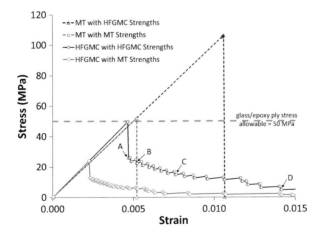

Figure 7.11 Simulated transverse (x_2-direction) stress-strain response of 0.55 fiber volume fraction glass/epoxy composite using MT and HFGMC (with a 46×80 hexagonal RUC), both using only the max stress failure criterion, and using the Table 7.1 and Table 7.2 constituent material strengths. A strain increment of 0.00005 was employed.

Table 7.2 Glass/epoxy constituent strengths backed out from HFGMC with a 46×80 RUC with hexagonal fiber packing ($V_f = 0.55$) with only the maximum stress criterion.

Mat	Material name	XT (MPa)	XC (MPa)	YT (MPa)	YC (MPa)	Q (MPa)	R (MPa)	S (MPa)
18	Glass - HFGMC	2445	–1653	2445	–1653	1000	1000	1000
15	Epoxy - HFGMC	86	–425	86	–425	184	184	184

(denoted by red triangular symbols) and HFGMC with a 46×80 hexagonally packed RUC (denoted by red circular symbols) have been used along with the maximum stress criterion. In addition, predictions from both micromechanics theories are shown when the adjusted constituent strengths given in Table 7.2 are employed. These strengths were backed out following a similar procedure as described in Chapter 4 to match the ply level glass/epoxy allowables (Table 4.2) using HFGMC, the maximum stress criterion, and the 46×80 RUC with hexagonal fiber packing. Here, however, progressive damage simulations have been used rather than simply projecting based on the minimum MoS, as in Chapter 4. Note that, in Table 7.2, the fiber tensile strengths (XT and YT) are slightly increased as compared to Table 7.1 due to the nonlinear behavior in the progressive damage simulations associated with matrix damage. Further, the matrix normal and shear strengths are increased, while the matrix axial strengths (XT and XC) are no longer set to artificially high values. In Fig. 7.11, the black lines correspond to the stress-strain predictions using these new HFGMC-based strengths from Table 7.2.

These progressive damage simulation problems can be defined as usual in `MicroProblemDef.m` as follows:

```
% -- 7.3 - Influence of Micromechanics Theory on Consistent Constituent Strength
MicroMat{NP + 1} = 176; % -- MT with MT props
MicroMat{NP + 2} = 177; % -- HF with MT props
MicroMat{NP + 3} = 178; % -- MT with HF props
MicroMat{NP + 4} = 179; % -- HF with HF props

for NP1 = 1:4
    NP = NP + 1;
    OutInfo.Name(NP) = "Section 7.3";
    Loads{NP}.DT = 0.0;
    Loads{NP}.Type  = [S,    E,    S,   S,   S,   S];
    Loads{NP}.Value = [0,    0.04,  0,  0,   0,    0];
    Loads{NP}.NINC = 800;
    Loads{NP}.CriteriaOn = [1,0,0,0];
end
```

where the composite materials are defined in `GetEffProps.m`, the additional constituent materials (see Table 7.2) are specified in `GetConstitProps.m`, the hexagonally packed RUC is specified in `GetRUC.m` (`Target_NG = 80`), and the problems are executed using `MicroSimDamage.m`.

As shown in Fig. 7.11, the stress-strain curves predicted by MT drop immediately following damage initiation since there is only a single average matrix stress to evaluate. Thus, once the matrix fails, there is no way for the transverse stress to reach the fiber, and therefore the composite cannot carry any additional load. In contrast, in the case of HFGMC (solid lines with circular symbols), damage initiation occurs significantly below that of the MT prediction, and the stress does not immediately drop to zero. Rather, after an initial significant drop in stress, a more gradual softening occurs over a wide range of strains. This gradual stress decrease results as the damage progresses through the RUC and more and more subcells are "eliminated" as the strain loading increases.

Clearly, when using constituent strengths that are consistent with the micromechanics theory used to back them out, both theories are able to predict a composite UTS close to the ply level allowable. Neither theory matches the ply allowable if inconsistent strengths are used. For both sets of constituent strengths, the stress at which damage initiates in HFGMC is much lower than in MT due to HFGMC's ability to capture stress concentrations that are missed by MT.

This example highlights the fundamental concept of idealization consistency from characterization to prediction. One should be as consistent as possible in terms of theory (specific constituent deformation/damage models, micromechanics theory, failure theory, etc.), numerics (grid density, increment size, microstructure or RUC (fiber packing, number of fibers, number of phases, etc.)), loading/boundary conditions, and included mechanisms (damage, delamination, etc.) for a given use case. Clearly, inaccurate predictions can result if idealization consistency is not preserved throughout the simulation process (e.g., utilizing constitutive properties and failure strengths without regard to how they were obtained).

The damage progression from the HFGMC simulation with the consistent, HFGMC-based, constituent strengths is depicted in Fig. 7.12. Four snap shots are shown, labeled A, B, C, and D, with the corresponding points from the stress-strain curve labeled in Fig. 7.11. Recalling that the transverse strain-based loading is in the x_2-direction, Fig. 7.12A shows where damage initiates (dark red subcells indicate failure) with matrix subcells failing between fibers at the fiber/matrix interface near the top and bottom of the fibers. As the loading continues (Fig. 7.12B, C, and D), the damage that initiated near the fiber/matrix interface propagates forming horizontal "crack-like" damage regions. These are then deflected as they approach adjacent fibers, and the damage begins to progress around the fibers (Fig. 7.12D). The left side of Fig. 7.12 depicts the von Mises stress states corresponding to each damage snap shot. These show that the stress in the failed subcells is very low, while a stress concentration arises at the tip of the horizontal crack-like damage zones. The intensities of these concentrations are dependent on the density of the subcell grid (as discussed previously and shown in Section 7.4.1.6). Areas between closely spaced crack-like damage zones are also at very low stress levels. For snap shot D, most of the RUC is at a low stress (blue areas) with very little of the geometry transferring load. The ability to model this type of damage, where it localizes and deflects in a crack-like manner is a capability that HFGMC owes to its inclusion of shear coupling. In the next section, this capability will be directly compared with that of GMC.

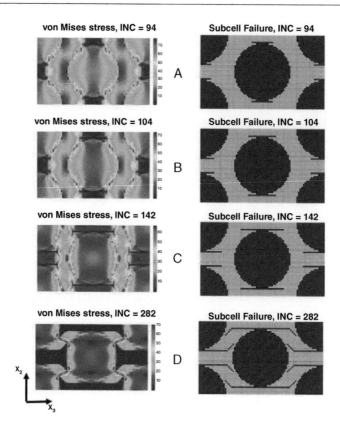

Figure 7.12 Damage progression and von Mises stress (MPa) snap shots associated with the HFGMC (with HFGMC strengths) stress-strain response in Fig. 7.11, with point locations indicated by letters.

7.4   PMC example problems

7.4.1   *Unidirectional PMC progressive damage*

In this section, progressive damage simulations of unidirectional glass/epoxy composites are considered, wherein the MT, GMC, and HFGMC micromechanics theories are compared. The effects of the applied loading orientation, fiber packing arrangement (ordered and disordered), number of fibers in the RUC representation, local failure theory, loading increment size, and geometric discretization are examined. For these and the remaining example problems, the MATLAB code used to specify the problem definition will not be given (as readers should be familiar with this by now). However, all code needed to execute the progressive damage example problems in this chapter is given in `MicroProblemDef.m` and `LamProblemDef.m`.

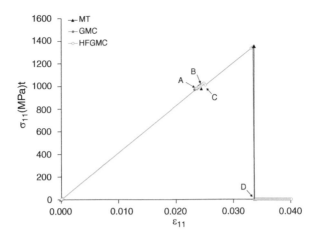

Figure 7.13 MT, GMC, and HFGMC simulated stress-strain curves for a 0.55 fiber volume fraction unidirectional glass/epoxy composite represented by a 32×32 RUC with square fiber packing (in the case of GMC and HFGMC), subjected to axial tensile strain, using the max stress criterion, with an applied strain increment of 0.0002.

7.4.1.1   Influence of micromechanics theory and applied loading

In this example, MT, GMC, and HFGMC progressive damage simulations are compared for a 0.55 fiber volume fraction unidirectional glass/epoxy composite represented by a 32×32 RUC with square fiber packing (in the case of GMC and HFGMC), subjected to strain controlled axial tension, transverse tension, and axial and transverse shear loading. This strain controlled loading involves applying a single component of strain, with all other stress components set equal to zero. The maximum stress criterion has been used along with the constituent material strengths given in Table 7.2.

The progressive damage results for axial tensile loading are shown in Fig. 7.13, while damage and von Mises stress snapshots (for GMC and HFGMC) are shown in Fig. 7.14. MT damage progression is not plotted as MT predicts only a single average stress state for the matrix, thus, once matrix failure initiates, all matrix has failed. In response to axial tensile loading, all three models predict damage initiation in the matrix, but at slightly different applied loading levels (see deviation from linearity in Fig. 7.13). Fig. 7.14 shows that GMC and HFGMC both predict some variation in the matrix σ_{11} field, leading to some progression of the matrix failure from point A to point C, when all the matrix has failed. From this point forward, in all three models, the fiber carries all of the load until the fiber strength is reached and it fails at an applied composite σ_{11} value of approximately 1340 MPa.

The progressive damage results for transverse tensile strain loading in the x_2-direction are shown in Fig. 7.15, while damage and von Mises stress snapshots (for GMC and HFGMC) are shown in Fig. 7.16. Note that results for transverse tensile loading in the x_3-direction would be identical (but rotated) due to the RUC's symmetry. As before, MT damage progression is not plotted since, once matrix failure initiates, all

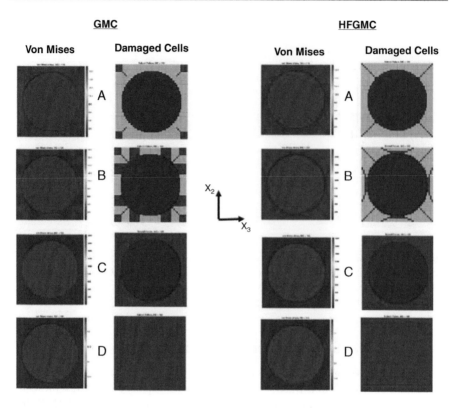

Figure 7.14 Damage progression and von Mises stress (MPa) snap shots as predicted by GMC and HFGMC associated with the stress-strain curves shown in Fig. 7.13.

the matrix has failed. In response to transverse tensile loading, all three models predict damage initiation in the matrix, but at significantly different global stress levels: MT (106 MPa), GMC (52.5 MPa) and HFGMC (50.4 MPa). The MT response is the same as that plotted in Fig. 7.11 for the MT with HFGMC strengths. Although GMC and HFGMC both predict damage progression in the matrix, the damage progression patterns are quite different because their local stress fields are dissimilar. Fig. 7.16 shows that, for HFGMC, the damage initiates between the fibers in the loading direction and progresses in a self-similar manner across the RUC. In contrast, although GMC's damage also progresses across the RUC, entire columns of matrix subcells fail all at once as the damage progresses. This is directly related to the traction continuity across subcell boundaries and lack of normal-shear coupling in the GMC formulation, see Chapter 5. Clearly GMC's local damage propagation pattern is distinctly different from HFGMC's, however, it does still allow for damage propagation within the RUC and provides an ultimate transverse strength that is within 5% of HFGMC's. It should be noted that the strain at which the stress reaches zero (point D) for GMC is more than three times that of HFGMC.

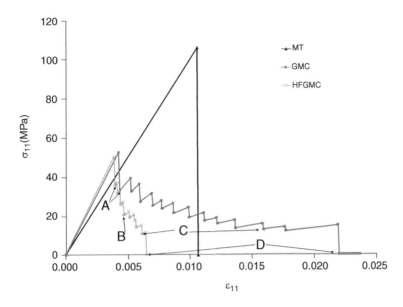

Figure 7.15 MT, GMC, and HFGMC simulated stress-strain curves for a 0.55 fiber volume fraction unidirectional glass/epoxy composite represented by a 32×32 RUC with square fiber packing (in the case of GMC and HFGMC), subjected to transverse tensile strain (x_2-direction), using the max stress criterion, with an applied strain increment of 0.00005.

Fig. 7.17 shows the evolution of the transverse Young's moduli, E_{22} and E_{33}, and axial shear moduli, G_{12} and G_{13}, as damage progresses during the transverse (x_2-direction) tensile simulation. MT predicts that, once damage initiates in the matrix, all of the matrix fails and E_{22}, E_{33}, G_{12}, and G_{13} all instantly fall to very low values (see black lines in Fig. 7.17). In GMC, as the damage progresses, as one would expect, the moduli associated with the x_2- loading direction (E_{22} and G_{12}, solid blue lines) drop gradually, but significantly. Because entire columns of matrix subcells fail as the damage progresses (Fig. 7.16), the moduli associated with the x_3-direction (E_{33} and G_{13}, dashed blue lines) drop significantly as well. In fact, once the E_{22} and G_{12} moduli have dropped to close to zero, the E_{33} and G_{13} moduli are also very close to zero (because all matrix subcells have failed, Fig. 7.16D). In contrast to both MT and GMC, HFGMC's ability to predict a more realistic crack-like damage pattern enables the theory to maintain most of its E_{33} and G_{13} stiffness (dashed orange lines), even after the damage has progressed completely through the RUC in the x_2-direction (Fig. 7.16D). This is clearly much more realistic as cracks oriented normal to the x_2-direction should have little or no effect on the x_3-direction moduli. This also shows that damage induces additional anisotropy in composite effective properties when using HFGMC (and during damage progression when using GMC). Initially $E_{22} = E_{33}$, but once damage develops, this is no longer true.

Damage progression simulation results for applied axial shear strain (γ_{12}) loading are given in Figs. 7.18 and 7.19, while results for applied transverse shear strain (γ_{23})

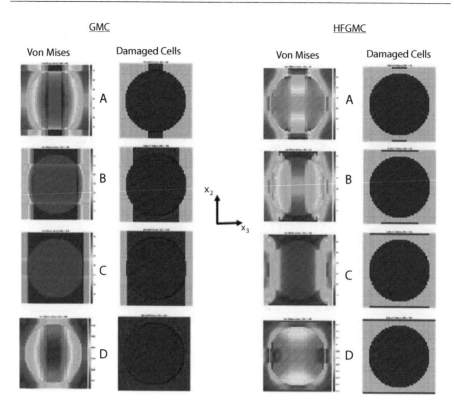

Figure 7.16 Damage progression and von Mises stress (MPa) snap shots as predicted by GMC and HFGMC associated with the stress-strain curves shown in Fig. 7.15.

loading are given in Figs. 7.20 and 7.21. The axial shear results are similar to the transverse tensile results, aside from the fact that, in the HFGMC axial shear simulation, the damage initiates in the matrix at the fiber matrix interface rather than between the fibers (see Fig. 7.16). The stress concentration at the tip of the damage zone in the HFGMC axial shear simulation, shown in Fig. 7.19C, is also more pronounced than that in the transverse tensile simulations, shown in Fig. 7.16C.

More noteworthy is the difference between GMC and HFGMC in the transverse shear progressive damage simulations (Figs. 7.20 and 7.21). In this case, the GMC results resemble the MT results (although with a difference in stiffness), with no progression after damage initiation. This is because GMC predicts constant transverse shear stress, σ_{23}, throughout the RUC and, like MT, no other local stress components arise due to applied transverse shear loading. Thus, once the matrix shear stress reaches the matrix shear strength, all of the matrix fails. HFGMC, on the other hand, captures realistic stress concentrations when loaded in transverse shear, leading to damage initiation at much lower load levels than GMC and MT. The damage pattern predicted by HFGMC is also quite interesting, with damage initiating and progressing on two opposite sides of the fiber in quadrants I and III (see HFGMC results in Fig. 7.21B). The

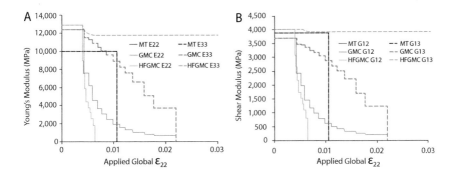

Figure 7.17 Evolution of (A) transverse Young's modulus and (B) axial shear modulus, associated with the stress-strain curves shown in Fig. 7.15.

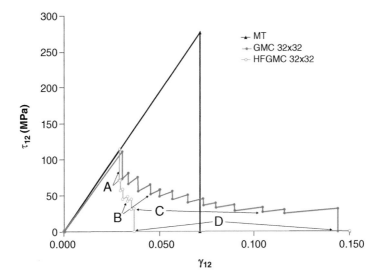

Figure 7.18 MT, GMC, and HFGMC simulated stress-strain curves for a 0.55 fiber volume fraction unidirectional glass/epoxy composite represented by a 32×32 RUC with square fiber packing (in the case of GMC and HFGMC), subjected to axial shear strain, using the max stress criterion, with an applied strain increment of 0.00005.

local transverse shear stress state prior to failure initiation is symmetric with respect to the x_2- and x_3-axes (Fig. 7.21A), thus it might be expected that damage initiation and growth would occur on all four sides of the fiber in all four quadrants. However, in the case considered, because the matrix tensile strength is so low, the damage initiation and growth are due to local tensile σ_{22} and σ_{33} stresses in the matrix (not local σ_{23}). This was also observed in Section 6.10.7. These stress components are compressive near the fiber/matrix interface in quadrants II and IV, so no damage initiation or growth occurs in these regions (see also Figure 6.17). Obviously, if the matrix shear strength were lower,

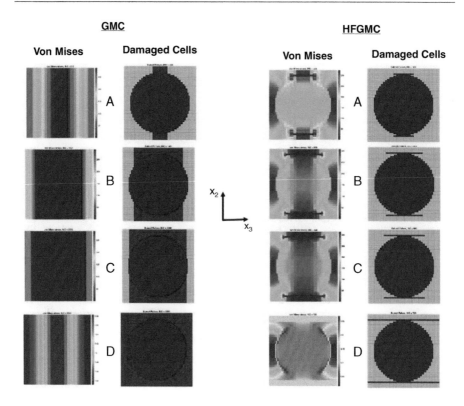

Figure 7.19 Damage progression and von Mises stress (MPa) snap shots as predicted by GMC and HFGMC associated with the stress-strain curves shown in Fig. 7.18.

or the matrix tensile strength were higher, the damage pattern could be much different. Clearly, in general, the progressive damage behavior can be significantly affected by the constituent material strengths and their relations to each other, opening the possibility of further tailoring composite performance by manipulating these relations based on HFGMC predictions.

7.4.1.2   Influence of fiber packing arrangement

In this example, HFGMC progressive damage simulations using different fiber packing arrangements are compared for a 0.55 fiber volume fraction unidirectional glass/epoxy composite, subjected to strain controlled transverse tension. The maximum stress criterion has been used along with the constituent material strengths given in Table 7.2.

Fig. 7.22 illustrates a significant influence of the fiber packing arrangement (square, hexagonal, and random) on both the transverse ultimate tensile strength (UTS) and the damage progression behavior. Note that results from only one representative random RUC, containing 10 fibers, has been plotted, whereas, for hexagonal packing, loading

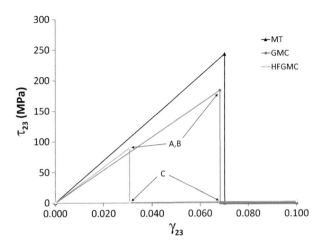

Figure 7.20 MT, GMC, and HFGMC simulated stress-strain curves for a 0.55 fiber volume fraction unidirectional glass/epoxy composite represented by a 32×32 RUC with square fiber packing (in the case of GMC and HFGMC), subjected to transverse shear strain, using the max stress criterion, with an applied strain increment of 0.00005.

has been applied in both the x_2- and x_3-directions. A .json file containing the Random-3 RUC definition is provided with the MATLAB code in the Output folder. Square fiber packing results in a higher transverse Young's modulus compared to hexagonal and random packing (as shown previously in Table 6.4), with a predicted UTS close to that associated with hexagonal packing loaded in the x_2-direction. In the results for hexagonal fiber packing, even though the Young's moduli are nearly the same, there is a stark difference in the damage initiation and progression between the two directions. As discussed in Section 6.10.4, this is because the local stress fields are quite different based on the loading direction, even though the effective material properties are close to transversely isotropic (see Figure 6.12A for the local fields in a CMC case). Therefore, materials that exhibit effective elastic transverse isotropy are not necessarily transversely isotropic when damage (or other nonlinearity) is involved. Similarly, as also shown in Chapter 6, the hexagonal RUC Young's moduli agree well with that from the random RUC. However, once again, (for the present random instance) the UTS and damage progression are vastly different. Effective elastic properties are based on averaging over the entire RUC, whereas damage initiation and propagation are based on the local fields.

Damage progression and local von Mises stress field snapshots for the hexagonal packed RUC (loaded in the x_3-direction) are shown in Fig. 7.23 and for the random fiber packing in Fig. 7.24. These can be compared to the hexagonal packed case loaded in the x_2-direction (Fig. 7.12) and the square packed case (Fig. 7.16). Figs. 7.12A and 7.16A show that, for hexagonal packing (x_2-loading) and square packing, damage initiates between the fibers in the x_2-loading direction, whereas, Fig. 7.23A shows that for hexagonal packing (x_3-loading), damage initiates near the

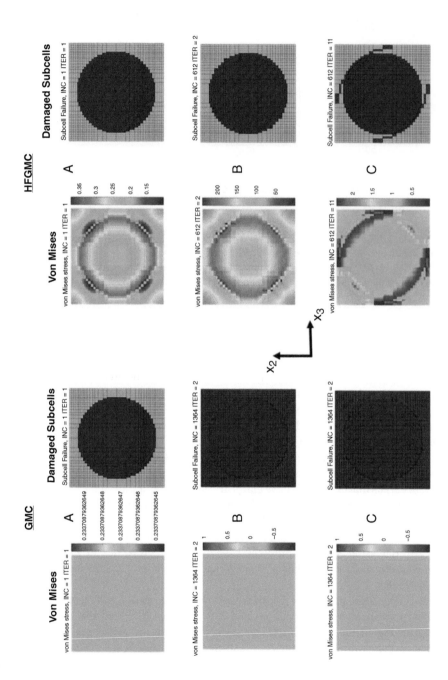

Figure 7.21 Damage progression and von Mises stress (MPa) snap shots as predicted by GMC and HFGMC associated with the stress-strain curves shown in Fig. 7.20.

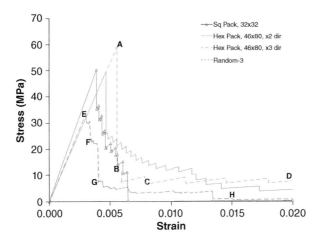

Figure 7.22 Influence of fiber packing arrangement on the transverse tensile response of a 0.55 fiber volume fraction unidirectional glass/epoxy composite with an applied strain increment of 0.00005 using the max stress criterion. The letter labels refer to the snapshots shown in Fig. 7.23 and Fig. 7.24. Random-3 is one particular realization of a random RUC.

fiber matrix interface at an angle of approximately 45° from the x_3-loading direction. In all three cases, damage initiates between closely spaced fibers. However, while for hexagonal packing (x_2-loading) and square packing the closest fiber spacing is aligned with the applied load, for hexagonal packing (x_3-loading), this is not the case. This explains why the hexagonal packing (x_2-loading) and square packing have similar UTS values (see Fig. 7.22), while the hexagonal packing (x_3-loading) predicts a much higher UTS. Examining the damage propagation in Figs. 7.12, 7.16, and 7.23, in the case with square packing (Fig. 7.16) the damage zones grow in a self-similar manner through a continuous matrix region and link up readily, leading to more rapid stress unloading in Fig. 7.22. In contrast the damage zones for hexagonal packing interact with the fibers, introducing tortuosity, and are greatly affected by the RUC geometry and the details of the evolving stress fields.

For the case with random fiber packing (Fig. 7.24), damage initiates between closely spaced fibers at a much lower global stress value compared to the other packing arrangements because more closely spaced fibers are present in the RUC, leading to higher stress concentrations. Several damage zones form and link up as the damage progresses, and the damage zones are deflected by the fibers, leading to the progressive response shown in Fig. 7.22. Obviously, this is just one single instance of random fiber packing. 200 additional realizations of random RUCs have been considered in Fig. 7.25. As shown, there is significant variation in the responses, with the maximum and minimum UTS cases highlighted. The case with hexagonal packing (x_2-loading) is also shown for comparison (blue line in Fig. 7.25) and produces a higher UTS than any of the random cases. This is because the hexagonally packed fibers are equally spaced, thus any random packing will almost definitely have some more closely spaced fibers oriented appropriately relative to the loading direction. GMC and MT, with their less

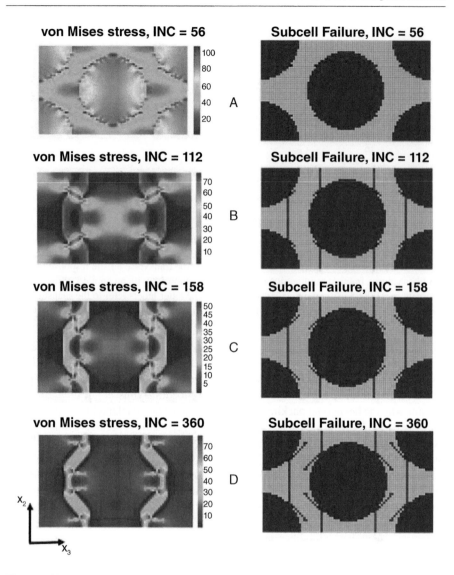

Figure 7.23 HFGMC damage progression and von Mises stress (MPa) snapshots with hexagonal packing (x_3-loading), associated with Fig. 7.22.

realistic local fields, are not able to capture these concentrations from closely spaced fibers and thus cannot be used to model random packing explicitly.

The 200 random cases in Fig. 7.25 have a mean transverse Young's modulus of 11.1 GPa with a coefficient of variation (CoV) of 2.4%, and a mean UTS of 35.0 MPa, with a CoV of 8.7%. Note that more variation in the predictions would be expected for lower fiber volume fractions (e.g., for CMCs) as there would be more space available for greater variations in the fiber locations. A histogram of the predicted UTS values is

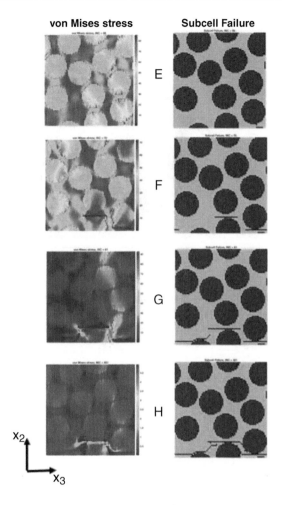

Figure 7.24 HFGMC damage progression and von Mises stress (MPa) snapshots with random packing, associated with Fig. 7.22.

shown in Fig. 7.26, while the best fitting distribution (of those available in MATLAB), a Rayleigh distribution, is plotted in Fig. 7.27. Furthermore, while the mean of the 200 realizations of the 10 fiber RUC responses may be representative, the variations are certainly dependent on the number of fibers in the RUC. As the number of fibers is increased, the variations in the progressive damage responses per realization should decrease as, on average, the RUC encapsulates more and more possible localized variations. The subcell grid refinement also has an impact, where, as has been shown, a more refined grid leads to damage progression requiring less energy. The influence of both of these effects are addressed in subsequent examples (Sections 7.4.1.3 and 7.4.1.6, respectively).

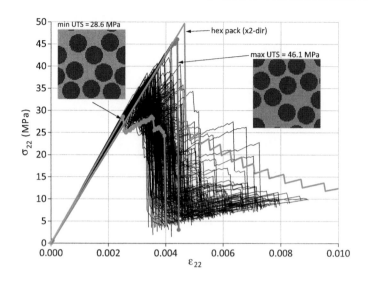

Figure 7.25 Comparison of the progressive damage response of 200 realizations of a random 10 fiber RUC to hexagonal packing (x_2-loading), all subjected to transverse tensile loading. The composite is 0.55 fiber volume fraction unidirectional glass/epoxy with an applied strain increment of 0.00005, and the max stress criterion was used.

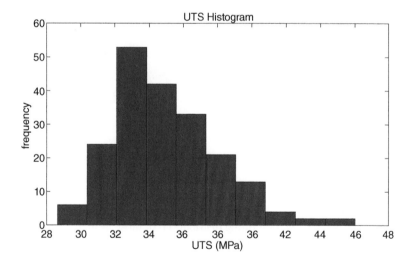

Figure 7.26 Histogram of the UTS predicted by HFGMC for 200 realizations of a random glass/epoxy RUC with 10 fibers subjected to transverse tensile loading, whose responses are plotted in Fig. 7.25.

From a practical standpoint, one must recognize that modeling many realizations of random (disordered) microstructures becomes quite expensive computationally. In contrast, ordered RUCs (e.g., square pack and hexagonal pack) provide a relatively inexpensive approach to obtaining composite effective properties and response curves.

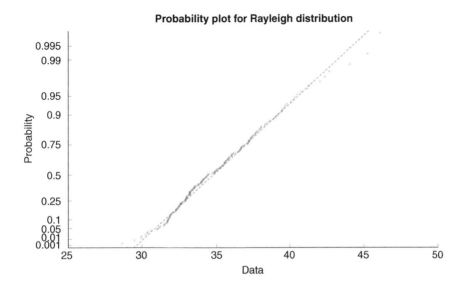

Figure 7.27 Rayleigh probability plot of the UTS values predicted by HFGMC for 200 realizations of a random glass/epoxy RUC with 10 fibers subjected to transverse tensile loading, whose responses are plotted in Fig. 7.25.

Consequently, it is more practical to calibrate in situ constituent properties within ordered RUC simulations to match either the mean response of multiple disordered simulations or higher scale experimental results.

7.4.1.3   Influence of number of fibers in RUC

As mentioned, in the case of random fiber packing, the number of fibers in the RUC influences the predictions. In this example, HFGMC progressive damage simulations using random RUCs containing 4, 8, and 16 fibers are compared for a 0.55 fiber volume fraction unidirectional glass/epoxy composite, subjected to strain controlled transverse tension. The maximum stress criterion has been used along with the constituent material strengths given in Table 7.2, and six instances for each random RUC have been considered.

Fig. 7.28 shows the transverse tensile stress-strain responses in the x_2-direction, where only the cases with the maximum and minimum predicted UTS have been plotted for each RUC. Figs. 7.29 and 7.30 show the damage initiation locations and final damage patterns, respectively, for two of the six instances for each RUC. Fig. 7.28 shows that the spread between the stress-strain curve with the maximum UTS and the minimum UTS is decreasing as the number of fibers within the RUC is increasing. This is expected as adding additional fibers in the RUC allows the RUC to approach a statistical representative volume element (RVE), which would produce a converged response curve. Regarding damage propagation, the different subcell grid sizes among the three different fiber counts also should have an influence. The 4 fiber

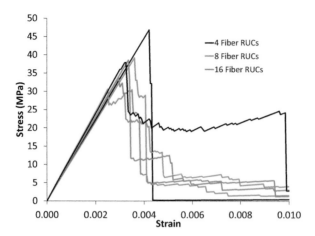

Figure 7.28 Influence of number of fibers within the RUC on the HFGMC transverse stress-strain response in x_2-direction, under incremental strain loading (0.00005 strain increment). The maximum stress failure criterion was used. The instances with the maximum and minimum UTS (per fiber count) are plotted.

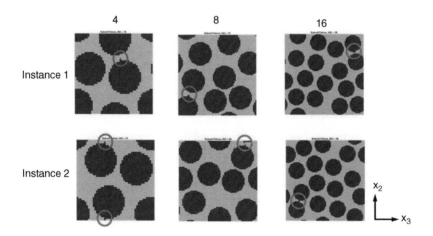

Figure 7.29 Initial damage state in two of the random instances for each RUC with a fixed number of fibers per RUC. The red circles highlight the damage initiation locations.

RUC consists of 48×48 subcells, the 8 fiber RUC consists of 68×68 subcells, and the 16 subcell RUC consists of 96×96 subcells, which means the 16 subcell RUC requires less energy density to propagate damage for each subcell that fails, and pathological mesh dependence is present (see Section 7.2). However, more realizations for each fiber count would be required to draw more definitive conclusions.

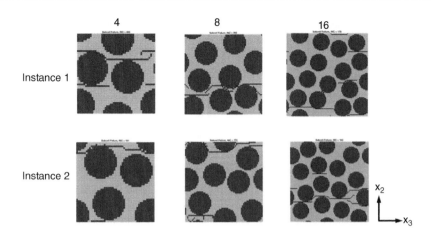

Figure 7.30 Final damage state in two of the random instances for each RUC with a fixed number of fibers per RUC.

Table 7.3 Glass/epoxy constituent failure strains determined from linear relationship with material strengths in Table 7.2.

Mat	Material name	$X_{\varepsilon T}$	$X_{\varepsilon C}$	$Y_{\varepsilon T}$	$Y_{\varepsilon C}$	Q_{ε}	R_{ε}	S_{ε}
18	Glass - HFGMC	0.0335	−0.0226	0.0335	−0.0226	0.0334	0.0334	0.0334
15	Epoxy - HFGMC	0.0249	−0.1232	0.0249	−0.1232	0.144	0.144	0.144

Examining the damage patterns in Figs. 7.29 and 7.30, it is clear that variability is present in both the damage initiation and the damage propagation. Both are influenced by the number of fibers in the random RUC and by the particular random instance.

7.4.1.4 Influence of failure criterion

In this example, HFGMC progressive damage simulations using the max stress, max strain, Tsai-Hill, and Tsai-Wu failure criteria are compared for a 0.55 fiber volume fraction unidirectional glass/epoxy composite, subjected to strain controlled transverse tension. The constituent material strengths are given in Table 7.2, the constituent material failure strains are given in Table 7.3, and two 10 fiber random RUCs (designated as "Random-1" and "Random-3") have been used. Note that .json files containing these random RUC definitions are provided with the MATLAB code in the Output folder.

The predicted stress-strain curves are plotted in Fig. 7.31, whereas damage initiation and final damage state snapshots are shown in Fig. 7.32. Fig. 7.31 shows that, for both random RUCs, the choice of failure criterion has a major impact on the damage initiation and ultimate tensile strength (with Tsai-Hill having the highest followed by max strain, then max stress, and with Tsai-Wu having the lowest). This is expected as the local stress fields in the random RUC are highly multiaxial, and interactive failure criteria (Tsai-Hill and Tsai-Wu) are quite different from each other and from

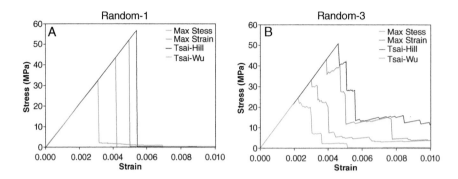

Figure 7.31 Influence of failure theory on the transverse (x_2-direction) tensile response of a 0.55 fiber volume fraction unidirectional glass/epoxy composite with an applied strain increment of 0.00005, using the (A) Random-1 and (B) Random-3 RUC with 10 randomly packed fibers.

max stress. The max strain results are different from the max stress results because the local strain fields are affected by the multiaxial stress fields and Poisson effects. It is also clear from comparing Fig. 7.31A and B, that the Random-1 microstructure results in a much more brittle response than does the Random-3 microstructure. The reason for this is clear in Fig. 7.32, which shows that the Random-1 damage paths are quite straight through the RUC (aside from Tsai-Wu), while the Random-3 paths are more tortuous. In the case of Random-1 Tsai-Wu, the damage is not straight, but it still proceeds rapidly across the RUC. An additional factor increasing the brittle response of Random-1 compared to Random-3 is the fact that damage initiates at higher global stress levels, thus the local stresses are higher when damage initiates, driving rapid damage propagation. Fig. 7.32 also shows that, for Random-1, all failure criteria predict initiation in approximately the same location even though the applied stress at which that initiation occurs is quite different. In contrast, for Random-3, Tsai-Hill predicts a damage initiation location different than the other criteria, and this initial damage does not propagate.

7.4.1.5 Influence of loading increment size

In this example, the influence of the size of the loading increment on HFGMC progressive damage simulations is examined for a 0.55 fiber volume fraction unidirectional glass/epoxy composite, subjected to strain controlled transverse tension using the max stress and Tsai-Wu criteria. The constituent material strengths are given in Table 7.2, and a 32×32 subcell RUC with square fiber packing has been employed.

Fig. 7.33 compares the HFGMC transverse tensile response for four different applied strain increment sizes for the two failure criteria. The smaller increment sizes are able to capture the details of the damage as it grows and then is briefly arrested, resulting in the zigzag pattern in the stress-strain curves. If the loading increment is large, the simulation will tend to step over these details, however, the damage

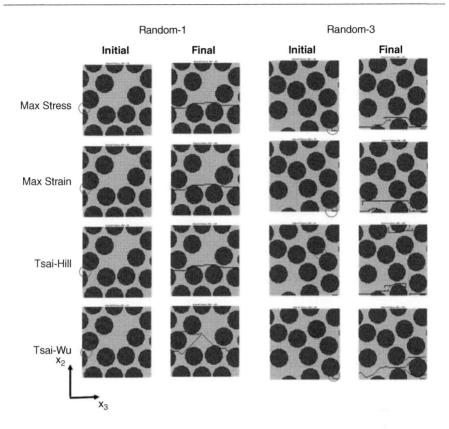

Figure 7.32 Initiation of damage and the final damage state predicted for the Random-1 and Random-3 RUCs loaded in transverse strain control in the x_2-direction. The Max Stress, Max Strain, Tsai-Hill, and Tsai-Wu failure criteria results are compared. The corresponding stress-strain curves are plotted in Fig. 7.31. The red circles highlight the damage initiation locations.

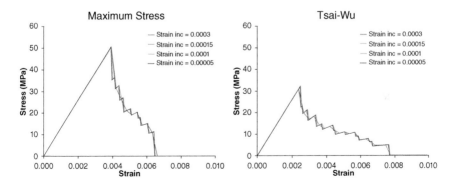

Figure 7.33 Influence of strain increment (inc) size on the x_2-direction HFGMC tensile response under incremental strain loading, using a 32×32 RUC and either the maximum stress or Tsai-Wu failure criterion.

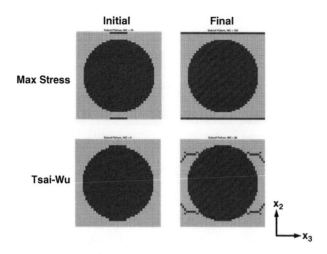

Figure 7.34 Damage initiation and final failure damage state snapshots as a function of applied strain increment associated with Fig. 7.33.

progression is very similar as, at each increment, the iteration procedure determines a converged damage state. In addition, the larger loading increments will tend to under predict peaks in the stress strain curve (see Strain inc = 0.0003 in the Tsai-Wu results).

The initial and final damage patterns (shown in Fig. 7.34) are not affected by the increment size, again, because the iteration procedure leads to a converged damage state at a given increment. As expected from the previous example, the two failure theories considered lead to different final failure patterns.

7.4.1.6   Influence of geometric discretization

In this example, the influence of the geometric discretization on HFGMC progressive damage simulations is examined for a (nominally) 0.55 fiber volume fraction unidirectional glass/epoxy composite, subjected to strain controlled transverse tension using the Tsai-Wu failure criterion and a strain increment of 0.00005. The constituent material strengths are given in Table 7.2, and a 32×32 subcell RUC with square fiber packing has been used.

As shown in Fig. 7.35, five geometric discretizations of an RUC containing a single fiber have been considered. These RUCs have been generated using the random RUC capability (RUCid = 300), specifying one fiber (Nfibers = 1), and altering the number of subcells across the fiber radius (Radf in GetRUC.m) from 5 to 80. Although the fiber center position varies among the five RUCs, the fiber packing arrangement of all five is square due to the periodicity. Furthermore, the variation of the geometric distribution is not simply refining the subcell grid while keeping the geometry constant, rather, as was done in Section 6.10.5, the circular fiber shape is also being refined. It should also be noted that the fiber volume fractions are not identical, but they are all close to 0.55.

The subcell grid density affects the results in two ways. First, the failure initiation is earlier for the more refined grids because of slight stress concentrations at the corners

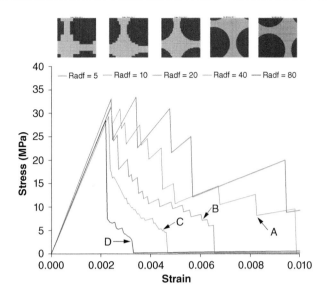

Figure 7.35 Influence of RUC geometric refinement (as controlled by the fiber radius, `Radf`) on the HFGMC transverse tensile response of nominally 0.55 fiber volume fraction glass/epoxy, under incremental strain loading (0.00005 strain increment) in the x_2-direction. The Tsai-Wu failure criterion has been utilized. The fiber volume fractions of all cases are close to 0.55, but they are not identical.

of the stair-stepped circular-fiber representation using square subcells. Second, as the mesh is refined, faster (more brittle) damage progression occurs, due to the pathological mesh dependence of HFGMC with softening damage (as described in detail in Section 7.2). Fig. 7.36 shows the initial and final damage patterns for the five different RUC subcell grid densities. Regardless of the grid refinement, damage initiates in the matrix at the fiber/matrix interface at the top and bottom of the fiber. The damage progression is also similar, but with the more refined grids, the damage zone propagates away from the fiber-matrix interface to a greater extent. Fig. 7.37 shows von Mises stress and damage snapshots for four of the five grid refinement cases corresponding to the labeled points in Fig. 7.35. This figure highlights the stress concentration at the tip of the growing damage zone and how it is more localized for the more refined grids. Note that the HFGMC failure pattern in Fig 7.16 is different than that in Fig. 7.36 (despite both having square fiber packing) because the failure criterion was maximum stress rather than Tsai-Wu.

7.4.2 PMC laminate progressive damage

In this section, progressive-damage simulations of glass/epoxy laminates are considered, wherein the MT and HFGMC micromechanics theories are used to model the behavior of the plies within the laminates. Cross-ply, angle-ply, and two types of quasi-isotropic laminates are examined.

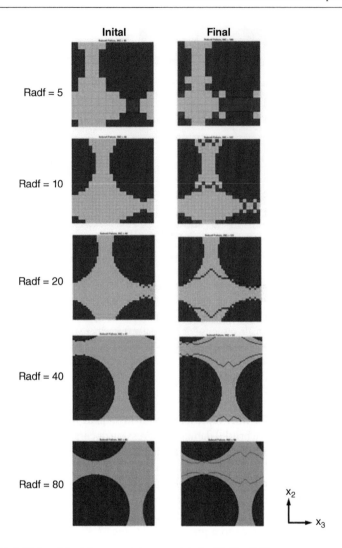

Figure 7.36 Initial and final damage states associated with Fig. 7.35.

7.4.2.1   Cross-ply laminate [0/90]$_s$

In this example, the progressive damage behavior of a 0.55 fiber volume fraction glass/epoxy cross-ply [0/90]$_s$ laminate, subjected to strain controlled axial midplane tension, using the max stress failure criterion and a strain increment of 0.00005 is examined. The MT theory is compared to HFGMC with square fiber packing and both 7×7 and 32×32 subcell RUCs. As in Section 7.3, two sets of allowables are also considered for consistency with the MT method. The constituent material strengths are given in Table 7.1 and Table 7.2. Note that a 0.25 mm ply thickness has been

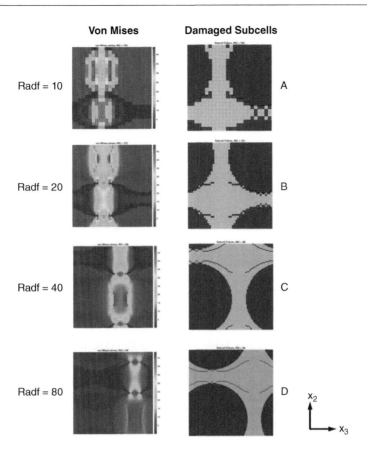

Figure 7.37 von Mises stress (MPa) and damage state snapshots associated with the points indicated in Fig. 7.35.

chosen (total laminate thickness of 1 mm) such that the force resultant, N_x, is equal in magnitude to the average stress, $\bar{\sigma}_{xx}$ (which has been used in the plots).

Comparing the two MT stress-strain curves plotted in Fig. 7.38, because the HF-Allowables are considerably higher than the MT-Allowables (Table 7.1 and Table 7.2), damage initiation occurs at a much higher stress in the MT HF-Allowables case. This damage initiation corresponds to failure of the matrix in the 90 plies. Next, failure of the matrix in the 0 plies occurs, but now the case with MT-Allowables is at the higher stress. This is because axial matrix failure in the 0 plies has been suppressed since the MT-Allowables in the axial direction (XT and XC) of the epoxy are taken to be very high (Table 7.1). Therefore the matrix in the 0 plies actually fails in the transverse direction rather than the axial direction. In contrast, for the MT HF-Allowables case, the 0 plies are able to fail in the axial direction, thus leading to the observed failure at the lower stress compared to the MT with MT-Allowables case. The UTS predictions of the two cases are different due to the difference in the fiber strength between the two

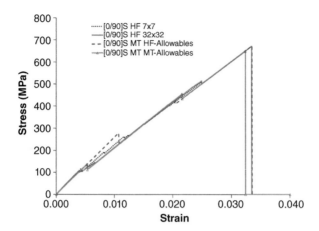

Figure 7.38 Influence of micromechanics theory (HFGMC and MT) and RUC (7×7 and 32×32) on the average stress vs. midplane strain response in the x-direction, under incremental strain loading (0.00005 strain increment) of a [0/90]$_s$ glass/epoxy laminate, $v_f = 0.55$. The maximum stress failure criterion was used.

sets of allowables (Table 7.1 and Table 7.2). In the HFGMC results, the stress-strain responses are more progressive due to the theory's ability to capture spatial variations in the constituent fields. The differences between the 7×7 and 32×32 RUC stress-strain responses are minor for this cross-ply laminate. Snapshots of the damage state at various applied midplane strain levels for the 7×7 and 32×32 RUC HFGMC cases are shown in Fig. 7.39A and B. Note that, even though these RUCs are within a laminate, they are plotted using the GMC/HFGMC orientation convention, with the x_2-direction vertical and the x_3-direction horizontal, even though the x_3-direction is the through-thickness direction of the laminate.

Comparing the MT theory using MT-Allowables with the HFGMC theory in Fig. 7.38, it is clear that the agreement is quite good, both in damage initiation and final failure. As should be obvious, the 0 plies in PMC laminates dominate the response and suppress the impact of lower length scale effects, such as microstructure, that are captured better by HFGMC. As such, the computational efficiency of each approach should be considered. The MT theory is the most efficient (by far) followed by the 7×7 HFGMC (which is still quite fast) and finally by the 32×32 HFGMC (whose execution takes on the order of minutes at the time of the writing of this text).

7.4.2.2 Quasi-isotropic laminates [60/-60/0]$_s$ and [45/0/-45/90]$_s$

In this example, the progressive damage behavior of two 0.55 fiber volume fraction glass/epoxy quasi-isotropic [60/-60/0]$_s$ and [45/0/-45/90]$_s$ laminates, subjected to midplane strain controlled tension, using the max stress failure criterion and a strain increment of 0.00005 is examined. The MT theory is compared to HFGMC with square fiber packing and both 7×7 and 32×32 subcell RUCs. As in Section 7.3, two sets of

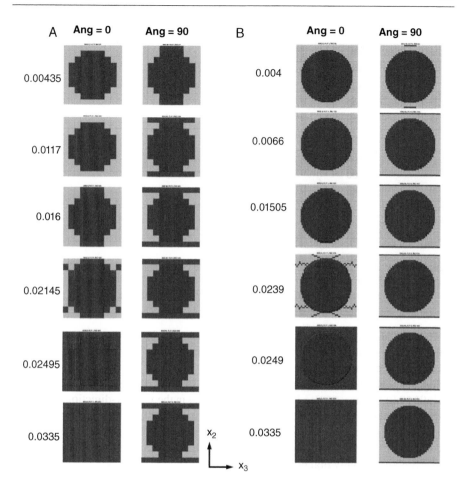

Figure 7.39 Incremental damage state, for various applied midplane strain levels in the x-direction, for the A) 7×7 RUC and B) 32×32 RUC, using the HFGMC micromechanics theory, see Fig. 7.38. The maximum stress failure criterion was used.

allowables are also considered for consistency with the MT method. The constituent material strengths are given in Table 7.1 and Table 7.2. Total laminate thicknesses have again been taken to be 1 mm such that the force resultants equal the average laminate stresses.

The average stress vs. midplane strain curves for the $[60/-60/0]_s$ laminate are shown in Fig. 7.40. The effects of the employed allowables, micromechanics theory, and subcell discretization are all quite similar to those discussed previously for the $[0/90]_s$ laminate. However, for the $[60/-60/0]_s$ laminate, only one third of the plies are 0 plies, compared to one half in the case of the $[0/90]_s$ laminate. As such, there is less suppression of the details of the progressive damage by the 0 plies, although the ultimate failure is still dominated by them. The ply damage progression in the HFGMC

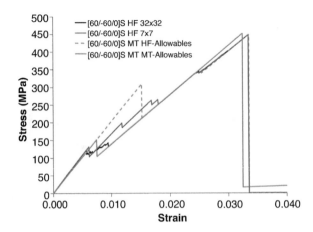

Figure 7.40 Average laminate stress vs. midplane strain response, under incremental strain loading (0.00005 strain increment) in the *x*-direction of a [60/-60/0]$_s$ glass/epoxy laminate, $v_f = 0.55$, analyzed using HFGMC and MT. The maximum stress failure criterion was used.

32×32 RUC simulation, shown in Fig. 7.41, indicates that the 60 plies behave similarly to the 90 plies in the [0/90]$_s$ laminate (although this behavior may be dependent on the failure criterion employed).

Fig. 7.42 compares simulated average laminate stress vs. midplane strain curves for the [60/-60/0]$_s$ laminate using HFGMC with loading in the *x*- and *y*-directions. Clearly, despite the fact that the laminate is quasi-isotropic in terms of effective laminate elastic properties, the damage initiation and progression is vastly different for loading in the two directions. Interestingly, although damage initiates earlier for *y*-direction loading, the UTS is significantly higher.

The [45/0/-45/90]$_s$ laminate results using HFGMC are plotted in Fig. 7.43. Results from the 32×32 RUC are compared with those from the 7×7 RUC, and, for the 7×7 RUC, results for loading in the two normal directions are compared. The *x*- and *y*-direction loading produces identical results for the [45/0/-45/90]$_s$ laminate, which was not the case for the prior [60/-60/0]$_s$ laminate. As in the laminates considered previously, there is some difference in the damage progression between the 7×7 and 32×32 RUCs, but the ultimate strengths are very similar. The post peak behavior, wherein stress increases again, is caused by the fact that the fibers in the 45 and 90 plies remain intact until after the fibers in the 0 plies fail, as shown in Fig. 7.44. Specifically, damage progression of the 32×32 RUCs is shown in Fig. 7.44, where it is clear that damage first initiates in the 90 plies in the matrix between the fibers at a global strain in the *x*-direction of 0.0042. Then it progresses in a self-similar manner through the 90 plies. This is followed by the initiation and propagation of damage in the 45 plies until, at a global strain of 0.02425, the matrix damage in the 0 plies initiates, and then all matrix in the 0 plies is failed at a global strain of 0.03. The fiber fails in the 0 plies at a global strain of 0.0335. Finally, at a global strain of 0.0679, the 90 ply fibers fail, reducing the global stress to a low value.

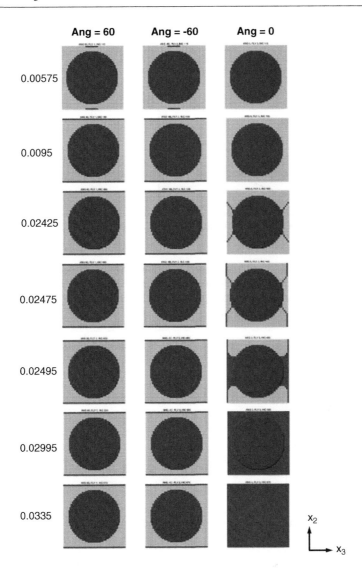

Figure 7.41 Incremental damage state, for various midplane strain levels in the x-direction, wherein a 32×32 HFGMC RUC has been used within a $[60/-60/0]_s$ glass/epoxy laminate, $v_f = 0.55$, see Fig. 7.40 . The maximum stress failure criterion was used.

Fig. 7.45 examines the influence of the choice of micromechanics theory, RUC discretization, and constituent allowables on the $[45/0/-45/90]_s$ stress-strain response. Clearly, both the MT and HFGMC theories provide reasonably close response histories, including the predicted UTS, due to the presence of 0 plies. The choice of allowables in the MT results, as well as the choice of RUC for HFGMC, affects damage initiation and pre-peak progression. Again, the MT theory is the most efficient (by far) followed by

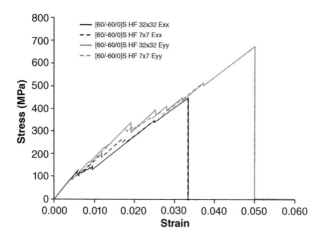

Figure 7.42 Influence of RUC discretization and loading direction on the average laminate stress vs. midplane strain response, under incremental strain loading (0.00005 strain increment) in either the x-direction or y-direction of a [60/-60/0]$_s$ glass/epoxy laminate, $v_f = 0.55$. The maximum stress failure criterion was used.

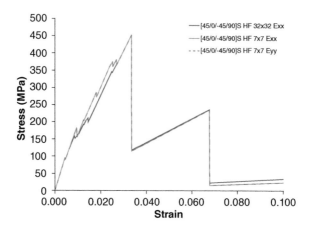

Figure 7.43 Average laminate stress vs. midplane strain response, under incremental strain loading (0.00005 strain increment) in the x-direction and y-direction, for a [45/0/-45/90]$_s$ glass/epoxy, $V_f = 0.55$, laminate analyzed using HFGMC for 32×32 and 7×7 RUCs. The maximum stress failure criterion was used.

the 7×7 HFGMC (which is still quite fast) and finally by the 32×32 HFGMC (whose execution takes on the order of minutes at the time of the writing of this text).

7.4.2.3   Angle-ply laminate, [45/-45]$_s$

Fig. 7.46 compares the stress-strain responses predicted employing the MT theory with those of the HFGMC theory using a 7×7 and a 32×32 RUC discretization, but now

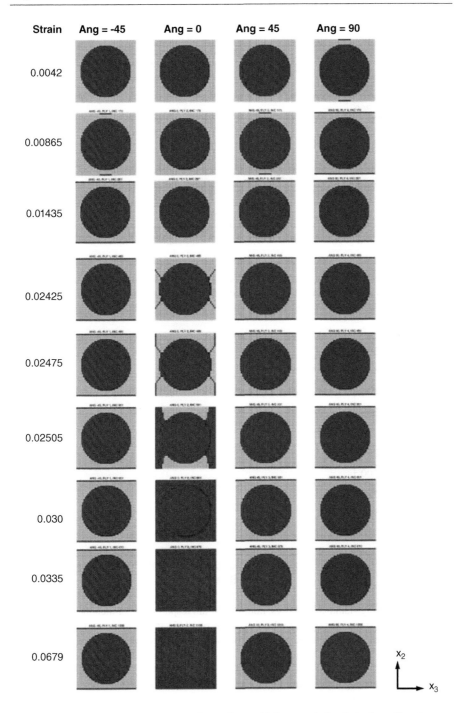

Figure 7.44 Incremental damage state, for various midplane strain levels in the *x*-direction, wherein a 32×32 HFGMC RUC has been used within a [45/0/-45/90]ₛ glass/epoxy laminate, $v_f = 0.55$, see Fig. 7.43. The maximum stress failure criterion was used.

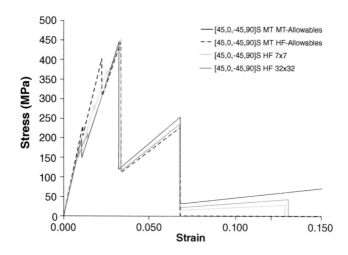

Figure 7.45 Average laminate stress vs. midplane strain response, under incremental strain loading (0.00005 strain increment) in the x-direction of a [45/0/-45/90]$_s$ glass/epoxy laminate, $v_f = 0.55$, analyzed using HFGMC and MT. The maximum stress failure criterion was used.

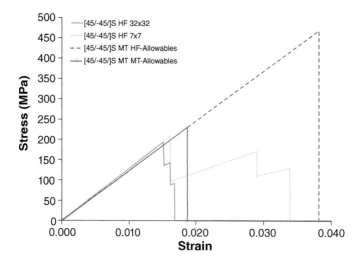

Figure 7.46 Average laminate stress vs. midplane strain response, under incremental strain loading (0.00005 strain increment) in the x-direction of a [45/-45]$_s$ glass/epoxy laminate, $v_f = 0.55$, analyzed using HFGMC and MT. The maximum stress failure criterion was used.

for the case of an angle ply laminate, i.e., [45/-45]$_s$. Now, since no 0 plies are present (which suppress the effects of the localized features), a larger difference between the MT and HFGMC theories is observed. The MT HF-Allowables case significantly over predicts (105%) the UTS of the laminate response relative to the HFGMC 32×32

RUC case. In the case of using the MT theory with consistent allowables (MT MT-Allowables case), the predicted UTS is only approximately 15% higher than the HFGMC 32×32 RUC case. Note, in both MT cases no progressive damage is apparent, as expected, since only ±45 plies are present. Finally, in comparing the two HFGMC RUC discretization cases, the more refined 32×32 RUC case behaves in a more brittle manner due to the pathological mesh dependence discussed earlier. Snapshots of the initial and final damage states for both HFGMC RUC discretizations are shown in Fig. 7.47. Note that both [45] plies identically initiate and propagate damage, as expected.

7.5 CMC example problems

7.5.1 Unidirectional CMC progressive damage

In this section, progressive damage simulations of unidirectional 0.28 fiber volume fraction SiC/SiC composites, with a 0.13 volume fraction BN fiber/matrix interfacial material, are considered, wherein the HFGMC micromechanics theory is used in conjunction with the max stress failure criterion. The effects of the applied loading orientation and ordered vs. disordered microstructures (hexagonal vs. random RUC fiber packing) are examined. First, the constituent properties are given in Table 5.2 are employed, then, for comparison, as done in the CMC failure envelopes shown in Section 5.7.5, results are presented wherein the interface is not permitted to fail.

7.5.1.1 Influence of loading orientation

Fig. 7.48 shows the stress-strain response for loading in the axial fiber direction of the unidirectional CMC as represented by a 32×58 hexagonal RUC. Note that the actual fiber and interface volume fractions produced by GetRUC.m are 0.276 and 0.129 for the hexagonal fiber packing RUC results shown in this section. Here the matrix first fails abruptly, then loading continues until the fiber fails. The interface stress does not rise to a high enough level to cause any failure. In reality, axial SiC/SiC behavior is more progressive, with less abrupt matrix failure (see Dunn, 2010). Bednarcyk et al. (2015) and Nemeth et al., (2016a,b) demonstrated that using a progressive damage model (rather than subcell elimination) for the matrix, or including stochastic effects, such as property variations and flaws, within micromechanics, enables the realistic gradual damage response of the matrix in the fiber direction prior to fiber failure to be captured.

In contrast, the transverse stress-strain response of the CMC is shown in Fig. 7.49, where loading in both the x_2- and x_3-directions has been considered. As in the PMC considered earlier, the hexagonal fiber packing leads to differences based on which transverse loading direction is considered (although elastically the two directions are the same). Comparing the two directions for the CMC, the damage initiates earlier for loading in the x_3-direction. As shown in Fig. 7.50A, for loading in both directions, failure initiation is in the interface material and is aligned with the loading direction.

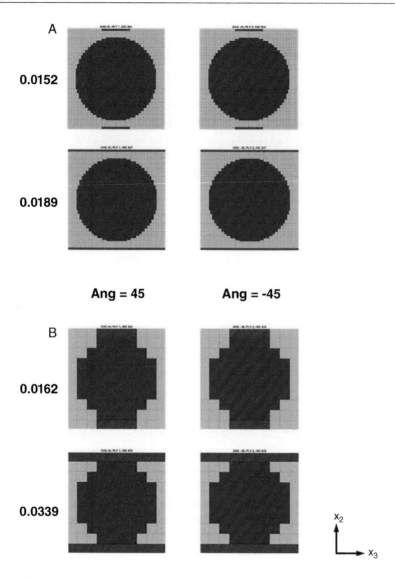

Figure 7.47 Incremental damage state, for damage initiation and final strain levels (labeled on the left), wherein A) 32×32 and B) 7×7 HFGMC RUCs have been used within a $[45/-45]_s$ glass/epoxy laminate, $v_f = 0.55$, see Fig. 7.46. The maximum stress failure criterion was used.

Due to the difference in fiber spacing in the two directions, the interfacial stress is higher for loading in the x_3-direction, leading to the earlier damage initiation evident in Fig. 7.49. Also, in both cases the stress continues to rise after initiation as long as the damage is confined to the interface, with the x_3-loading case exhibiting a much higher UTS prior to any damage in the matrix (see Fig. 7.50B). After matrix damage has initiated, however, the damage progression is more rapid for loading in the

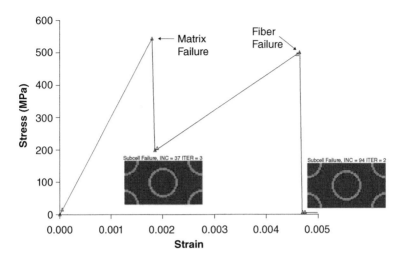

Figure 7.48 Simulated HFGMC stress-strain curves for a 0.28 fiber volume fraction, 0.13 interface volume fraction, unidirectional SiC/SiC composite represented by a 32×58 hexagonal RUC subjected to axial tensile strain, using the max stress criterion, with an applied strain increment of 0.00005.

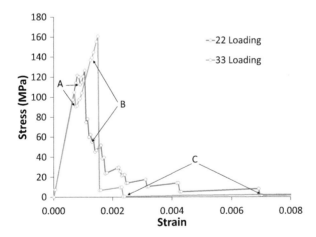

Figure 7.49 Simulated HFGMC stress-strain curves for a 0.28 fiber volume fraction, 0.13 interface volume fraction, unidirectional SiC/SiC composite represented by a 32×58 hexagonal RUC subjected to transverse strain, using the max stress criterion, with an applied strain increment of 0.00005.

x_3-direction as the damage grows into the high stress regions of matrix (as shown in Fig. 7.50B). Then the damage grows through the RUC, resulting in the largest stress drop in Fig. 7.49. The post-peak damage progression is more gradual for x_2-loading as the matrix damage zone does not proceed straight through the RUC, but rather it is deflected toward adjacent fibers (Fig. 7.50C).

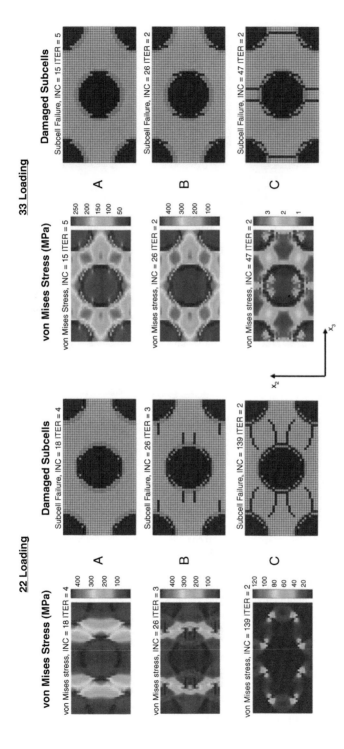

Figure 7.50 von Mises stress (MPa) and damage state (indicated by red subcells) in the RUC with fiber (dark blue), coating (light blue) and hexagonal packing, at the designated points A,B,C along the stress-strain response curve (see Fig. 7.49) under incremental transverse strain loading (0.00005 strain increment).

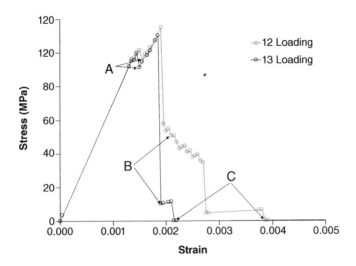

Figure 7.51 Simulated HFGMC stress-strain curves for a 0.28 fiber volume fraction, 0.13 interface volume fraction, unidirectional SiC/SiC composite represented by a 32×58 hexagonal pack RUC subjected to axial shear strain, using the max stress criterion, with an applied strain increment of 0.00005.

Results for axial shear loading (γ_{12} and γ_{13} applied) are shown in Figs. 7.51 and 7.52. Clearly, the damage initiation stress and ultimate strength are in much better agreement for the two applied strain components compared to the two transverse tensile loading component cases considered above. This is because the character of the local stresses between the simulations is much more similar for axial shear (see Fig. 7.52A), with high stress regions between the most closely spaced fibers. The γ_{13}-loading case exhibits a greater post-peak stress drop as the damage zones grow directly through the highest stress region of the matrix to link up the damaged regions of the interface. In contrast, for γ_{12}-loading, the matrix damage zones take a more tortuous path as they link up the interfacial damage.

Figs. 7.53 and 7.54 show the stress-strain responses and snapshots for applied γ_{23}-loading. As in the previous loading cases, damage initiates in the interface and then progresses into the matrix. There is quite a bit of damage progression restricted to the interface, resulting in two peaks in the stress-strain response, before the damage grows into the matrix (Fig. 7.54B), causing the large stress drop in Fig. 7.53. It is also interesting to note that the damage initiation is caused by shear (max σ_{23}) failure, but subsequent to point A, the failure switches to max σ_{33} until final failure. This is consistent with the orientation of the matrix damage zones in Fig. 7.54C. Of course, this is dependent on the relative magnitudes of the matrix strength components.

7.5.1.2 Influence of ordered vs. disordered microstructure

In this example, the same SiC/SiC CMC is considered, but now three RUCs with random fiber packing are simulated and compared with the hexagonal fiber packing case for transverse (x_2-direction) loading. The random cases have been labeled

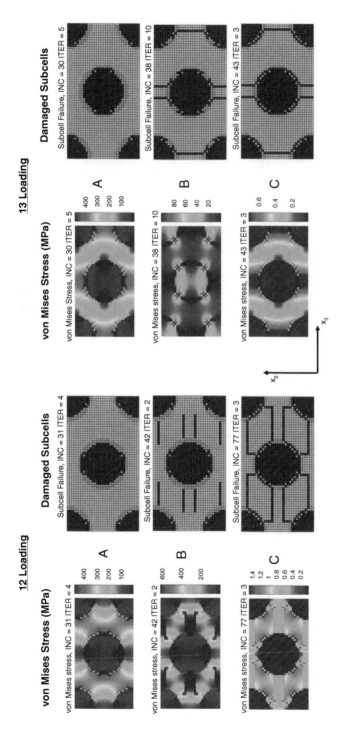

Figure 7.52 von Mises stress (MPa) and damage state (indicated by red subcells) in the RUC with fiber (dark blue), coating (light blue) and hexagonal fiber packing, at the designated points A,B,C along the stress–strain response curve (see Fig. 7.51) under incremental axial shear strain loading (0.00005 strain increment).

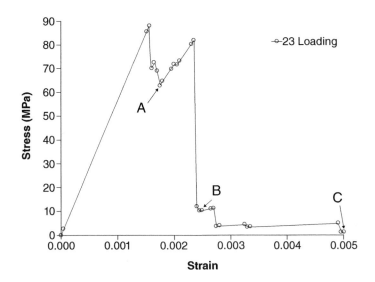

Figure 7.53 Simulated HFGMC stress-strain curves for a 0.28 fiber volume fraction, 0.13 interface volume fraction, unidirectional SiC/SiC composite represented by a 32×58 hexagonal pack RUC subjected to transverse shear strain, using the max stress criterion, with an applied strain increment of 0.00005.

Random-A, B, and C, and the associated .json files are provided with the MATLAB code in the Output folder. As in the PMC examples presented earlier, 10 fiber random RUCs, wherein the fiber radius is 10 subcells, have been generated using GetRUC.m. An interface that is two subcells thick has been chosen, which results in an actual fiber volume fraction of 0.281 and an actual interface volume fraction of 0.117, although nominal values of 0.28 and 0.13, respectively, were specified.

Fig. 7.55 compares the transverse tensile stress-strain curves of the three random cases with the hexagonal fiber packing case (plotted previously in Fig. 7.49). As observed in the PMC, the damage initiation in the random cases is earlier than for the ordered hexagonal packing case as hexagonal packing maximizes the distance between fibers and thus minimizes the induced local stress concentrations. The initial and final damage patterns, shown in Fig. 7.56, indicate that, as was the case with hexagonal packing, damage initiation tends to occur in the interface material in the random cases.

The final damage patterns shown in Fig. 7.56 indicate that Random-A and Random-C have more tortuous damage paths compared to Random-B, in which 4 fibers are reasonably well-aligned across the top of the RUC. The ramifications can be seen in the Fig. 7.55 random RUC stress-strain curves, wherein the Random-B curve indicates failure in the most brittle fashion and has the lowest strain energy density (area under the curve). Of course, these results are merely examples as other instances of a 10 fiber random RUC will give different results. As discussed earlier, the hexagonal packing RUC response is quite progressive, even as the matrix begins to fail, because the matrix damage zones progress towards low stress regions between the fibers (see Fig. 7.50) and are thus deflected toward adjacent fiber interfaces.

23 Loading

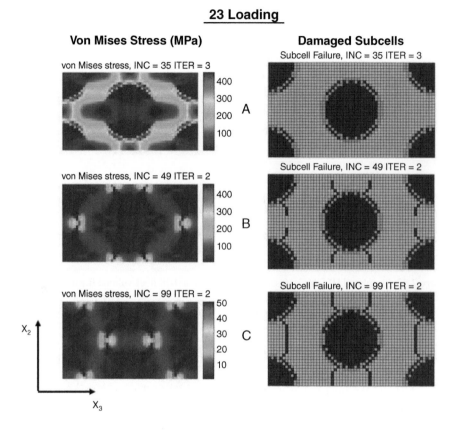

Von Mises Stress (MPa)   **Damaged Subcells**

Figure 7.54 von Mises stress (MPa) and damage state (indicated by red subcells) in the RUC with fiber (dark blue), coating (light blue) and hexagonal packing, at the designated points A,B,C along the stress-strain response curve (see Fig. 7.53) under incremental transverse shear strain loading (0.00005 strain increment).

7.5.1.3   Influence of interfacial failure

As discussed above, all of the previous CMC examples exhibited initiation of damage in the BN interface material. For the employed modeling approach, this characteristic is wholly dependent on the assigned strength values of the interface and their relationship to the matrix strength values. That is, if the interface material strengths were high enough, damage would initiate in the matrix rather than the interface. To investigate this quite different damage initiation scenario, the possibility of failure in the interface has been turned off by setting the interfacial strengths to very high values (a new material, MAT = 17, was added to the MATLAB code in GetConstitProps.m).

Fig. 7.57 compares the transverse tensile stress-strain curves of the three random cases with the hexagonal fiber packing case (considered in Section 7.5.1.2) where, now, no failure has been permitted in the BN interface material. The associated

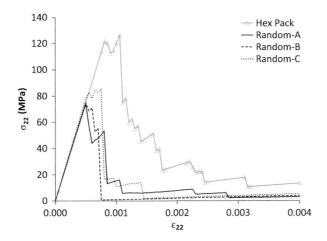

Figure 7.55 Simulated HFGMC stress-strain curves for a unidirectional SiC/SiC (nominal 0.28 fiber volume fraction and 0.13 interface volume fraction) composite subjected to transverse strain, using the max stress criterion, with an applied strain increment of 0.00005. Three 10 fiber random RUCs, and a 32×58 hexagonal RUC (x_2-direction tension) were employed.

initial and final damage snapshots are shown in Fig. 7.58. For the random fiber packing cases, the damage initiates in the matrix between closely spaced fibers in regions of high local stress. Additional damage initiates and propagates, linking the undamaged interfaces of other fibers. This is vastly different than the previous case wherein interfacial failure was permitted (Fig. 7.56) in which the matrix damage tends to link up existing interfacial damage. The impact of interfacial failure on the composite stress-strain response is shown in Fig. 7.59, wherein a direct comparison is made for the Random-A case and the hexagonal case with and without interfacial damage permitted. The presence of interfacial damage reduces the damage initiation stress and also significantly decreases the composite's toughness (ability to dissipate energy, area under curve). This should be expected because, when the interface is not permitted to fail, it is now an infinitely tough material. This toughening mechanism is precisely the objective of reinforcing brittle ceramic matrices with brittle fibers in CMCs.

For hexagonal packing with no interfacial damage allowed, damage initiates in the matrix on the left and right of each fiber (see Fig. 7.58) and then progresses in a self-similar manner until approaching the low stress matrix regions between fibers. Subsequently, the damage growth starts and stops, resulting in the progressive nature of the post-peak stress-strain response in Fig. 7.57. By the time a strain of 0.0027 is reached, the damage zones have been slightly deflected toward adjacent fiber interfaces (see Fig. 7.58). It is interesting to note that, when interfacial failure is not permitted, the hexagonal packing idealization transverse response is a much better approximation of the random RUC behavior.

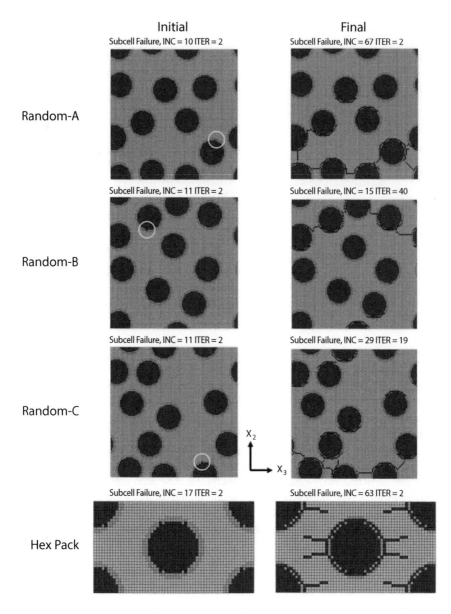

Figure 7.56 Initial and final damage states for three random microstructures and hexagonal packing associated with Fig. 7.55. Damage is indicated by red subcells in the RUC, with fiber (dark blue) and BN coating (light blue).

Obviously it is not possible to achieve a compliant and infinitely tough interface in a CMC, as simulated here. In reality, the BN interface behavior is likely somewhere in between the two cases examined here (brittle with a strength of 70 MPa vs. infinitely tough).

7.5.2   CMC cross-ply [0/90]_s laminate progressive damage

In this section the same nominally 0.28 fiber volume fraction SiC/SiC CMC (with a nominal 0.13 volume fraction BN interface) is examined, but now in the more technologically relevant [0/90]_s laminate configuration, where HFGMC is used to model the ply level response. The influences of the RUC microstructure and interfacial failure on the cross-ply laminate progressive damage response are studied. The following RUCs are considered: 13×13 with square fiber packing, 32×58 with hexagonal fiber packing, and two with random fiber packing, one with four fibers and one with eight fibers (the two random RUCs are provided in the Output folder as .json files). The 13×13 RUC was introduced in Exercise 5.6 as RUCid = 113, and must be present within GetRUC.m as follows:

```
% -- 13x13 with interface (hardwired Vf and Vi)
elseif RUCid == 113
    RUC.Vf_max = 0.6;
    RUC.NB = 13;
    RUC.NG = 13;
    RUC.h = [0.15555, 0.050931, 0.022449, 0.050931, 0.022449, ...
             0.050931, 0.29352, 0.050931, 0.022449, 0.050931, ...
             0.022449, 0.050931, 0.15555];
    RUC.l = RUC.h;
    RUC.mats = [M, M, M, M, M, M, M, M, M, M, M, M, M; ...
                M, M, M, M, M, I, I, I, M, M, M, M, M; ...
                M, M, M, M, M, I, F, I, M, M, M, M, M; ...
                M, M, M, I, I, I, F, I, I, I, M, M, M; ...
                M, M, M, I, F, F, F, F, F, I, M, M, M; ...
                M, I, I, I, F, F, F, F, F, I, I, I, M; ...
                M, I, F, F, F, F, F, F, F, F, F, I, M; ...
                M, I, I, I, F, F, F, F, F, I, I, I, M; ...
                M, M, M, I, F, F, F, F, F, I, M, M, M; ...
                M, M, M, I, I, I, F, I, I, I, M, M, M; ...
                M, M, M, M, M, I, F, I, M, M, M, M, M; ...
                M, M, M, M, M, I, I, I, M, M, M, M, M; ...
                M, M, M, M, M, M, M, M, M, M, M, M, M];
% -- Calculate Vf and Vi
RUC.Vf = 0;
RUC.Vi = 0;
for B = 1: RUC.NB
    for G = 1: RUC.NG
        if RUC.mats(B,G) == F
            RUC.Vf = RUC.Vf + RUC.h(B)*RUC.l(G);
        elseif RUC.mats(B,G) == I
            RUC.Vi = RUC.Vi + RUC.h(B)*RUC.l(G);
        end
    end
end
RUC.Vf = RUC.Vf/(sum(RUC.h)*sum(RUC.l));
RUC.Vi = RUC.Vi/(sum(RUC.h)*sum(RUC.l));
```

Both random RUCs use a fiber radius of 10 subcells and an interface that is 2 subcells thick, resulting in fiber and interface volume fractions of 0.281 and 0.117, respectively. Note these random microstructure simulations will take a long time to complete. In all of these cases, the interface material has been allowed to fail (constituent material

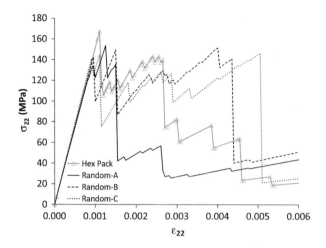

Figure 7.57 Simulated HFGMC stress-strain curves for a unidirectional SiC/SiC composite (nominal 0.28 fiber volume fraction and 0.13 interface volume fraction) subjected to transverse strain, using the max stress criterion, with an applied strain increment of 0.00005 and no interface material failure allowed. Three 10 fiber random RUCs and a 32×58 hexagonal RUC have been employed.

properties given in Table 5.2). Furthermore, the laminate total thickness was taken as 1 mm such that the force resultants are equal in magnitude to the average laminate stresses.

Fig. 7.60 shows the laminate x-direction tensile stress-strain responses for the four RUCs. In contrast to the PMC laminate cases examined previously, in the case of CMC laminates, the 0 plies suppress the influence of local features to a much lesser extent. This is because, in the CMC, the 90 plies carry considerably higher loads than in the PMC due to the high stiffness of the SiC matrix (see Table 5.2). Note that, in the CMC laminate, there is still very little difference between the 4 fiber and 8 fiber random cases in Fig. 7.60.

Damage snapshots for the 0 and 90 plies at points A and B in Fig. 7.60 are shown in Fig. 7.61. For all RUCs, damage initiates in the interface material of the 90 plies. However, unlike the unidirectional transverse tensile CMC results with interface failure, the stress-strain curves continue to rise significantly after interfacial damage initiation. This is because, as shown in Fig. 7.61, at point A, damage has not yet initiated in the 0 plies. During this rise in the stress-strain curves, damage continues to progress in the 90 ply interface and matrix materials. In the coarse 13×13 subcell RUC with square packing, the damage is actually limited to the interface until the stress-strain curve reaches its absolute peak, at which point the matrix subcells in the 90 plies between the fibers fail. This lack of pre-peak matrix failure explains why the UTS of the square fiber packing case is the highest. After the peak, the 13×13 subcell RUC case drops to coincide with the two random RUC curves. It then follows the random RUC curves closely as, in all three cases, the 90 plies have failed all the way through the RUC and are contributing little to the laminate stiffness. The hexagonal packing RUC

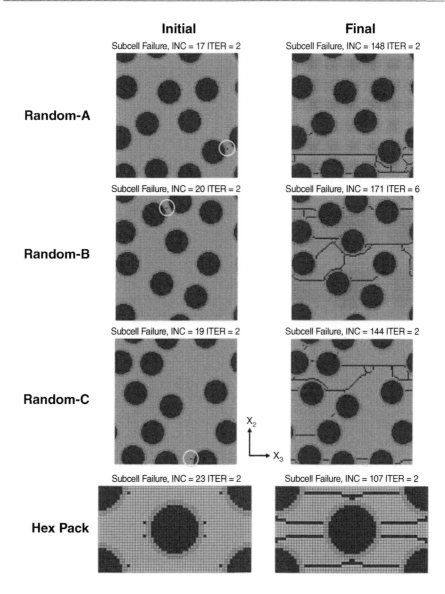

Figure 7.58 Damage snapshots from points A and B in the Fig. 7.57 stress-strain curves. Damage is indicated by red subcells in the RUC, with fiber (dark blue) and BN coating (light blue).

90 plies have not completely failed at the stress-strain curve peak, but, along with the other three cases, the stress-strain curve exhibits a large drop to point B associated with the failure of the matrix in the 0 plies, as shown in Fig. 7.61. Subsequent to this large stress drop, the laminate is still able to carry load, albeit with significantly reduced stiffness, as the fiber in the 0 plies is still intact. This is similar to the axial response

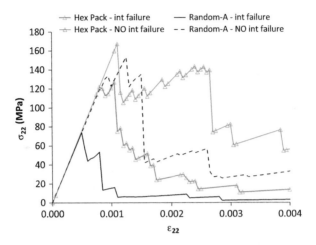

Figure 7.59 Comparison of simulated HFGMC stress-strain curves for a unidirectional SiC/SiC composite (nominal 0.28 fiber volume fraction and 0.13 interface volume fraction) subjected to transverse strain, using the max stress criterion, with an applied strain increment of 0.00005 with and without interface failure active. A 10 fiber random RUC and a 32×58 hexagonal RUC have been employed.

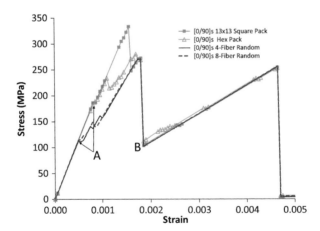

Figure 7.60 Influence of RUC on the HFGMC stress-strain response of a $[0/90]_s$ laminate loaded in the x-direction under incremental strain loading (0.00005 strain increment). The maximum stress failure criterion was used.

of the unidirectional CMC, as shown in Fig. 7.48. The final stress drop in Fig. 7.60 is associated with fiber failure in the 0 plies.

Fig. 7.62 shows cross-ply laminate results when the interface is not permitted to fail, which is in sharp contrast to the previous results with interface failure (see Fig. 7.60). Specifically, the damage initiation stress and UTS are significantly increased and much

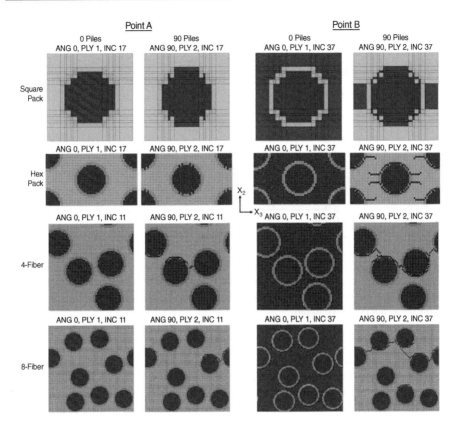

Figure 7.61 Damage state at points A (left side) and B (right side) associated with the Fig. 7.60 stress-strain curves for each type of RUC.

less sensitive to the RUC microstructure when interfacial failure is disallowed. In addition, previously, in Fig. 7.60, the hexagonal packing RUC stress-strain curve was above the random case curves, but now, in Fig. 7.62, this curve is predominantly below the random RUC curves (although the damage initiation point is still higher). As shown in Fig. 7.63, the 90 plies in the random fiber packing cases behave similarly to the unidirectional composite loaded in the transverse direction (Fig. 7.58) with matrix damage initiating and growing until encountering the interface of another fiber. As in the case with interfacial failure (Fig. 7.60), the large stress drop to point B in Fig. 7.62 is associated with matrix failure in the 0 plies, whereas the second large stress drop is associate with fiber failure in the 0 plies.

7.6 Concluding remarks

This chapter introduced micromechanics-based progressive damage simulations, wherein local failures occur and accumulate as damage progresses through a

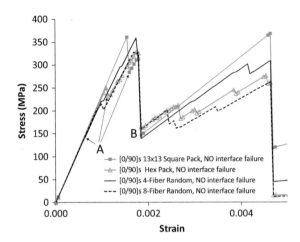

Figure 7.62 Influence of RUC on the HFGMC stress-strain response of a [0/90]$_s$ laminate loaded in the x-direction under incremental strain loading (0.00005 strain increment). The maximum stress failure criterion was used for the fiber and matrix materials, while the interface was not permitted to fail.

composite material, resulting in a nonlinear composite stress-strain response. In contrast, Chapter 4 considered only damage initiation, which may be sufficient for composites that behave linearly and fail in a brittle manner. A subvolume elimination method was employed, where subcell or constituent material stiffnesses are instantaneously reduced to a very low value once a specified failure criterion is reached within the subvolume. For laminates, this is done at the microscale within each ply. The pathological mesh dependence introduced by this method (when used within HFGMC or the finite element analysis), was illustrated and discussed.

Although significant portions of the MATLAB code from previous chapters have been preserved, due to the added complexities associated with progressive damage, new driver scripts that include the ability to apply loading incrementally and perform iterations within a given loading increment are required. Because of the nonlinearity present in progressive damage, which is controlled by the local fields and their redistribution as damage progresses, the impact of the chosen micromechanics theory, local failure criterion, and microstructural representation is significantly amplified compared to linear elastic analysis. Consequently, example problems involving PMC and CMC composite materials and laminates were presented, illustrating the following influences on the composite progressive damage response:

- Subcell grid and RUC geometry
- Consistency and interaction of the constituent failure strengths
- Chosen micromechanical theory
- Number of fibers included within the RUC

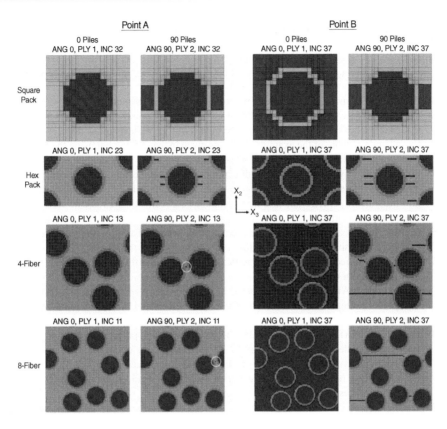

Figure 7.63 Damage state at points A (left side) and B (right side) associated with Fig. 7.62 for each type of RUC. The interface was not permitted to fail.

- Employed failure criterion
- Loading orientation and number of loading increments

It was demonstrated that HFGMC, with its normal-shear coupling and accurate local fields, enables capture of the details of damage progression, which are particularly influential for disordered (random) microstructures. Of course, since detailed progressive damage simulations are computationally expensive, maximizing computational efficiency is highly desirable. If only damage initiation is of interest, then the margin of safety predictions from Chapter 4 can be employed without the need for more expensive progressive damage analysis. Furthermore, in fiber-dominated situations in PMCs, the present chapter demonstrated that simpler, more efficient, micromechanics theories like MOC and MT are sufficient to predict the composite ultimate strength. This is because the effect of local fields is suppressed in fiber-dominated situations. Finally, for CMCs, it has been shown that the progressive damage behavior is highly influenced by damage initiation location (i.e., interface vs. matrix), which is controlled by their relative strengths.

7.7   Progressive damage self-learning

While Exercises are not provided for this chapter, suggestions are provided here that can help readers to better understand progressive damage analysis using HFGMC.

1) Execute one or more of the example problems given in the chapter.
2) Determine 0.55 volume fraction unidirectional IM7/8552 composite constituent allowables that enable a hexagonal RUC to reproduce the IM7/8552 ply-level allowables.
3) Execute one or more of the example problems given in the chapter, but using the IM7/8552 constituent allowables determined in #2 above.
4) Determine 0.55 volume fraction unidirectional glass/epoxy composite constituent allowables that enable a random RUC to reproduce the glass/epoxy ply-level allowables. Consider how many fibers per RUC and how many random RUC realizations should be used to obtain a good approximation.
5) Consider progressive damage under compressive loading. Examine the interplay of the matrix compressive and shear strengths and how they impact the results.
6) Examine the effect of interface properties, allowables, and thickness on the progressive damage behavior of a SiC/SiC CMC.
7) Enhance the MATLAB scripts to enable multiple loading steps, wherein the loading can change (for example, first load in the axial direction, followed by loading in the transverse direction).
8) Enhance the MATLAB scripts to enable thermal loading with progressive damage analysis. Using this and the enhancement from #7 above, examine the effect of a thermal pre-load on the progressive damage response of a composite or laminate.
9) Use progressive damage analysis to generate final failure surfaces and compare to damage initiation surfaces given in previous chapters.
10) Enhance the MATLAB scripts to enable, instead of subvolume elimination, progressive damage at the micro scale, wherein the stiffness decrease of subcells (or materials for MT) occurs gradually rather than all at once. This requires equations for gradually decreasing the stiffness. Implement the (relatively simple) damage equations given by Liu and Arnold (2011, 2013) (see Eqs. 14 and 21 in Liu and Arnold, 2011).
11) Use the enhancement from #10 above to examine the effect of the parameter n in the damage equations (the damage normalized secant modulus) on the progressive damage behavior of unidirectional composites or laminates. Note that the damage equations of Liu and Arnold (2011, 2013) are isotopic, and thus they should not be applied to anisotropic materials such as IM7.
12) Implement a version of the MT theory that includes an interface material (see Barral et al., 2020). Use this version of MT to model CMC properties and progressive damage and compare with GMC and HFGMC.

References

Aboudi, J., 2011. The effect of anisotropic damage evolution on the behavior of ductile and brittle matrix composites. Int. J. Solids Struct. 48, 2102–2119.
Barral, M., Chatzigeorgiou, G., Meraghni, F., Leon, R., 2020. homogenization using modified mori-tanaka and tfa framework for elastoplastic-viscoelastic-viscoplastic composites: theory and numerical validation. Int. J. Plast. 127, 102632.

Bažant, Z.P., 1994. Nonlocal damage theory based on micromechanics of crack interactions. J. Eng. Mech. - ASCE 120 (3), 593–617.

Bažant, Z., Cedolin, L., 1979. Blunt crack band propagation in finite element analysis. J. Eng. Mech. Div.-ASCE 105, 297–315.

Bažant, Z.P., Cedolin, L., 1991. Stability of Structures: Elastic, Inelastic, Fracture and Damage Theories. Oxford University Press, New York, Oxford.

Bažant, Z., Oh, B.H., 1983. Crack band theory for fracture of concrete. Mater. Struct. 16, 155–177.

Belytschko, T., Moës, N., Usui, S., Parimi, C., 2001. Arbitrary discontinuities in finite elements. Int. J. Numer. Meth. Eng. 50, 993–1013.

Bednarcyk, B.A., Aboudi, J, Arnold, S.M, 2010. Micromechanics modeling of composites subjected to multiaxial progressive damage in the constituents. AIAA Journal 48 (7), 1367–1378.

Bednarcyk, B.A, Mital, S.M., Pineda, E.J., Arnold, S.M, 2015. Multiscale modeling of ceramic matrix composites. In: 56th AIAA/ASME/ASCE/AHS/ASC Structures, Structural Dynamics, and Materials Conference, AIAA Science and Technology Forum 2015, January 5-9. AIAA 2015-1191.

Dunn, D.G., 2010. "The Effect of Fiber Volume Fraction in Hipercomp SiC-SiC Composites", Ph.D. Thesis, Alfred University, Alfred, NY.

Eringen, A.C., 1966. A unified theory of thermomechanical materials. Int. J. Eng. Sci. 4, 179–202.

Garikipati, K., 2002. A variational multiscale method to embed micromechanical surface laqs into the macromechanical continuum formulation. Comput. Model. Eng. 3, 175–184.

Garikipati, K., Hughes, T.J.R., 1998. A study of strain localization in a multiple scale framework the one dimensional problem. Comput. Meth. Appl. Mech. Eng. 159, 193–222.

Jirásek, M., 1998. Nonlocal models for damage and fracture: comparison of approaches. Int. J. Solids Struct. 35, 31–32.

Kachanov, L.M., 1958. On the creep fracture time, Izvestiya Akademii Nauk SSSR. Otdeleniya Tekhnika Nauk 8 (1), 26–31.

Krajcinovic, D., 1996. Damage Mechanics. Elsevier, New York, United States.

Liu, K.C., Arnold, S.M., 2011. Impact of material and architecture model parameters on the failure of woven Ceramic Matrix Composites (CMCs) Via the Multiscale Generalized Method of Cells. NASA-TM-2011-217011.

Liu, K.C., Arnold, S.M., 2013. Influence of scale specific features on the progressive damage of woven ceramic matrix composites". Comput. Mater. Continua 35 (1), 35–65.

Lemaitre, J., Desmorat, R., 2005. Engineering Damage Mechanics. Springer, Berlin.

Lemaitre, J., Chaboche, J.L., 1990. Mechanics of Solid Materials. Cambridge University Press, Cambridge, United Kingdom.

Nemeth, N.N., Bednarcyk, B.A., Pineda, E.J., Walton, O.J., Arnold, S.M, 2016a. Stochastic-Strength-Based Damage Simulation Tool for Ceramic Matrix and Polymer Matrix Composite Structures. NASA Technical Memorandum 219113.

Nemeth, N.N., Mital, S.K, Murthy, P.L.N, Bednarcyk, B.A., Pineda, E.J., Bhatt, R.T., Arnold, S.M, 2016b. Stochastic-Strength-Based Damage Simulation of Ceramic Matrix Composite Laminates. NASA Technical Memorandum 219115.

Ortiz, M., Pandolfi, A., 1999. Finite-deformation irreversible cohesive elements for three-dimensional crack-propagation analysis. Int. J. Numer. Meth. Engng. 44, 1267–1282.

Pineda, E.J., Waas, A.M., 2012. Modelling Progressive Failure of Fiber Reinforced Laminated Composites: Mesh Objective Calculations. Aeronaut. J. 116 (1186), 1221–1246.

Pineda, E.J., Bednarcyk, B.A., Waas, A.M., Arnold, S.M, 2012. Progresive Failure of a Unidirectional Fiber-Reinforced Composite Using the Method of Cells: Discretization Objective Computational Results. NASA TM 2012-217649, 2012.

Pineda, E.J., Bednarcyk, B.A., Waas, A.M., Arnold, S.M, 2013. On multiscale modeling using the generalized method of cells: preserving energy dissipation across disparate length scales". Comput. Mater. Continua 35, 119–154.

Soden, P.D., Hinton, M.J., Kaddour, A.S., 2004. Biaxial test results for strength and deformation of a range of E-glass and carbon fibre reinforced composite laminates: Failure exercise benchmark data. In: Hinton, M.J., Kaddour, A.S., Soden, P.D. (Eds.), Failure Criteria in Fibre Reinforced Polymer Composites: The World-Wide Failure Exercise. Elsevier, New York, United States.

Sukumar, N., Moës, N., Moran, B., Belytschko, T., 2000. An extended finite element method (x-FEM) for two- and three-dimensional crack modeling. Int. J. Numer. Methods Eng. 48 (11), 1741–1760.

Rabotnov, Y.N., 1969. Creep Problems of Structural Members. North-Holland, Amsterdam, The Netherlands.

Rudraraju, S.S., Salvi, A., Garikipati, K., Waas, A.M., 2010. In-plane fracture of laminated fiber reinforced composites with varying fracture resistance: Experimental observations and numerical crack propagation simulations. Int. J. Solids Struct. 47, 901–911.

Voyiadjis, G.Z., Ju, J.W., Chaboche, J.L. (Eds.), 1998. Damage Mechanics in Engineering Materials, Studies in Applied Mechanics, Vol. 46. Elsevier.

Xie, D., Waas, A.M., 2006. Discrete cohesive zone model for mixed-mode fracture using finite element analysis. Eng. Fract. Mech. 73 (13), 1783–1796.

Index

Page numbers followed by "*f*" and "*t*" indicate, figures and tables respectively.

Material properties

Ply material properties

Thermoelastic properties

Mat #	Material name	Density (g/cm^3)	E_1 (GPa)	E_2 (GPa)	G_{12} (GPa)	ν_{12}	α_1 ($10^{-6}/°C$)	α_2 ($10^{-6}/°C$)
1	IM7/8552	1.58	146.8	8.69	5.16	0.32	–0.1	31
2	glass/epoxy	2.0	43.5	11.5	3.45	0.27	6.84	29
3	SCS-6/Ti-15-3	3.86	221	145	53.2	0.27	6.15	7.90
4	SiC/SiC	3.21	300	152.4	69.5	0.17	3.0	3.2
5	Al	2.7	70	70	26.923	0.3	23	23

Stress allowables

Mat #	Material name	X_T (MPa)	X_C (MPa)	Y_T (MPa)	Y_C (MPa)	S (MPa)
1	IM7/8552	2323	–1531	52.3	–235	88
2	Glass/epoxy	1346	–944	50	–245	122

Constituent material properties

Thermoelastic properties

Mat #	Material name	E_1 (GPa)	E_2 (GPa)	G_{12} (GPa)	ν_{12}	ν_{23}	α_1 ($10^{-6}/°C$)	α_2 ($10^{-6}/°C$)
1	IM7	262.2	11.8	18.9	0.17	0.21	–0.9	9.0
2, 18	Glass	73.0	73.0	29.918	0.22	0.22	5.0	5.0
3	8552	4.67	4.67	1.610	0.45	0.45	42.0	42.0
4, 15	Epoxy	3.45	3.45	1.278	0.35	0.35	54.0	54.0
5	Tsai&Hahn Glass	113.4	113.4	46.475	0.22	0.22	5.0	5.0
6	Tsai&Hahn Epoxy	5.35	5.35	1.981	0.35	0.35	54.0	54.0
7	Dean & Turner Carbon	232	15	24	0.279	0.49	–	–
8	Contour Glass	69	69	28.75	0.2	0.2	5.0	5.0
9	Contour Epoxy	3.42	3.42	1.276	0.34	0.34	54.0	54.0
10	SiC Fiber	385	385	164.53	0.17	0.17	3.2	3.2
11	SiC Matrix	327	327	134.02	0.22	0.22	3.1	3.2
12	BN Coating	10	10	4.065	0.23	0.23	4.0	4.0
13	Sun & Vaidya Carbon	235	14	28	0.20	0.25	–0.9	9.0
14	Sun & Vaidya Epoxy	4.8	4.8	1.8	0.34	0.34	54	54

Note all constituents are assumed transversely isotropic so $E_3 = E_2$, $G_{13} = G_{12}$, $G_{23} = E_2/(2(1-\nu_{23}))$, $\nu_{13} = \nu_{12}$, and $\alpha_3 = \alpha_2$.

Stress allowables

Mat #	Material name	X_T (MPa)	X_C (MPa)	Y_T (MPa)	Y_C (MPa)	Q (MPa)	R (MPa)	S (MPa)
1	IM7	4335	−2608	113	−354	128	138	138
2	Glass (MT)	2358	−1653	2358	−1653	1000	1000	1000
3	8552	59.4	−259	59.4	−259	112	112	112
4	Epoxy (MT)	1×10^6	-1×10^6	42	−181	81	81	81
10	SiC Fiber	1800	−1800	1800	−1800	700	700	700
11	SiC Matrix	600	−600	600	−600	350	350	350
12	BN Coating	70	−70	70	−70	45	45	45
15	Epoxy (HFGMC)	86	−425	86	−425	184	184	184
18	Glass (HFGMC)	2445	−1653	2445	−1653	1000	1000	1000

Note all constituents are assumed transversely isotropic so ZT = YT and ZC = YC.

Repeating Unit Cells (RUCs)

GMC and HFGMC RUCs provided in GetRUC.m

RUCid	Options in GetRUC.m	Schematic
2 (2x2)	-	
3 (3x3)	-	
7 (7x7)	-	
8 (24x24)	Continuous, discontinuous, and overlap	
26 (26x26)	Extra	
30 – 33 (25x25 – 201x201)	-	

RUCid	Options in GetRUC.m	Schematic
99 (user defined)	Full RUC definition	?
105 (5x5)	-	
200 (hexagonal fiber packing)	Target_Ni and Target_NG	
300 (random fiber packing)	Nfibers, Radf, Tint and touching	
1000 (JSON file)	Read from JSON file	?

9780128206379